How Life Increases
Biodiversity

How Life Increases Biodiversity

Biodiversity

An Autocatalytic Hypothesis

David Seaborg

CRC Press
Taylor & Francis Group
Boca Raton London New York

CRC Press is an imprint of the
Taylor & Francis Group, an **informa** business

First edition published 2022
by CRC Press
6000 Broken Sound Parkway NW, Suite 300, Boca Raton, FL 33487-2742

and by CRC Press
2 Park Square, Milton Park, Abingdon, Oxon, OX14 4RN

© 2022 Taylor & Francis Group, LLC

CRC Press is an imprint of Taylor & Francis Group, LLC

ISBN: 978-1-138-34140-1 (hbk)
ISBN: 978-0-367-63134-5 (pbk)
ISBN: 978-0-429-44013-7 (ebk)

Typeset in Times
by codeMantra

"If I have seen further than others, it is by standing on the shoulders of giants."

I dedicate this book to the following giants who have inspired me with their love, wisdom, or insight:

Glenn Seaborg, Helen Seaborg, Peter Seaborg
Lynne Cobb, Steve Seaborg Eric Seaborg, Ellen Dudley
Lela, Todd, Cole, and Lucy Arthur
Molly, Brad, and Matilda Cobb-Holderness
Adele Seaborg
Megan, Nikhil, Anish, and Myra Kalyankar
Alex Loijos; Eddie and Joe Glaeser
Hal Segelstad, Bill Ryan, Mark Koperweis, Harold and
Janet Wood, Jim Saunders, Sonoko Masui Rooney
"Limitless undying love which shines around me like a million suns
And calls me on and on across the universe
Jai guru deva om"

Sterling Bunnell, Paul Segall, Mamade Kadreebux
"Dreaming dreams no mortal ever dared to dream before"

John Muir and David Brower
"Great spirits have always encountered
opposition from mediocre minds."

The rainforests and coral reefs of the world
"Degged with dew, dappled with dew,

Are the groins of the braes that the brook treads through,
Wiry heathpacks, flitches of fern,
And the beadbonny ash that sits over the burn."

What would the world be, once bereft
Of wet and of wildness? Let them be left,
O let them be left, wildness and wet;
Long live the weeds and the wilderness yet."

The first quote is from Isaac Newton.
The second quote is from the song, "Across the Universe",
recorded by The Beatles, written by John Lennon, attributed
to John Lennon and Paul McCartney. The song first appeared
on the 1969 various artists' charity compilation album,
"No One's Gonna Change Our World". It appeared later, in
a different form, on The Beatles' 1970 album, "Let It Be".
The third quote is from the poem, "The Raven", by Edgar Allan Poe.
The fourth quote is from Albert Einstein.
The fifth quote is from the poem, "Inversnaid",
by Gerard Manley Hopkins.

Contents

Acknowledgments

Sterling Bunnell encouraged and inspired me and had several valuable discussions with me about the Autocatalytic Biodiversity Hypothesis/Pachamama Hypothesis and my ideas before I began writing this book or gave the hypothesis its name. He had similar ideas to mine, although he never published them.

The following scientists discussed the Autocatalytic Biodiversity Hypothesis with me, encouraged me, and gave me confidence and inspiration – the latter three before I began writing the book or gave the theory its name:

James Lovelock, Lynn Margulis, Eugene Odum, and Edward O. Wilson.

The following scientists helped me with ensuring that the science and logic are correct:

Frank Almeda, Roni Avissar, Terry Erwin, Sergey Gavrilets, Egbert Giles Leigh, Jr., Aditee Mitra, Mary Power, and Geerjat Vermeij.

Sylka Perez, Roni Avissar's assistant, was very helpful in conveying my messages.

Of those above, the following scientists gave extraordinary help and a great deal of their time, discussing ideas and/or reading and commenting on and/or editing parts of the manuscript, and giving invaluable feedback to assure scientific accuracy:

Frank Almeda and Aditee Mitra.

The following laypeople read chapters of the manuscript, doing valuable editing, improving the English and presentation of my ideas, including making the wording more succinct, comprehensible, interesting, and exciting:

Lynne Cobb, Kevin Langdon, Eric Seaborg, and Steve Seaborg. Lynne Cobb was especially helpful, editing several chapters, with exceptionally helpful suggestions.

The following laypeople sent me articles supporting the hypothesis that were helpful because they addressed aspects I was at least partially unaware of:

Lynne Cobb, Bob Jansen, Adele Seaborg, and Steve Seaborg.

Lynne Cobb sent me the book, *The Medea Hypothesis*, by Peter Ward, making me aware of a hypothesis that is in opposition to the Autocatalytic Biodiversity Hypothesis.

Mamade Kadreebux encouraged me constantly and provided me a forum to speak to an audience about the hypothesis.

Adele Seaborg solved computer issues that were beyond my limited knowledge in that area.

Chuck Crumly and Ana Lucia Eberhart of Taylor & Francis, the publisher of this book, kindly provided me with guidance through the publishing process.

Countless teachers, extended family members, friends, and scientists I knew, but did not speak with specifically about this book, made this book better. So did many scientists I never met, but whose works I am familiar with, and gurus and teachers of philosophical systems I met or only know through their teachings. So, too, did innumerable living organisms, especially snakes, museums, and beautiful ecosystems in nature. It would not be in keeping with the spirit of this book to fail to acknowledge these.

Any errors of any sort in this book are solely my fault and not the fault of my commensal helpers.

Author

David Seaborg is an evolutionary biologist. His undergraduate degree is from the University of California at Davis in zoology, and his graduate degree is from the University of California at Berkeley, also in zoology. He originated the concept that organisms act as feedback systems in their evolution, and that they thus play an important role in their evolution. This concept is a mechanism for punctuated equilibrium. He showed that the canonical genetic code is on an adaptive peak, and how populations cross over maladaptive valleys from one adaptive peak to another. He published a hypothesis to explain how homosexuality evolved, even though it theoretically reduces the number of offspring produced.

He has taught biology at all levels from kindergarten to the university level. He taught the basic biology course at the University of California at Berkeley, university extension courses, courses at museums, field courses for all ages, at Burton Academic High School in San Francisco, and elementary students. He currently teaches various life science courses at the Osher Lifelong Learning Institute, which is part of the University of California at Berkeley extension program, and at the Fromm Institute for Lifelong Learning, at the University of San Francisco.

David is an environmental leader. He is Founder and President of the World Rainforest Fund, a nonprofit, tax-exempt foundation dedicated to saving the Earth's tropical rainforests and biodiversity. This organization has saved rainforests in Ecuador, Brazil, Columbia, the Democratic Republic of the Congo, and Borneo. It set the record for the most species saved per dollar when it helped stop a road that if built would have resulted in the destruction of a 10,000-acre rainforest in Ecuador that has the highest biodiversity of any ecosystem on Earth according to scientists at the Missouri Botanical Garden. Had the road been built, exploiters would have used it to access the rainforest and destroy it. The World Rainforest Fund spent only $3,500.00 to stop the road and save this rainforest.

He raised $20,000 in less than a year to successfully help save Acalanes Ridge, a pristine hillside oak and grassland habitat in Lafayette, California.

He wrote an article that is a summary of the scientific research on the effects of high atmospheric levels of carbon dioxide other than global warming. Unlike the climatic effects, these effects are not well known to the general public. They are very serious, and have the potential to cause high levels of extinction of species and greatly disrupt ecosystems and our food supply.

He was on the General Plan Advisory Committee of the city of Lafayette, California, which he guided to producing a ten-year General Plan for that city that emphasized environmental sustainability, preserving open space, combating global warming, and energy conservation.

In the 1990s and part of the first decade of this century, he served on the Board of Directors and as Vice President of the Club of Rome of the United States, the environmental think tank that published the Limits to Growth in the 1970s. This is a computer simulation study that showed that continued growth and consumption of resources will lead society to disaster. He is currently on the nominating committee for the Goldman Environmental Prize, the most prestigious grassroots environmental prize in the world.

He was on the Board of Directors of the East Bay Chapter of the United Nations Association of the United States from 2006 to 2009, where he was the lead environmental person. He gave the keynote address at their 2006 annual meeting, and helped secure the passage of key resolutions on biodiversity and global warming and the Kyoto Protocol, at the local, state, and national levels of the United Nations Association/USA. These resolutions call for action on these issues by the UN and US government.

He conceived and helped secure passage by the Berkeley City Council an ordinance banning the use of old-growth rainforest and redwood in all products used by the city of Berkeley. This ordinance also required all businesses contracting with Berkeley to stop using old-growth rainforest and redwood in any products or services Berkeley hires them to use or perform, or in any product they sell this city.

David carried the Ten Commandments for the Earth, a version of the original Ten Commandments rewritten to focus on saving the Earth's environment, while riding on a camel down Mount Sinai, the mountain in Egypt down which Moses allegedly carried the original Ten Commandments. Then, in a brief ceremony, he presented these Ten Commandments to a Bedouin youth, who represented the indigenous people and the youth of the planet, the generation inheriting the Earth for its stewardship. After completing this act, which was captured on video camera, David swam for over an hour with a dolphin in the Red Sea.

David conceived the idea for and was the head organizer for a press conference of Nobel Prize winners on global environmental issues that was held at the time of the 100th Nobel Prize ceremonies in Stockholm, Sweden, in December, 2001.

David has been to over 30 countries, observing various natural ecosystems and wildlife. He is an award-winning nature and wildlife photographer and an award-winning poet. He wrote a popular and acclaimed poetry book called *Honor Thy Sow Bug*. He is listed in Who's Who in America. An excellent public speaker, he lectures to various scientific, environmental, civic, business, and other organizations on evolutionary biology, the philosophical implications of science, and environmental issues.

1 Introduction

Evolution, ecology, and population biology have given us tremendous understanding and insights, but have for the most part missed one of the most important, interesting, exciting, explanatory, and unifying aspects of evolution, ecology, and how ecosystems function. This is that organisms, species, and life itself generate the vast majority of the biodiversity on Earth, much more than would be present from physical and chemical factors alone. The subject and thesis of this book, the Autocatalytic Biodiversity Hypothesis (ABH), proposes that organisms create most of the biodiversity on Earth, and that ecosystems maximize biodiversity. It proposes that the number of species on Earth is far greater because of the presence, actions, and interactions of its organisms. The hypothesis has autocatalytic in its name because it proposes that life itself and organisms themselves create more life, more species, and more biodiversity. Life causes far greater biodiversity than can be explained solely by physical and chemical processes. This is accomplished by two or three mechanisms. The first underlying mechanism uniting all the seemingly unrelated mechanisms is that every species is an ecosystem engineer that has a net beneficial impact on other species, its ecosystem, and biodiversity in a natural ecosystem over a sufficient time. Oxygenic photosynthesis caused a great loss of diversity because oxygen was toxic to many prokaryotes when it first became abundant in the atmosphere. It also caused the Earth to become very cold. Sulfate-reducing microbes produced toxic hydrogen sulfide that decreased diversity for a long time. But these phenomena were in time changed by life, making them not very harmful to life. In time, they both were essentially permanently converted to being factors that increase diversity. This is why the words "over a sufficient time" are in the above statement that every species is beneficial to biodiversity. This is discussed in more detail in Chapter 15. This statement is also the major underlying mechanism uniting most of the seemingly unrelated mechanisms by which organisms cause biodiversity to increase. Crucial to this theory is the assertion that mutualism is much more important than commonly thought, fundamental, an organizer of ecosystems, and a great promoter of biodiversity. The ABH and its mechanisms are entirely scientific, naturalistic, and mechanistic, and any attempt to interpret this hypothesis in teleological, pseudoscientific, or "Aquarian Age" terms, as some incorrectly did with the Gaia Hypothesis, is simply wrong. The presentation of this theory and the evidence for it are the theme and subject of this book.

The second mechanism by which organisms increase biodiversity is the behavior of the genome. Sexual recombination, transposable elements, viral transduction, repurposing of viruses as genetic material by the host, polyploidy, hybridization, epistasis and changes in regulatory genes, and other such mechanisms cause macro-evolutionary innovations and great increases in biodiversity. This can be considered a form of ecosystem engineering in the sense that viral transduction, transposable elements, sexual recombination, and so on are molding (engineering) the genome, and the genome is part of the ecosystem. If one accepts this, there is no need for this additional unifying mechanism for the ABH, since all species are ecosystem engineers that tend to have a positive effect on diversity in natural systems over time covers genomic behavior. Genomic behavior can be incorporated in the ecosystem engineering mechanism of the ABH. I think this is legitimate. But some might differ. If one thinks calling the behavior of the genome a form of ecosystem engineering to be too much of a stretch, then a second unifying mechanism of the ABH can be added. That mechanism is the behavior of the genome, examples of which are listed in the second sentence of this paragraph. This mechanism is not discussed in this volume of book because of space limitations. However, I will discuss it in a second volume of this book that will be published next year or sooner.

The other mechanism by which the ABH works and organisms generate biodiversity is natural selection.

Lovelock [1] and Lovelock and Margulis [2] proposed the Gaia Hypothesis followed by Lovelock's [3] popularizing, seminal book, *Gaia: A New Look at Life on Earth* [3]. The hypothesis proposed that organisms interact with their inorganic surroundings to form a cybernetic and self-regulating system that helps maintain and perpetuate the conditions for life. The hypothesis sees the Earth as analogous to an organism because it is homeostatic due to its life. Specifically, the hypothesis states that life regulates Earth's atmospheric composition, and makes it favorable to life, largely by negative feedback. They both provided evidence for this. They are responsible for this insightful idea. Margulis presented evidence that mitochondria and chloroplasts originated from bacteria that were ingested by prokaryotic cells, but this was never incorporated as part of the hypothesis. They rejected uninformed, pseudoscientific interpretations of the hypothesis that claimed Earth is literally an organism with consciousness. My ABH was partly inspired by the Gaia Hypothesis, and I owe a debt to Lovelock and Margulis. The Gaia and Autocatalytic Biodiversity Hypotheses are in agreement on the point that life created an atmosphere favorable to life.

It is important to note that the ABH is not the same as the Gaia Hypothesis. There are three major differences between them. The first is that the Gaia Hypothesis is a subset of the ABH. The ABH is more general and broad than the Gaia Hypothesis. The Gaia Hypothesis is limited to life's interaction with nonbiological aspects of the Earth system, focusing mainly on the atmosphere. It does address other factors than the atmosphere, such as oceanic salinity. The Gaia Hypothesis can be summarized as follows: Life made the Earth, mainly the atmosphere, more favorable to life, and this helped life thrive. It does recognize a coevolution between life and the nonbiological environment. It does not include interactions between species. Margulis did emphasize the importance of mutualism, especially with respect to the origin of the mitochondrion and chloroplast (see, for example, Margulis [4]), but mutualism was never incorporated into the Gaia Hypothesis. The Autocatalytic Hypothesis includes life's coevolution with the nonbiological environment, and expands on the idea. For example, one way the ABH expands on this idea is that it includes life's coevolution with and formation of the soil, which the Gaia Hypothesis does not include. But it also addresses interactions between species, such as predators and their prey, competition, and symbiosis; altruistic behavior between organisms of the same species; behavior of genomes; effects of viruses on the biosphere; and other diversity-enhancing actions of organisms.

Second, the only mechanism proposed by Gaia is negative feedback. The ABH accepts the importance of negative feedback in some cases, but proposes many other mechanisms that life employs to maintain and increase biodiversity, including the unifying concept that all species are ecosystem engineers. These mechanisms are summarized a few paragraphs below this one.

Finally, the Gaia Hypothesis does not propose that any variable in particular in an ecosystem is maximized by life. The ABH claims that life's activities maximize an important variable of the ecosystem: biodiversity. Thus, the ABH can be briefly stated as follows: Organisms create and increase biodiversity, all species are ecosystem engineers that maintain or increase biodiversity under natural conditions over sufficient time and ecosystems maximize biodiversity. I hope to demonstrate these claims in this book.

The ABH may also be called the Pachamama Hypothesis, following the lead established by Lovelock [3], who named his idea the Gaia Hypothesis after the Greek primal Mother Earth goddess. Ward's [5] *Medea Hypothesis*, named after the destructive Greek goddess, Medea, proposes that life is self-destructive, claiming that life has the opposite effect of the Gaia Hypothesis. Ward thus continued the idea of naming hypotheses about life's effect on life after Greek goddesses. Pachamama is the ever-present Earth and time mother goddess who has her own self-sufficient and creative power to sustain life on Earth. She is also the prime origin of her four cosmological Quechua principles—Water, Earth, Sun, and Moon. She is revered by indigenous peoples of the Andes, such as the Quechua and Aymara. Pachamama is usually translated as Mother Earth, but a more literal translation is World Mother in Quechua and Aymara. This departs from the use of Greek goddesses, but continues the use of goddesses that are appropriate, powerful metaphors for these hypotheses. And like Gaia, Pachamama is an Earth goddess. Thus, I consider the name appropriate to the ABH.

A summary this volume and the various mechanisms by which life creates and increases biodiversity follows.

Evolution by natural selection is one several mechanisms by the ABH works. Life has extraordinary abilities to adapt to environmental challenges, resulting in extreme and incredible adaptations. Mutualism is far more important and common than commonly accepted; it is fundamental; it organizes ecosystems; and it has increased biodiversity tremendously, both directly and indirectly. Mutualism results from the coevolution of two species. It often creates a new adaptive zone with a multitude of new niches. When it does so, the mutualists radiate into a tremendous number of new mutualist species. Other species then evolve to exploit the many niches provided by the new species of mutualists, further increasing diversity. Commensalism is common, and maintains and increases diversity. Ecosystem engineers greatly increase biodiversity by creating habitat and niches for many species commensal and mutualistic with them. Examples are beavers and termites. Competition does not significantly reduce species richness because competing species partition niches and evolve divergently. Plants are ecosystem engineers that sequester carbon; produce oxygen; make and improve soil; liberate soil nutrients; aerate the soil; and provide food and a three-dimensional habitat for many animals, other plants, fungi, and prokaryotes. Rivers would not exist without plants stabilizing their banks. Phytoplankton are the basis of oceanic food webs, major climate regulators, and produce most of the planet's oxygen. They often produce oxygen in great amounts by using nutrients provided by rainforests, the two ecosystems working together. Herbivores increase plant diversity, fertilize soil and marine and freshwater ecosystems, maintain mosaic habitats, and provide food to predators. They till and aerate the soil when they run from predators. Predators are mutualists with their prey if one considers prey at the levels of population and species. Predators regulate prey populations, increase prey diversity by consuming more of the prey that is the best competitor, prevent catastrophic trophic cascades, and select for novel adaptations in their prey. Parasites enhance the diversity of their hosts by similar mechanisms. Decomposers remove carcasses and feces, recycle nutrients, enrich soil, and can even reduce the amount of the powerful greenhouse gas methane (CH_4) in the atmosphere. This book proposes eight principles of ecology and evolution. The first states that in a natural ecosystem, situation, and conditions, over a time interval sufficiently long to accurately represent the effect of a species on its ecosystem, every species has a net positive effect on other species, its ecosystem, and biodiversity. The second states that all species are ecosystem engineers.

I am asking the reader to keep an open mind to my thesis that we need a fundamental paradigm shift. That is central to the ABH. The new paradigm states that biology is the major generator of biodiversity, and has an inherent tendency to do this. Ecosystems maximize biodiversity. Life is autocatalytic. Mutualism is a major source of ecosystem structure, a driver of high biodiversity, and a fundamental feature of evolution. All biological evolution is coevolution. No gene, organism, or species evolves in isolation. Genes, genomes, organisms, and species cannot be isolated from the network they evolve in. All of these entities evolve in an interconnected web. No species can be understood without understanding its connections, species it interacts with, and its environment. The relationships between species are as important in affecting ecosystems and evolution as the species themselves. Organisms do not exist as individuals, but are both chimeras and interconnected parts of systems. There is a tendency toward order when certain conditions are met that does not violate the Second Law of Thermodynamics.

This book attempts a comprehensive presentation of the evidence for the autocatalytic, coevolutionary, symbiotic generation of biodiversity by life called the ABH, drawing from a large number of disciplines. This presents the problem that the specialized scientist will be quite knowledgeable of one or perhaps two fields discussed in this book, while not being very familiar with the literature and terminology of the other fields. I have done my best to solve this dilemma with a compromise, whereby I use language less technical than in a scientific journal, but more rigorous than a popular science book for the educated layman. My hope is to make all chapters comprehensible to all readers, but not boring to the experts in the area being discussed at any given time.

I have resisted the temptation of setting up an easy-to-falsify straw man hypothesis that claims that the standard theory of evolution has no recognition of ideas similar to the ABH in the literature and that it opposes any suggestion that the actions of organisms can increase diversity. Also, I recognize that other authors have proposed similar ideas. Scientific integrity requires these two actions. There is the problem that many of the ideas in this volume are floating in the literature, and some may question the originality of some ideas in this book. I have done my best to credit previous authors and state when ideas presented are not novel, and when they are. I have been as diligent as I could about this. Sometimes this is obvious or implied.

Also, the reiteration of already existing concepts such as mutualism to support arguments being made could make my ideas appear unoriginal. It would be unfair and unscientific to judge the ABH unoriginal by taking it piecemeal; one must consider the whole picture, the entire gestalt. To describe the mutualism between angiosperms and their pollinators is not original; demonstrating how it fits contextually into a larger theory of how organisms generate diversity is.

I will state what I consider to be original in this book and concerning the ABH, to avoid any confusion on this issue. It is as follows.

First is the generality of the idea that organisms and life create biodiversity. The Gaia Hypothesis already pointed out that life regulates atmospheric gases, but the ABH adds many more mechanisms, including species interactions, sexual selection, and the behavior of DNA. This is the first compilation of the evidence supporting, logical arguments for, and mechanisms by which, life generates diversity and is auto-catalytic in one book. The idea has certainly not yet achieved consensus or proof. The central importance of life's effect on diversity and an attempt to focus people's attention on it is a novel emphasis. Darwin and Wallace showed environment molds life through selection; this book asserts that life also molds the environment in a coevolutionary feedback. Next is the synthesis of several mechanisms and pieces of evidence to the effect that life creates diversity. It is not original to say predators increase the diversity of their prey by eating the best competitor among their prey species. But it is original to state this in the context of a synthesis that results in a model stating that life generates higher diversity by a number of different mechanisms. As far as I am aware, I am the first to say that ecosystems maximize biodiversity. Mutualism and commensalism are elevated to a much more important position than commonly accepted, and they are major drivers and maintainers of diversity. Coevolution is more pervasive than recognized. For the first time, two unifying mechanisms are presented by which organisms create and increase diversity. Eight Principles of Ecology and Evolution are proposed for the first time as general phenomena. Concerning the idea that life generates diversity, the eight principles and the other ideas presented in this book, I concede that it is hard to fully and precisely judge their originality. The ideas are in the literature in various forms and are accepted to varying degrees by different researchers. A goal of this book is to make a large contribution by demonstrating the general validity of these ideas.

There are a few places in this book where a topic is discussed a second time in a different context, resulting in some repetition. This is unavoidable because sometimes the same phenomenon acts as an especially important example illustrating two different topics. However, this occurs only seldomly and is not problematic.

I capitalize all common names of species and subspecies, but not common names that do not refer to a specific species, or when I refer to a species without using its full name. For example, I call *Castor canadensis* the North American Beaver, but when I refer to this animal in general and no specific species of it, I use the lowercase and spell it beaver. I am consistent concerning this throughout the book.

REFERENCES

1. Lovelock, J. E. (1972). Gaia as seen through the atmosphere. *Atmospheric Environment* 6 (8): 579–80. doi: 10.1016/0004-6981(72)90076-5.
2. Lovelock, J. E. & Margulis, L. (1974). Atmospheric homeostasis by and for the biosphere: the Gaia hypothesis. *Tellus. Series A. Stockholm: International Meteorological Institute* 26 (1–2): 2–10. doi: 10.1111/j.2153-3490.1974.tb01946.x. ISSN 1600-0870.
3. Lovelock, J. (1979). *Gaia. A New Look at Life on Earth*. Oxford University Press: Oxford, London.
4. Margulis, L. (1998). *Symbiotic Planet*. Basic Books. A member of the Perseus Books Group: New York, NY.
5. Ward, P. (2009). *The Medea Hypothesis. Is Life on Earth Ultimately Self-Destructive?* Princeton University Press: Princeton, NJ.

2 Biodiversity
Organisms Create It, Ecosystems Maximize It

WHAT IS BIODIVERSITY?

Because my thesis claims that organisms increase biodiversity, it is important to precisely define biodiversity and state what I mean by it. It is the variability among living organisms, including diversity within species, between species, and of ecosystems. It is the information in a system. A greater variability of DNA represents more information. As a first approximation, biodiversity is species richness, which is defined as the number of species in a community. I am using community as it is used in ecology to mean a group or association of populations of two or more, but usually several, different species occupying the same geographical area at the same time. Enhancing this definition of biodiversity, it can be considered to be species diversity, which is both species richness and the evenness of distribution of organisms within different species in a community. Thus, a system with ten species with 100 organisms in each species has higher species diversity than one with ten species and 910 organisms in one species and ten organisms in the other nine species, even though each has the same number of species and organisms. Further, biodiversity is more than species diversity. It also includes variation within species and populations. This includes genetic variability between individuals in the same population or species, as well as between different populations, races, and subspecies of the same species. This includes number of subspecies, amount of polymorphism, and similar quantities. All of these measures include morphology, behavior, karyotype, and gene sequence—all aspects of the phenotype and genotype. Another measure is the number of species in a higher taxon, such as within a genus or family; number of genera within a family; families within a class; or any lower taxonomic level within a higher one. The term can also be used to refer to the relative number of species or other measures of diversity within different ecosystems. Biodiversity can be defined as the variation of life forms within a given ecosystem, within a given biome, or on the entire Earth. In these cases, species number could be considered, but so could numbers of higher taxa. For example, a coral reef is in a sense more diverse than a rainforest because it has at least one entire phylum that a rainforest lacks, Echinodermata. Biodiversity can also be viewed as the diversity or sum and nature of relationships among species (predator-prey, mutualism, etc.). The definition is determined by the questions being asked. All of these definitions are relevant to this book. Different aspects of biodiversity will be emphasized depending on the question I am discussing. Most of the time I will be using species richness and species diversity as my working definitions.

NUMBERS OF ORGANISMS AND BIODIVERSITY
WERE BOTH SPECTACULARLY HIGH BEFORE
HUMAN IMPACTS BECAME DOMINANT

Both population sizes and numbers of species were extremely high before the industrial revolution. There are reports of the sky being blackened by flocks of birds. Even the now-extinct Passenger Pigeons darkened the sky with their numbers. People reported salmon so dense that one could not see the stream bottom and could walk on them. Dasmann [1] documented the amazing abundance of Californian animals and plants. California Poppies once covered hillsides in continuous orange. The journalist Mackinnon [2] cited records from recent centuries that show the abundance of wildlife in those times:

> In the North Atlantic, a school of cod stalls a tall ship in mid-ocean; off Sydney, Australia, a ship's captain sails from noon until sunset through pods of sperm whales as far as the eye can see... Pacific pioneers complain to the authorities that splashing salmon threaten to swamp their canoes.

There were flocks of birds that took days to fly by in the sky, lions in the south of France, walruses at the mouth of Britain's Thames River, and 100 Blue Whales in the Southern Ocean for every one that's there today. The world's largest King Penguin colony shrank by 88% in 35 years, and more than 97% of the bluefin tuna that once lived in the ocean are gone, meaning these two species were recently many times more abundant than they are today.

A study found thriving wildlife at the site of the Chernobyl nuclear accident, including White-tailed Eagles, American Mink, and river otters, showing that wildlife becomes abundant and diverse when humans are excluded [3].

HOW MANY SPECIES ARE THERE?

It is impossible to say how "high" the biodiversity of the entire planet is without a relative scale for comparison, and none exists. We do not know what a "low" or "high" number of total global species is. However, we can estimate the global species number and thus get a feeling for how biodiverse the planet is.

Estimates of the present global macroscopic species diversity vary from 2 million to 100 million, likely representing the outer bounds of the total number of species, with a reasonable estimate of 9 million. However, one of the best estimates of the total global species number is that by Erwin [4], extrapolated from insects in a Panamanian tropical rainforest canopy collected by fogging with a biodegradable pesticide together with estimates of tropical plant host specificity. This indicated that one hectare of not very rich seasonal forest may have over 41,000 species of arthropods. Further extrapolation of available data based on known relative richness of insect orders and canopy richness led him to conclude that there could be as many as 30 million tropical, terrestrial arthropod species extant globally, far exceeding the usual estimate of 1.5 million at the time of the study (ibid). Even with an estimate that 75% of terrestrial arthropods are tropical, this calculates to a conservative estimate of 40 million terrestrial arthropods worldwide. Approximately 85% of all animal species are terrestrial or live in freshwater. Assuming the same percentage of arthropod species as all animal species live on land or in freshwater, this gives a global total of about 47 million arthropod species. Researchers have no consensus, but the Encyclopedia Britannica says that arthropods comprise about 84% of all species, which is a good rough estimate. This yields an estimated total of approximately 56 million animal species on Earth. Without a standard for comparison, we cannot say whether this is a large number, but it is far higher than was previously thought, and much larger than expected by most. If human effects lowered the species count in Erwin's study, the estimate for total species on the planet before human impacts would be even higher than 56 million. However, Gaston [5] asserts that, for insects, specialist knowledge of the taxonomic community at large has largely been ignored and a figure under 10 million species is more tenable, and one of about 5 million feasible.

May [6] stated estimates of the number of species (he seems to be referring to metazoans) range from 3 million to 30 million, and that the number is unknown. Storks [7] stated that most methods of answering how many species there are estimate global totals of 5–15 million. Mora et al. [8] did a study resulting in an estimate that there are about 8.7 million (±1.3 million standard error) eukaryotic species on Earth, of which about 2.2 million (±0.18 million standard error) are marine. Over 1.2 million species are cataloged in a central database (ibid). Their results suggest that 86% of the existing species on Earth and 91% of species in the sea are as yet undescribed (ibid). Quentin Wheeler, an entomologist who directs the International Institute for Species Exploration said: "Our best guess is that all species discovered since 1758 represent less than 20 percent of the kinds of plants and animals inhabiting planet Earth". He also said, "A reasonable estimate is that 10 million species remain to be described, named, and classified before the diversity and complexity of the biosphere is understood". Many discovered species have not been described yet. It is estimated that nearly 90% of all discovered arthropod species have not been described or classified. Only 1.8 million animal species have been given scientific names. Between 5 and 10,000 new species are found each year, most of them are insects. Many species have gone extinct, and many more will meet the same fate before being discovered.

For insects, approximate numbers of described species are 400,000 beetle, 170,000 lepidopteran, 120,000 dipteran, 110,000 hymenoptera, 82,000 hemiptera, 20,000 grasshopper, 5,000 dragonfly, and 2,000 praying mantis. It is estimated that only about 20% of insect species have been described; there are millions of undescribed insect species. Among arachnids, a square foot of soil 2 inches deep is estimated to contain 22 unique mite species.

A recent study estimated that there are about 18,000 bird species, twice the number previously thought to be the case [9].

The World Register of Marine Species records 230,000 deep sea species thus far [10]. But there are still millions more ocean species to be discovered.

Grassle and Maciolek [11] found that deep sea communities of 1,500–2,000 m depth off New Jersey contained up to 1,597 species in 171 families and 14 phyla. Species-area curves did not level off when sampling stations were added, indicating the number of species is higher, potentially much higher, than what they observed. They said species diversity there is high in both species richness and evenness.

For the plant kingdom, a rough estimate for tropical tree species number is 50,000 [12]. But Joppa et al. [13] showed that the current number of known Angiosperm species should grow by between 10% and 20%.

Locey and Lennon [14] combined a universal dominance scaling law with a lognormal model of biodiversity to predict that there are between 100 billion and upward of 1 trillion (10^{12}) microbial species on Earth, much more than ever anticipated. Since the genomes of only 100,000 microbial species have been sequenced and only 10,000 species have been grown in the lab, only a very small fraction of the total have been studied at all. Virus types (species) are at least ten times all cellular species combined, probably much more than this.

CRYPTIC SPECIES

The estimates of species numbers are almost surely far below the actual numbers. A major reason for this is cryptic species, which are species that appear identical but are actually distinct species. They potentially cause gross underestimates of species numbers. They have been shown in a number of taxa. An example is *Astraptes fulgerator*, a common neotropical skipper butterfly (Hesperiidae), which is in fact a complex of at least ten species [15]. DNA analysis showed the lichen, *Dictyonema glabratum*, found throughout the Americas and previously thought to be a single species, is at least 126 different species [16]. The researchers estimate it will consist of 452 when the research is finished. Many have strikingly different morphologies in nature, with habitats that range from rocks to shrubs to trees. Scientists have identified other groups of lichens that likely will show similar patterns of unrecognized species. Gomez et al. [17] showed that the marine bryozoan *Celleporella hyaline* is in reality over ten ecologically distinct species that had been diverging for several million years. Mitochondrial DNA of giraffes revealed at least 11 genetically distinct populations [18]. Knowlton [19] showed that cryptic species complexes are quite common in marine environments. Pfenninger and Schwenk [20] analyzed all known data on cryptic animal species and discovered that they are much more common than thought. Pfenninger said 30% of species could be cryptic.

Other examples of cryptic species groups include the *Brachionus plicatilis* rotifer complex (about ten species), the harvester ants *Pogonomyrmex barbatus* and *P. rugosus* (at least six lineages within the two morphospecies), Algerian Barb (several suspected), Asian glass catfishes (several known), the Lake Malawi cichlid genus *Placidochromis* (over 30 described in 2004), leopard frogs, *Rhacophorus* flying frogs (several suspected), legless lizards in California of genus *Anniella*, Audubon's shearwater (several suspected), common bush-tanager (several suspected), gray-cheeked thrush, and mouse lemurs (16 species).

LIFE ACHIEVED AMAZINGLY HIGH DIVERSITY
WITH VERY LITTLE SPACE AND NUTRIENTS

It is useful to bear in mind the small size of the biosphere, the thin layer of gases, soils, and liquids within which all life exists, to comprehend and visualize the magnificence and improbability of life's diversity. Imagine the planet, which is a mere

12,000 km in diameter, the size of a tennis ball. Then the thickness of the biosphere from the deepest abyss to the highest mountain amounts to no more than the thickness of a human hair (0.1 mm). The troposphere, the atmospheric layer where life and the weather are, varies between 17 km at the equator to 7 km at the poles. The deep biosphere extends to about 2.5 km, and the mean depth of the ocean is about 3.7 km. Using these values, and extending the top of the troposphere to 20 km and the deepest life to 5 km to answer any critics, the zone of life is only 25 km! Within this ultra-thin film, all life's processes occur, as well as the ocean currents and the weather, including the high jet streams. In preindustrial times, the amount of carbon dioxide (CO_2) in the atmosphere was only about 1.864×10^{12} metric tons, of which the carbon content was only about 5.08×10^{11} metric tons. It is incredible that photosynthesizers could power life in the area fueled by photosynthesis on such a small amount of carbon, and that life achieved such high diversity, complexity, and information content within such a small area.

NATURAL SELECTION FAVORS BIODIVERSITY

Leigh and Vermeij [21] pointed out that three types of evidence indicate that natural selection results in natural ecosystems being organized for high productivity and species diversity: (1) novel changes to natural ecosystems, such as human disturbances, tend to diminish their productivity and/or diversity; (2) humans must recreate the properties of natural ecosystems to enhance the productivity of artificial ones; and (3) productivity and diversity have increased during the Earth's history as a whole, and after every major extinction. They assert that natural selection results in ecosystems organized to maintain high species diversity and productivity. Energy that is poorly exploited attracts more efficient exploiters, favoring the creation of new niches and more efficient resource recycling. Ecological monopolies that reduce productivity are eliminated. Efficiency of predators and herbivores tends to increase, favoring faster turnover of resources. Lenton [22] addressed a concern that the Gaia Hypothesis was incompatible with the theory of natural selection by demonstrating that a model based on Daisyworld, which is a Gaia Hypothesis model, was strengthened by incorporating natural selection.

AFTER MASS EXTINCTIONS, SUBORDINATE TAXA OFTEN REPLACE DOMINANT ONES

Often a dominant taxon occupies most of an adaptive zone and most of the niches, and has more species, variety, and most or all of the large species, while a subordinate taxon with lower diversity occupies less niches and has generally smaller species. Yet the subordinate taxon persists and competition from the dominant one does not fully eliminate it. If a major catastrophe causes a vast majority or all of the species in the dominant taxon to go extinct, the subordinate taxon often survives and diversifies, and fills the niches the dominant taxon had occupied. This allows eventual recovery of diversity after the mass extinction. Without the subordinate taxon, a catastrophe that wiped out the dominant one would greatly decrease biodiversity. The subordinate taxon serves as a "redundancy" in the system that "safeguards" diversity.

The demise of the dominant taxon allows novel innovative, adaptive changes in the subordinate taxon as it diversifies and becomes the new dominant taxon, resulting in even higher biodiversity and complexity than before the catastrophe and mass extinction of the (formerly) dominant taxon. It is interesting that the formerly subordinate taxon emerges as more complex and diverse than the formerly dominant one it replaces. Specific examples will clarify this principle.

About two-thirds of marine animals were sessile and attached to the seafloor before the Permian-Triassic (P-Tr) extinction event. After recovery, half were sessile, and half were free-living. Sessile suspension feeders such as brachiopods and sea lilies decreased and were replaced by more complex, mobile species such as snails, sea urchins, and crabs, all of which increased [23]. Predation pressure selected for the increased mobility. Bivalves were rare before the event, but with reduced competition from the sessile organisms of the Permian, became common and diverse in the Triassic. One such group, rudist clams, became the Mesozoic's main reef builders, providing a habitat that greatly increased marine diversity. Simple and complex marine ecosystems were equally common before the event. After the recovery, complex ecosystems outnumbered the simple ones by nearly three to one (ibid). The ecosystems after the extinction were also more diverse. The increase in complexity was an increase in the information content of the system, as was the increase in diversity.

The late Devonian extinction of 358.9 mya led to the extinction of the ostracoderms, placoderms, and other fish, leaving a myriad of empty niches. Ray-finned fishes (Actinopterygians) appeared shortly after this at the start of the Carboniferous, about 360 mya [24]. Lacking competitors and equipped with novel adaptations, they radiated into the modern fish of today, the largest group of vertebrates, being about half of all vertebrates, with 30,000 species, including gars, salmon, and bowfin. They are almost 99% of all fish species today, and are ubiquitous in fresh and salt water from the deep sea to high mountain streams. They are more complex and diverse than their predecessors. Ray-finned fishes did not evolve until after most of their potential competitors were extinct, so were never a subordinate taxon, but still illustrate rapid diversification into empty niches with a lack of competitors after a mass extinction.

The P-Tr extinction killed off most Therapsids (mammal-like reptiles), which had been the dominant terrestrial vertebrates before the extinction. This allowed their subordinate competitors, the more complex Archosauria, the group containing dinosaurs and Pseudosuchians, which include crocodilians (the taxon containing today's crocodiles, caimans, and alligators) and their extinct relatives, to arise in the early Triassic period, about 245 mya. With many empty niches available and a lack of competitors, dinosaurs and Pseudosuchians underwent an adaptive radiation. The Archosauria became much more diverse than Therapsids. The P-Tr extinction also allowed the rise of the ancestors of today's lizards, frogs, and salamanders by eliminating their competitors and creating empty niches available to them. The modern forms of these are more complex and diverse than their pre-extinction competitors.

Brusatte et al. [25] studied the first 50 million years of dinosaur evolution, focusing on their ascent from a small, almost marginal group of reptiles in the Late Triassic to the preeminent terrestrial vertebrates of the Jurassic and Cretaceous, one of the most important evolutionary radiations in Earth's history. They showed that dinosaurs followed the pattern of being the subordinate group that diversified after the mass extinction of their dominant competitors at the end of the Triassic. The Archosauria branched into two main clades: Pseudosuchia and Avemetatarsalia, which includes birds, nonavian dinosaurs, and pterosaurs. The Pseudosuchia contained many taxa other than the crocodilians. The clade diversified more quickly than dinosaurs and became dominant, occupying the major niches of the time. Dinosaurs were a marginal group. Dinosaurs were losing to and being outcompeted by the Pseudosuchians until the Triassic-Jurassic extinction event, about 199.6 mya. This mass extinction happened in less than 10,000 years, and killed off all the Pseudosuchians except the crocodilians. It also eliminated the dinosauromorphs, competitors of dinosaurs that were quite similar to them. Shortly afterward, as a result of the lack of competition and unoccupied niches, dinosaurs diversified greatly. Some became extraordinarily large, and they became the dominant group, remaining so for 130 million years. The Avemetatarsalia were more complex and diverse than Pseudosuchians.

Competition with dinosaurs probably contributed to the evolution of the subordinate mammals by forcing the surviving therapsids and their mammaliform successors to live as small, mainly nocturnal insectivores. A nocturnal lifestyle likely caused selection on at least the mammaliforms to develop fur and higher metabolic rates [26] as well as endothermy and a keen sense of smell. So competition with dinosaurs provided the selective pressure to jump-start mammalian evolution in its early stages (ibid). But it also kept them small and nocturnal, denying them numerous niches.

The extinction of dinosaurs opened a novel adaptive zone with many new unoccupied niches that allowed mammals to diversify tremendously. Lyson et al. [27] collected fossils that documented the first million years of recovery after the Cretaceous-Paleogene (K-Pg) extinction in Corral Bluffs, Colorado. Ferns appeared quickly; they are pioneers that rapidly colonize disturbed habitats. In the first 1,000 years after the extinction, small, rat-like mammals lived among the ferns. Within about 100,000 years after the extinction, mammalian taxonomic richness doubled, and maximum mammalian body mass increased to near pre-extinction levels, which was the size of an average modern raccoon. Ferns built soil for palm trees, which appeared as the next stage of this evolutionary succession. By about 200,000 years after the extinction, mammals, benefiting from the palm trees and other new vegetation, had another increase in diversity, and the rat-like mammals obtained the size of beavers, eating nuts similar to walnuts. Palms built soil, and as a result were largely replaced by a more diverse community of trees, the next successional stage. Concomitant with diverse forests of large trees 300,000 years after the K-Pg event, mammals increased their maximum body mass threefold and underwent dietary niche specialization. The appearance of additional large mammals occurred by about 700,000 years after the event, coincident with the first appearance of legumes, which can fix nitrogen. Many mammals now ate seed pods to obtain protein. Seed pods of legumes provided a new, large source of protein that fueled mammalian evolution and allowed mammals to increase in size. The largest mammals now included the Eoconodon, which was the size of a large dog. Mammal populations were large. Forests had recovered fully. These coinciding plant and mammal originations and body-mass increases occurred during warming periods, suggesting that climate also influenced the biotic recovery. This elegant work shows how a subordinate group diversifies and becomes the dominant group after its dominant competitor is removed. It also shows evolutionary succession, the help of and coevolution with plants in animal recovery and diversification, and the role of climate in animal and plant recovery and diversification. The recovery period of 700,000 years is rapid. The researchers want to know if this timeline is normal or the exception.

Over a longer timespan, Smith et al. [28] showed that the demise of the dinosaurs paved the way for mammals to increase in size about a thousand-fold, demonstrating a near-exponential increase in the maximum size of mammals after the extinction of dinosaurs. When they coexisted with dinosaurs, mammals weighed a maximum of 10 kg, but after dinosaurs died off, the maximum weight of mammals was 170,000 kg. Within the geologically short time of 25 million years, ending about 40 mya, the system reset to a new maximum in body size and then the increase stopped. So evolution was rapid. There was remarkable congruence in the rate, trajectory, and upper limit of size across continents, orders, and trophic guilds. This study also showed that ecosystems can reset themselves relatively quickly. The demise of the dinosaurs led to mammals taking over the niches of large animals previously occupied by dinosaurs. This was a large diversification. And mammals are more complex than and at least as diverse as dinosaurs.

However, the story is more complex than evolutionary stagnation by mammals and then a sudden diversification after the extinction of the dinosaurs. Mammals actually started diversifying before the demise of the dinosaurs. Bininda-Emonds et al. [29] found that net per-lineage diversification rates barely changed immediately after the extinction of the dinosaurs. Instead, these rates spiked significantly with the origins of the currently recognized placental superorders and orders approximately 93 mya, well before the extinction of the dinosaurs, probably due to the appearance of the placenta. This key innovation put placental mammals in a new adaptive zone with many new open niches, leading to a big diversification. Mammals did not diversify at a very rapid rate right after the mass extinction cleared niche space, and their diversification rate remained low until accelerating again throughout the Eocene (56 to 33.9 mya) and Oligocene (33.9 to 23 mya) Epochs. So, the researchers found mammals did not start their rapid, tremendous diversification until 10–15 million years after the demise of the dinosaurs. A few mammal groups diversified right after the dinosaurs' extinction; most of these have since become extinct. Nevertheless, the extinction of the dinosaurs created unfilled niches for large organisms, and eventually mammals were able to fill these niches. So the principle of diversifying by filling open niches created by the extinction of competitors is illustrated. It is possible that in many cases the diversification of the subordinate taxon is more complex than a sudden radiation immediately after the extinction of the dominant taxon. Still, the general principle of reasonably rapid diversification after the extinction of a dominant competitor is valid.

Feng et al. [30] used a molecular dataset of unprecedented size to show that about 88% of living frogs, which comprise about 90% of all amphibian species, originated from three main lineages that arose coincident with the K-Pg extinction. Frog families and subfamilies containing arboreal species originated for the first time in all three lineages near or after this boundary. Many of the new frog species also started to lay their eggs on land, skipping the tadpole stage; about half of all frog species do this today. There are currently about 6,775 known frog species. The K-Pg extinction eliminated their competitors and opened up numerous niches that led to this diversification.

These more recently-evolved frogs are generally more complex than the ones that originated before them.

Archaic enantiornithine birds were dominant over the more modern neornithines, the ancestors of today's birds, until the K-Pg mass event, which eliminated the former entirely. The neornithine lineage then underwent one of the greatest radiations of all time; today, there are 10,000 bird species, all from neornithines, only surpassed in vertebrates by bony fishes. Neornithines are more complex and diverse than were enantiornithines.

The first major point of these examples of diversification of subordinate groups after their dominant competitors go extinct is that the new group radiates and fills niches rapidly, showing niches do not remain unfilled for long time periods, supporting the ABH. The second main point is that the new group is either about as diverse as or more diverse than the group it replaces. Thus, diversity stays the same or increases. The final point is that the new group and its ecosystem tend to be more complex than the group and the ecosystem it replaces. Complexity of a taxon is hard to define, but can be considered as having more DNA per species on the average, a more complex nervous system, more complex behavior, or more intelligence. A more complex taxon, like a more diverse community, has more information. So these replacements involve an increase in information, which can be viewed as a combination of diversity and complexity. This is interesting, for there is no easily seen reason why the taxon that replaces its dominant competitor should be more complex, be more diverse, and contain more information.

NICHES BECOME FILLED QUICKLY; THERE ARE NO EMPTY NICHES FOR LONG TIME PERIODS

Life tends to rapidly reach its maximum possible level of biodiversity. Ecosystems always quickly go to their "carrying capacity" for their number of species. Every empty niche becomes occupied in a short time; no niche remains empty for a long period of time. The best evidence for this is that natural ecosystems that are not disturbed by humans appear to be at carrying capacity with respect to the number of species in them. Other evidence is that diversity returns to at least its previous level quickly after mass extinctions, similar niches are filled by unrelated species, and metazoan diversity has increased since the Cambrian explosion.

Brayard et al. [31] found fossils showing a rapid recovery and diversification only about 1.3 million years after the P-Tr mass extinction event of about 251.9 mya, the largest mass extinction in Earth's history, which killed off 70% of land species. The fossils include top predators, primary producers, and scavengers, and bear characteristics of much later organisms. There was a relatively fast recovery of some amphibians, in spite of them nearly becoming extinct [32]. The fossils show a faunal transition into modern marine life, although the timing of a full transition remains unknown. The rapid diversification and recovery are consistent with the ABH, since it shows life's tendency to diversify. However, fully terrestrial vertebrates took an unusually long time to fully recover from the extinction. Benton [33] thinks it took 30 million years, not being complete until the Late Triassic, when dinosaurs, pterosaurs, crocodiles, archosaurs, amphibians, and mammaliforms were abundant and diverse. This is not fully consistent with the ABH.

Other evidence that empty niches are quickly filled is found in convergent evolution, whereby if a species that fills a specific niche is absent, an unrelated species evolves to fill that niche. For example, the Honey Possum (*Tarsipes rostratus*), a tiny marsupial that feeds on the nectar and pollen of a diverse range of angiosperms in southwest Australia, and an important pollinator of many such species, occupies the niche of hummingbirds, which do not occur in Australia.

In a broad sense, deer and kangaroos occupy similar niches, both consuming leaves, shrubs, grasses, fungi, and other items. There are a number of species of each and the comparison is general, with various specific kangaroo species each having more or less the same niche as a corresponding deer species. There are several marsupials that evolved by convergent evolution with placentals, filling in niches that placentals would normally occupy, but do not in Australia because placentals do not live there. There are marsupial equivalents of placental moles, mice, rats, rabbits (bandicoots), raccoons (wombats), flying squirrels, ground cats, and even a groundhog. In the past, but now extinct, there were marsupial wolves, lions, and even sabertoothed cats. This is clear evidence that if a taxon is absent from a large geographic area, another taxon will evolve to fill the niches it normally exploits by convergent evolution with this taxon. This is strong evidence that empty niches do not remain unfilled for long, supporting the idea of the tendency of organisms to generate diversity and thus the ABH.

Parrott et al. [34] found that native Australian Water Rats (*Hydromys chrysogaster*) have learned to flip over and eat the toxic Cane Toad (*Rhinella marina*), which was introduced by humans, avoiding its poison by removing the gall bladder, which contains toxic bile salts, with surgical precision, and eating the heart. They learned to avoid the toxic areas, eating only the nontoxic parts. They learned the technique by trial and error, or had previous experience from eating Australian native toxic frogs. The parents apparently teach their young how to attack the toads. They targeted the bigger toads; it was easier to avoid toxic organs in them. When they did eat medium-sized toads, they stripped off the toxic skin from the legs and ate the nontoxic thigh muscle. The Cane Toad is an invasive species whose populations have exploded. They have negatively impacted their prey, competitors, and predators tremendously. This shows that ecosystems in time control high-population problem species because they offer an abundant resource. The higher their numbers, the greater the resource and selection to adapt to exploit them. However, introduced species can greatly decrease diversity before this happens, and we should do everything possible to prevent introducing species to new areas. This also shows that when a novel niche (in this case, a species) is added to an ecosystem from the outside, it is in time exploited by a species that fills the open niche.

Except for three bat species, New Zealand has no native mammals. So mammalian niches have been filled by insects, gastropods, reptiles, and birds. The 70 species of weta, which resemble giant crickets, eat seeds and smaller invertebrates, playing the part that mice do almost everywhere else. Genus *Powelliphanta* are snails up to three and half inches across that forage earthworms, like moles, although they also overlap the niches of robins and some other birds. On some small islands, the lizard niche is occupied by the three species of tuatara, the only extant members of an ancient reptile order, Rhynchocephalia, which flourished 200 mya. The extinct

South Island Giant Moa was a 6-foot-6-inch flightless bird, which could reach 12 feet up into trees with its neck to eat leaves, occupying a similar niche to elephants and giraffes, which it underwent convergent evolution with.

The kiwi (genus *Apteryx*), a bird with five extant species, is the most interesting animal to have evolved into a mammalian niche due to lack of mammals in New Zealand. Like mammals, but not birds, they have a highly developed sense of smell. They are flightless, have tiny wings, have no keel on the sternum to anchor wing muscles, have hair-like feathers, and nostrils at the end of their long beaks. Adults birds typically have bones with hollow insides to minimize weight for flight, but adult kiwis have bone marrow, like mammals and the young of other birds. Their sight is so poorly developed that blind specimens have been observed in nature; birds generally rely on sight more than olfaction, and mammals generally are the opposite. They prefer subtropical and temperate podocarp and beech forests, but deforestation has forced them to adapt to other niches and habitats, including subalpine scrub, tussock grassland, and the mountains. This shows that species can adapt to novel niches and habitats under strong selection pressure. Their nocturnal habits may be yet another niche shift from habitat intrusion by introduced predators. Where such predators have been removed, they are often seen in daytime.

Unpublished anecdotal observations show two species of day gecko (genus *Phelsuma*) introduced to the Hawaiian Islands have adapted to eating introduced insect species there. This exploitation of an unfilled niche has aided the ecosystem by controlling exotic species that negatively impact ecosystem diversity.

An extinct crocodile-like genus (*Simosuchus*) evolved to fill a vegetarian niche on Madagascar.

The third piece of evidence that empty niches are quickly filled is that life rapidly diversified exponentially after the Cambrian explosion and after each of the five mass extinctions in Earth's history. This is discussed in detail later in this chapter under the section on patterns of diversity, because the unique pattern of the history of life's diversity provides yet further evidence for the ABH.

KEY INNOVATIONS LEAD TO MAJOR ADAPTIVE RADIATIONS

Every time a species undergoes a macroevolutionary breakthrough with a key inno-vation, it enters a novel adaptive zone, and this is followed by an adaptive radiation creating many new species that fill the many new niches made available by the key innovation. Then many additional species use the niches created by the species that evolved as a result of the radiation, further increasing diversity. Examples are the evolu-tion of chloroplasts, mitochondria, flowers, fruits and nuts, pharyngeal jaws in cichlid fish, adaptations associated with invasion of the land by invertebrates and vertebrates, the amniotic egg in reptiles, flight in birds, and the placenta in placental mammals.

QUASI-STABLE STATES SUPPORT THE ABH

Some biological systems exist in quasi-stable states. "Quasi-stable state" has meanings in other fields than ecology that differ somewhat from what I mean here. Quasi means "almost". By a quasi-stable state, I mean a state that is mostly stable and resistant to perturbation, but can be thrown into a different quasi-stable state by a phase shift if sufficient disruption, usually from outside of the system, is applied to it. A threshold or tipping point is crossed when the phase shift occurs. It is of note that the natural state found in nature is in all cases I am aware of the one of higher diversity and complexity, with the more complex species dominant, the less complex one subordinate. Thus, there is higher information and order in the natural state. Humans have inadvertently run experiments on these systems, driving them to less complex, less diverse, lower information states, providing strong empirical evidence for the ABH, by demonstrating the natural state is the more diverse, complex, and information-rich state.

For example, the natural state of coral reefs is the quasi-stable, high-diversity, high-information state of abundant coral and sparse macroalgae. Humans have added an excess of dissolved inorganic nutrients to some coral reefs, causing an increase in macroalgae. Though this contributes to reef degradation, it alone does not tend to cause a phase shift by allowing algal overgrowth of coral. If humans also cause the herbivorous reef fish that eat algae to be artificially low by overfishing, the two human-caused environmental disruptions allow macroalgae to grow over and cover the coral, smothering it. This causes a phase shift to an ecosystem of abundant macroalgae and sparse coral, with a resulting tremendous decrease in diversity of species of coral, fish, and invertebrates—a system of low diversity, complexity, and information [35].

Another coral reef example concerns the Crown of Thorns Sea Star (*Acanthaster planci*). In the natural situation of high-coral and low-sea-star abundance, the sea star eats, among other prey, the fast-growing corals, such as staghorn and stony coral species. This prevents the fast-growing, better competitors from outcompeting slower-growing coral species, maintaining coral reef diversity. The normal high-coral, low sea-star quasi-stable state is maintained by coral that eat the sea star larvae, and other predators of the sea star. The sea star is kept at population levels optimal to maintain a healthy, high-diversity reef that also protects other ecosystems.

Human activity that is destructive to coral reefs is one set of factors that weakens the reef and reduces coral predation on the small adult sea stars, causing a phase shift to an abundance of sea stars and low numbers of coral. Fertilizers from agriculture run into the ocean, causing phytoplankton populations to increase, which allows the populations of their zooplankton predators to increase. Crown of Thorns larvae are among these zooplankton. The increased numbers of zooplankton also provide alternative food for predators of the larvae of the Crown of Thorns. Humans often kill off the natural predators of the sea star's larvae, which are its most vulnerable stage. These include filters feeders, such as coral; shrimp, especially the Harlequin Shrimp (*Hymenocera picta*); worms; and some reef fish. Overfishing and destructive impacts on the reef negatively impact these predators. Humans sometimes offer alternative food sources to the sea star's predators. For example, snorkelers feed bread to certain species of triggerfish that eat the sea star's eggs, so they consume fewer eggs. Most importantly, the Pacific Triton, a beautiful, large marine snail, is a major predator of the full-grown adult Crown of Thorns. Excessive collecting of its shell has devastated its populations in various areas at various times. This is true in most areas of the Pacific, where it is endangered or locally extinct.

Since some species of coral eat the Crown of Thorns, it can increase its population size when coral are less abundant. A large single coral polyp of the genus

Pseudocorynactis was observed wholly ingesting a Crown of Thorns of like size. Normally the coral polyp eats the larvae of the Crown of Thorns as part of the zooplankton they prey on. They can eat a good number of the larvae. And removing its predators, especially the triton, is also important in the sea star's population explosion. And its population increases enormously when the conditions listed in the previous paragraph occur. Females produce 60 million eggs in one season. Under normal circumstances, the vast majority are eaten as eggs, larvae, or young sea stars. But when predators are lacking, this egg production can help cause rapid population explosions. It preys on other organisms, but it is a voracious predator of coral polyps. Damaged reefs attract it more, as the sea star and its swimming larvae are drawn to the metabolites released by damaged corals.

The sea star's preferred prey are the stony coral species, genus *Acropora*. But when Crown of Thorns becomes overpopulated, many of them become desperate for food, and will consume coral species that they normally would not prey on, including slow-growing ones. One Crown of Thorns can eat a volume of coral up to its size in a day, or up to $6\,m^2$ per year. Millions upon millions of them can be present when their population explodes. In some cases, they can cause the seafloor to decrease from 25% to 45% covered in coral to 1% in only 2–3 seasons. The coral help maintain the high coral state by feeding on sea star larvae, while sea stars help maintain the high sea star state by feeding on coral. The more abundant form in each of the two quasi-stable states maintains the state in which it is dominant, by its predation on the subordinate one. Thus, in both states, the more abundant form aids itself.

Human-induced climate change causes extended droughts followed by high rains. This causes low salinity and high sediment and nutrients at the Great Barrier Reef. Cutting of forest also causes high nutrient input to this reef. Low salinity increases sea star larval survival, and high nutrients increase microscopic algae, which are food for the larvae. Thus, these conditions also induce a high-sea-star, low-coral.

The Crown of Thorns contributes to coral diversity when the normal high-coral, low-sea-star quasi-stable state exists, and destroys coral reefs when the unnatural quasi-stable state of abundant Crown of Thorns occurs. As the sea star's population explodes, a new quasi-stable state with abundant Crown of Thorns Sea Stars and low numbers of many coral species sets in. This shift is also promoted because the sea star may help with transmission of some coral diseases. This phase shift is among the most illustrative of the ABH, since when the unnatural state takes hold, the Crown of Thorns can decimate coral reefs, with a loss of biodiversity far beyond most other phase shifts.

The natural, high-coral state is not only highly diverse, it also helps maintain high diversity in ecosystems beyond the coral reef. The profound negative effects on biodiversity from coral loss, which are not limited to coral reefs, are as follows. Coral reefs are the most diverse ecosystems in the sea, being home to one-fourth of all known species of marine fishes and perhaps about 1 million species of animals. In the high-sea-star state, the sea star greatly reduces coral diversity. Essentially all reef fish and invertebrates cannot exist without their coral habitat, so the great diversity of fish and invertebrates on the reef falls precipitously. To make a bad situation worse, the digestive juices of the Crown of Thorns are very toxic and contain saponin, a chemical that harms marine organisms. When their populations are as low as they are in the natural state, this is not a problem. But at abnormally high population levels, there is a potential for poisoning of large predators that attempt to eat them. After the Crown of Thorns devastates the reef, the dead coral skeletons become slowly covered with a gray-colored alga. The dead reefs are desolate and colorless in contrast to the beautiful, vibrant reefs full of bright colors from both corals and fish. Coral reefs protect an estimated 155 of the world's beaches and coastlines from storms and erosion by reducing the action of ocean waves. These habitats get destroyed if coral reefs are badly damaged or eliminated. Coral removes CO_2 from the atmosphere as part of the carbon cycle, stabilizing Earth's climate, aiding biodiversity. This also keeps the level of CO_2 in the ocean lower, so acidity is controlled, making the environment better for life. Coral reefs aid mangrove and seagrass ecosystems and the many species they harbor. These influences on biodiversity of coral reefs and their destruction are discussed in more detail elsewhere in Chapter 4.

There is a quasi-stable state in the tropics between tropical rainforest and savanna (see [36]). Rainforests have the highest biodiversity of any terrestrial ecosystem, with about half of all land species. Though savanna has moderately high diversity, it is less diverse than rainforest. The normal state is a patchwork of the two ecosystems that is higher diversity than pure rainforest, since this provides two different ecosystems, each with its own set of species. Higher rainfall and less fire favor rainforest. Rainforest keeps the quasi-stable state of high rainforest intact because the trees undergo evapotranspiration, which adds moisture and creates rain; this forest-generated rainfall limits forest fires. Forests typically have 80% tree cover, and savanna 20%. Intermediate states (for example, 40% trees) are very rare, and this is evidence that indicates they are unstable. When humans cut a rainforest and remove its trees beyond a threshold, there are too few trees to generate sufficient rain to aid tree growth and suppress fire. In the dryer condition, grass takes over and a phase shift to a savanna occurs. Because of the exceptional biodiversity of rainforests, the loss of species is tremendous. It is thought, but needs to be tested and proven, that a similar situation holds for temperate forests and prairies.

In the natural situation, the Nile Crocodile (*Crocodylus niloticus*) is more abundant, and the Nile Monitor (*Varanus niloticus*) is less abundant. The male Nile Crocodile averages 3.5–5 m and is 30% larger than the female, while the Nile Monitor averages 2.7 m. Although complexity is hard to measure, the best interpretation might be to consider the crocodile the more complex species due to its larger size. The crocodile certainly increases diversity and hence information of the system more than the monitor, by controlling prey, eating the more abundant prey, and fertilizing the water with its feces. The monitor increases diversity by the first two mechanisms, but less so, and not by the third. The crocodile preys on the monitor, while the monitor eats the crocodile's eggs. With higher crocodile populations, predation by the crocodile is more intense, and the high-crocodile state is stable and maintained. However, when human hunting reduced crocodile numbers beyond a threshold, crocodile predation on monitors was reduced, monitor populations increased, and monitor predation on crocodile eggs escalated. The result was a phase shift to a state with high Nile Monitor and low Nile Crocodile numbers, and a resulting decrease in diversity and complexity of the ecosystem.

In the normal state, the Dungeness Crab is in high numbers compared to its parasite. Of course, the parasite has a higher population size because parasites outnumber their hosts, but the relative number of infected crabs is normally low. Commercial crabbers are legally obligated to harvest only large crabs. Large adult crabs prey on smaller crabs. Before the legal requirement to spare smaller crabs, humans overharvested them. As a result, very few small crabs reached adult size, and there were very few adult crabs to eat small ones. The generation that was not preyed on by adult crabs matured to an overpopulation of adult crabs. They ate so many small crabs that the crab population declined precipitously. Furthermore, crab fishermen keep healthy crabs, and throw the ones with parasites back into the sea, selecting for high rates of parasitism. These two factors caused a phase shift to a new quasi-stable state of high parasite and low crab populations. This caused the ecosystem to decrease in diversity, because the crabs increase biodiversity by providing their predators with food and controlling prey populations. Also, the quasi-stable state of high crab and low parasite numbers is more complex, since the crab is a more complex organism than its parasite. The crab maintained the high crab state by regulating its numbers, ultimately keeping them relatively high, by cannibalism.

In all cases, the abundant species maintained the quasi-stable state that it was dominant in. This was true in both the more complex and less complex states. The more complex and diverse state is the natural state, and is maintained by the abundant species in that state. Thus, this is an example of organisms generating and maintaining high biodiversity—the ABH.

LIFE'S EVOLUTIONARY HISTORY HAS TRENDED IN THE DIRECTION OF BIODIVERSITY, NOT COMPLEXITY

There is a view held by some scientists as well as many laymen that there was a force in evolution that drove life inevitably to increasing complexity from the beginning of life until recent times. Gould [37] refuted this. There appears to be an increase in complexity through geologic time only if the most complex taxon at any given time is considered. For example, the most complex group in the Silurian period, 443.8–419.2 mya, was fish with jaws; in the Devonian, 419.2–358.9 mya, it was amphibians; and so on. But more forms reduced complexity than increased it in each transition from one geologic period to the next, largely because of the appearance of new parasites and viruses.

Throughout the entire history of life, the dominant cellular life forms (as opposed to viruses) have been prokaryotes. This holds for virtually all measures of dominance, whether one considers number of organisms, number of species, range of habitats occupied, importance to ecosystems, the range of chemical reactions the taxon can perform, and even biomass. Bacteria have been found as much as four miles beneath Earth's surface [38]. This and other findings have increased estimates of their biomass to levels exceeding all surface animals and plants. Throughout the entire history of life, the major cellular forms have been bacteria, and every age has been the age of bacteria; there has never been in reality an age of reptiles or of mammals.

Cellular life had to start in the simplest state because of the laws of physics and chemistry, for chemical evolution must produce bacteria, and cannot produce giraffes. Thus, cellular forms could not evolve to be less complex than the first prokaryotes;

the only niche space available to evolve into entailed more complexity. As life diversified, more complex life forms appeared because they filled empty niches, not as a result of a driving force for more complex forms. Complexity increased as a secondary effect of diversification and filling of empty niches. The niches for simple forms like bacteria had already been filled, but niches for more complex forms like eukaryotes were available. The many factors discussed in this book caused constant increases in biodiversity of the Earth, and that increase through time is the major trend that life displayed throughout its history, with increased complexity of some forms as a secondary effect, resulting in increased complexity of the most complex forms. This is not to say that complexity was never selected for; it was often favored. Complexity can increase size, intelligence, behavioral plasticity, and other traits that are often adaptive.

Figure 2.1 shows a graph of complexity plotted against its frequency of occurrence. It shows a left wall of minimum complexity that cellular life had to start at and cannot go beyond. There is only one direction available for expanding into, and that is toward more complexity. Figure 2.1a shows the situation at the beginning of the Precambrian, when there were only prokaryotes. Figure 2.1b shows the present day distribution, showing that life's complexity becomes increasingly right-skewed through time, but that prokaryotes remain the dominant form. Now imagine evolution moves randomly toward more complexity or less complexity, but cellular forms can never evolve past the left wall of minimum allowable complexity. When this wall is hit, life must make one move in the direction of more complexity, then can

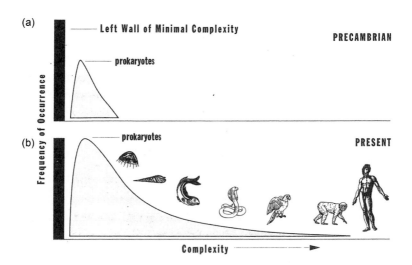

FIGURE 2.1 The frequency distribution of cellular life's complexity becomes increasingly right-skewed through time as the complexity of the most complex form increases. But prokaryotes never alter in complexity, and remain the dominant, most diverse form. Cellular life must start at the left wall of minimum possible complexity for cellular life, so evolves on the average toward more complexity because that is the area where empty niches are. (a) Frequency distribution of life's complexity in the Precambrian. (b) Frequency distribution of life's complexity in the present time. (Modified from Gould [37].)

continue moving randomly. This random walk alone, without the aid of selection, will result in an increase in complexity of the most complex forms. Further, consider life forms that are already far enough from the left wall that they have sufficient niche space to evolve to be simpler as well as more complex. McShea [39,40] tested if there was a bias in evolution toward complexity by looking at the vertebral column of a number of vertebrates (which are of course far from the left wall of minimum complexity of cellular life), and found that there is no tendency in evolution toward more complex forms in groups far from the left wall.

Most evolution is in fact toward less complexity because of the many times that parasites have evolved. Parasites that evolved from nonparasitic species are always or almost always less complex than their ancestors because they no longer have to provide for some functions provided by their hosts. It is no doubt fair to surmise that every nonparasitic species has more than one cellular parasitic species. So the vast majority of evolution is from the more complex to the simpler. And, as in the case of the more complex cellular forms evolving into unfilled niches, the viruses evolved into empty niches and occupied them. This also greatly increased biodiversity, for there are manyfold more types (species) of viruses than all cellular species combined. Essentially all cases of the evolution of new virus types involved a decrease in complexity. The decrease in complexity that resulted from the evolution of non-viral parasites likewise increased diversity, although much less than the evolution of viruses did. We can thus conclude that the primary variable that on the average over time increased throughout life's history is biodiversity, not complexity.

FOUR PATTERNS OF BIODIVERSITY STRONGLY SUPPORT THE ABH

The first of the four patterns of biodiversity supporting the ABH is that a graph of the number of species in any given taxon on the y-axis against the size of the individual organisms in a species on the x-axis yields an exponential curve of increasing species number as the size decreases. There are many more species with organisms of small size than large. There are at least ten species of virus for each species of cellular organism. Over 99% of all cellular species are prokaryotes. There are two reasons for this pattern. First, size of a species is generally correlated with its generation time. In general, the smaller the species, the shorter the generation time. Species with shorter generation times adapt more quickly to environmental challenges, and diversify more rapidly. So they have lower extinction rates and higher speciation rates. This is biology creating diversity. It supports the ABH because natural selection is a mechanism by which the ABH operates, as explained in Chapter 3. The second reason for the negative correlation of size to diversity is there are more niches available to smaller species. All else being equal, a given area provides more niches to a smaller species than a larger one. The relevance of this to the ABH is that smaller species use larger species as habitat and niches, living as parasites, commensals, and mutualists, and this is tremendously common.

The second pattern is that the number of species of almost all taxa, with few exceptions, increases at a rate that is greater than linear from the poles to the equator, for both land and sea habitats. There are many hypotheses to explain this. If physical and chemical factors alone accounted for this increase in diversity, one might expect a linear relationship. That the increase is greater than linear may indicate that life plays a role in increasing diversity by making the environment better for life and because the existence of species themselves provides niches for other species. This is far from proof. There could be other factors that account for it than life being autocatalytic and aiding life. However, it is suggestive of the validity of the ABH. It needs further exploration.

Third, the ocean occupies about three-quarters of Earth's surface; provides a vertical environment and a huge volume; and life as well as higher life started in the sea, giving diversification a longer time to occur there. Yet it is estimated that about 80% of species live on land, 15% in the sea, and 5% in freshwater, and this is not an artifact of sampling [41]. This is at least partially due to the effects of biology [41,42].

Vermeij and Grosberg [43] cite three reasons the land has more diversity than the sea: productivity, habitat heterogeneity, and differences in viscosity and density of the medium. As a general rule, high diversity is correlated with high productivity, although there are exceptions to this. Terrestrial environments tend to be far more productive and cover a larger area relative to marine ones. The most productive marine environments are confined to hydrothermal vents and shallow-water coastal zones. Much of the pelagic zone and the deep sea are relatively unproductive. On the other hand, terrestrial forests and grasslands tend to have high productivity and cover large areas. They are geologically relatively new. Fossil leaves strongly indicate that 100 mya, there was a dramatic increase in terrestrial productivity, when angiosperms, whose photosynthetic capacities far exceeded that of their predecessors, came to dominate terrestrial vegetation (ibid). Thus the productivity, though influenced by physics and chemistry, was enhanced by life.

Second, the pelagic zone is inhabited by comparatively few macroscopic species because it is one of the more homogeneous environments. It lacks variation because there is no substrate for sessile organisms such as coral and seagrass to provide heterogeneous habitats. The shallow seafloor provides environmental heterogeneity because of the corals, seaweeds, seagrasses, mussels, oysters, and others. But this is a small area compared to terrestrial habitats, which are generally far more physically complex, largely because of habitat provided by plants, both above and below ground. And the Earth transitioned from predominantly marine to terrestrial life when early angiosperms evolved to be extraordinarily successful on land 100 mya (ibid), indicating habitats and other ecosystem services provided by plants helped land pass the sea in diversity. Environmental heterogeneity on land is provided by the myriad forms of plants, and the actions of animals and other life forms.

Water is more viscous and dense than air. This allows smaller animals to move greater distances at higher speeds on land and in air than in water (ibid). This permits animal-mediated pollination in the air, a mutualism that caused great diversification of pollinators and plants. Animal-mediated gamete transfer and its associated coevolution, mutualism, and diversification are rare in the sea. Greater ease of movement, faster diffusion of chemical signals, and more effective transmission of visual signals in air mean that species attracting or choosing among mates at a distance can maintain populations at much lower densities than in the ocean. Visual and chemical signals involved in mate location and recognition can function over greater distances in the air than in water, though this is not true with sound. Thus, terrestrial environments are on the whole more conducive to speciation by sexual selection and to maintenance of species. Over the last 100 million years, factors permitting rarity and inducing life-generated diversification have allowed diversity to rise faster on land than in the sea, especially among such very diverse clades as flowering plants, fungi, and insects. Physical factors played a role in all three of these explanations, especially the third. But biological factors are crucially important in all of these explanations of why the land has more species than the sea.

The final pattern is the changes in diversity through geologic time. The bones and shells of skeletonized marine animals have left the most complete fossil record there is. Sepkoski [44] plotted the number of families of skeletonized marine animals throughout the Phanerozoic, from 542 mya to present. His famous graph is shown in Figure 2.2a. It shows a rapid increase in family number from the beginning of the Phanerozoic until a plateau is reached, and then family number remains approximately constant until the first mass extinction, the end-Ordovician extinction of about 443 mya. The first two mass extinctions decrease diversity significantly and precipitously, but after both of these, family number returns to approximately the same plateau. Then the P-Tr extinction, the largest in Earth's history, occurred 252 mya. Up to 96% of all marine and 70% of terrestrial vertebrate species went extinct. It is the only known mass extinction of insects. About 57% of all biological families and 83% of all genera went extinct. Terrestrial diversity took significantly longer to recover than after any other mass extinction event, possibly up to 10 million years. But then diversity rose past the old plateau, and kept rising steadily almost fourfold until human effects became prevalent, interrupted only by the two other mass extinctions. It recovered and continued to rise after each of these two mass extinctions. Family number rose at a very rapid rate and went way beyond the pre-Triassic plateaus. The fossil record indicates there were just over 900 marine animal families at the last measurement point. There are about 1,900 families alive today, including those rarely or never preserved as fossils. So the family number increased greatly even since the last measurement point.

Figure 2.2b shows a chart by Foote [45] of the number of genera of skeletonized marine mammals through time. It shows a similar pattern to Figure 2.2a, with more fluctuation before the post-Permian rise. That rise gets steeper after the K-Pg mass extinction. Diversity increased steadily on land after the P-Tr mass extinction as well, especially after the mid-Cretaceous [46].

Figure 2.3 shows the numbers of skeletonized marine animal genera through geologic time, corrected for biases such as temporal differences in rock volume and the fact that more recent fossils are better preserved. The smooth curve is a running average from the late Ordovician to the mid-Cenozoic. Significantly, this graph shows the steady, steep increase in marine diversity from the P-Tr mass extinction until recent times holds, even when one corrects for these biases. It indicates that, aside from mass extinctions and recoveries, biodiversity had a stronger increasing trend after this extinction than in the Paleozoic.

Figure 2.4 shows the number of insect families, vascular land plant species, and nonmarine tetrapod vertebrate families from their origins until recent times. Insects

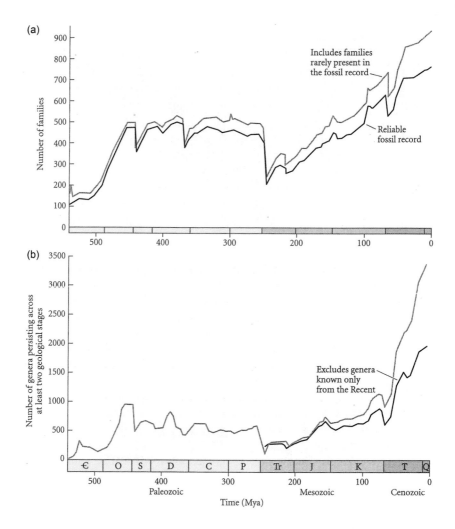

FIGURE 2.2 Taxonomic diversity of skeletonized marine animals from the beginning of the Phanerozoic to pre-industrial times. The number of taxa for each geological stage includes all those whose known temporal extent includes that stage. (a) Diversity of families. Black curve represents only those families with a reliable fossil record. Blue curve includes families rarely preserved. Black and blue curves concur well, so any bias from including or excluding rare forms is not important. (From Sepkoski [44].) (b) Diversity of genera of these animals, counting only those that crossed boundaries between two or more stages. Black curve excludes genera known only from the Recent, to avoid bias created because older fossils do not preserve until the present as readily. The fairly good agreement of the two curves shows this bias is not very important. (From Foote [45].)

and plants show the basic increase and plateaus, and exponential increase without a plateau since the P-Tr extinction, giving further evidence this pattern holds for a wide variety of taxa. Nonmarine tetrapods do not show an exponential increase until after the K-Pg extinction, but, interestingly, the increase is exceptionally steep after

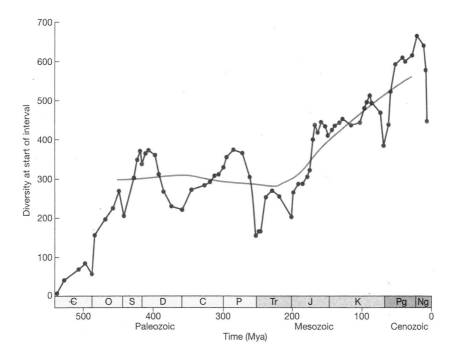

FIGURE 2.3 Numbers of skeletonized marine animal genera through geologic time, corrected for biases such as temporal differences in rock volume and the fact that more recent fossils are better preserved. The smooth curve is a running average from the late Ordovician to the mid-Cenozoic. The graph shows a steady increase in diversity from the P-Tr extinction until recent times even when one corrects for biases. In fact, the increasing trend is even stronger after this extinction than before it. (From Foote [50].)

this extinction. And it appears that nonmarine tetrapods have not even come close to filling all the niche space available to them on the planet. This does not mean niches remain unfilled for long time periods, because the increase and filling of niches since the K-Pg extinction is rapid and exponential. And the general increase in terrestrial diversity since the P-Tr extinction would have created many new niches that nonmarine tetrapods could have diversified into and filled. The basic pattern of an exponential increase in diversity without a plateau after the P-Tr extinction until the present holds in all three of these groups, though the nonmarine tetrapod increase is delayed compared to the others.

McGuire et al. [48] showed in a thorough analysis that hummingbird diversification began about 22 mya, was rapid, occurred by heterogeneous clade-specific processes, and involved radiation into a diverse assemblage of specialized nectarivores comprising 338 species. Above all, although diversification is slowing in all taxa, several major clades are still diversifying rapidly on a par with classical examples of rapid adaptive radiation. The researchers estimated that, if undisturbed, hummingbirds would radiate until they reached a speciation/extinction equilibrium of perhaps as many as 767 species, twice the number of species currently extant. This is a major

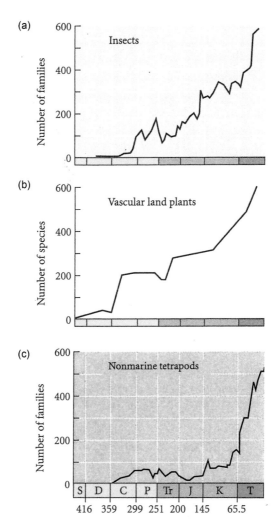

FIGURE 2.4   Since their origination, changes in the number of known: (a) insect families. (From Labandeira and Sepkoski [47].) (b) Vascular land plant species. (From Benton [46].) (c) Nonmarine tetrapod vertebrate families. (From Benton [46].)

diversity increase for the last 22 million years with no plateau, showing the same basic pattern as the other groups just discussed, though starting much later.

There are several possible sampling errors in graphs of diversity through time. These include rare species which are detected more easily in large than small samples, geologic stages vary in duration, more recent times have more preserved fossils, and fossils preserve more readily in some time periods than others regardless of how recent the period. The most relevant sampling bias to this discussion is that more recent fossils have less time to be destroyed and so are better preserved. While some researchers believe modern biodiversity is about equal to that of 300 mya when corrected for sampling errors, others consider the fossil record to be a fairly accurate representation of diversity's history. Some, but not all, of the sampling errors are corrected for in Figure 2.2. Figure 2.3 shows what I consider the most important corrections to minimize sampling errors for skeletonized marine genera [49,50]. It shows a less steady increase since the P-Tr extinction, but its diversity increase since this mass extinction is still apparent. Thus, the graphs that are corrected for sampling errors indicate the continued exponential increase in diversity since the end-Permian extinction is real.

The figures all show that life diversified exponentially after the beginning of higher life at the Cambrian explosion and after each mass extinction. Exponential growth of species (or other taxa) implies that changes in diversity are guided by a first-order positive feedback, whereby more species (taxa) create more descendant species (taxa). There may also be second-order positive feedback whereby species number increases with increasing complexity of community structure. This complexity is created by organisms. An example is trees of a rainforest providing a three-dimensional habitat for animals, plants, and fungi. Both first- and second-order positive feedbacks with respect to species increases are examples of the ABH.

But the most important observation, which seems to hold true even when one corrects for sampling errors, is as follows. After the Cambrian explosion, diversity increased exponentially, reached a plateau, and stayed there until it declined during the first mass extinction, the Ordovician–Silurian Extinction about 439 mya. Then it increased exponentially again, recovering fully, staying more or less constant at a second plateau that had about the same diversity as the first, until the second mass extinction, the late Devonian extinction of about 369 mya. Then there is a third exponential increase that then levels off at a third plateau at about the same diversity as the first two followed by a third mass extinction, the largest in life's history, the P-Tr extinction of about 252 mya, followed by an exponential increase in diversity. But this time there was a continuous exponential increase in diversity, temporarily interrupted twice by the final two mass extinctions, but never leveling off. There was no plateau. Except for interruptions by the two mass extinctions, biodiversity on Earth increased steadily and exponentially from the P-Tr extinction until the time of human impacts. Why did diversity plateau after the first two mass extinctions, but increase exponentially without ever leveling off after the third one? One would expect it to plateau after all mass extinctions. We do not know the answer, and research is needed to answer this important, fascinating question. However, we have some good, plausible explanations.

There is no evidence for a revolutionary change in the fundamental genetics of organisms as a result of the P-Tr extinction. Sexual recombination had evolved by the time of the event, and it is likely transposable elements, polyploidy, and all other mechanisms of genetic change were in place at that time. The continents started to separate after the extinction, and continued to drift apart to their present-day positions ever since. This created large isolated areas both on land and in the sea for speciation, and is a nonbiological mechanism. But once the continents had separated a sufficient distance, further distance between them would not increase speciation. A new carrying capacity for the number of species should be reached, where diversity should remain more or less constant. Thus, diversity should have leveled off long ago if continental drift were the only driving force.

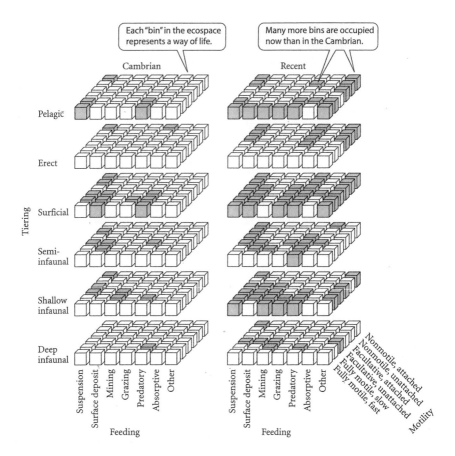

FIGURE 2.5 Employment of ecological space by marine animals in the Cambrian compared with the Recent (present time). Each layer represents the vertical space used by the animal groups, from pelagic to deep infaunal (deep in the sediment). Cubes from left to right within each layer represent different feeding niches, and those from front to rear represent different motilities. Green indicates that the modes of life are used; white indicates that they are not used. Far more modes of life are employed now than in the Cambrian. (From Bush and Bambach [51].)

In the case of skeletonized marine organisms, an increase in use of ecological space provides a partial explanation. Figure 2.5 shows that the variety of different feeding niches, habitat types, and motility in marine animals is significantly greater in the present time than it was in the Cambrian. This would have been accomplished by active colonization of new habitats, selection, and use of resources and habitat created by other species. The origination of new forms and key adaptations, allowing employment of more modes of life, would have played a role. This may have kept increasing at a more or less constant rate since the Cambrian, possibly accounting for some of the observed diversity increase since the P-Tr extinction in the sea. Part of the explanation is that when new forms arise, they provide niches and/or habitat for several other species; this is true both on land and in the sea.

This gives a plausible explanation of why diversity increased exponentially since the P-Tr extinction, but it does not explain why diversity plateaued after the first two mass extinctions, but not the third. In my view, the best explanations for that involve diversifications after the P-Tr extinction. Gymnosperms became dominant and diversified after the extinction, providing three-dimensional habitat for animals. Gymnosperms are also the ancestors of angiosperms, so aided diversity via their evolutionary breakthrough in giving rise to flowering plants. The ancestors of angiosperms diverged from gymnosperms in the Triassic, about 245 to 202 mya, and the first angiosperms appeared about 140 mya. They diversified extensively until they were widespread by 120 mya, and have continued to diversify since, and are now 64 orders, 416 families, about 13,000 known genera, and about 300,000 known species. They provided three-dimensional habitat to animals, and created and enriched soil. Angiosperm ancestors coevolved with insects until angiosperms appeared, then insects and angiosperms greatly diversified. Then angiosperms coevolved with other pollinators, and the angiosperms and these pollinators radiated tremendously as a result. Then other species were able to evolve to exploit the numerous niches created by the flowering plants and their pollinators. Insects aided diversification of numerous animal taxa by being pollinators, seed dispersers, decomposers, aerators of the soil, food sources, regulators of populations by being herbivores and disease vectors, and other mechanisms. The radiation of angiosperms and insects, as well as other factors, helped birds and mammals diversify tremendously. Birds and mammals then provided many novel niches to many other species, allowing a large further increase in the number of species and some taxa above the species level.

Angiosperms released nutrients to the sea and aided the four major groups of phytoplankton of today (coccolithophores, dinoflagellates, diatoms, and foraminiferans), helping them replace less complex forms. This added complexity to the sea. These four phytoplankton taxa greatly diversified. This helped diversification of all levels of the marine food webs, from zooplankton to large fish, sharks, and whales. Angiosperm pollen also went into the sea, providing nutrients to organisms there. A good deal of pollen ends up in the sea, including on the seafloor, as food. Pollen was found in a deep-sea trench 35,000 feet below the surface of the Pacific Ocean. These contributions of nutrients to the sea by flowering plants only account for some of the steady increase in marine diversity since the end of the Permian.

Grasslands released silica to freshwater systems, and this increased nonmarine diatom diversity in the Miocene [52]. Miocene volcanism also likely contributed directly to diatoms' thriving as well as to grassland's success and release of silica to freshwater ecosystems.

Cazzolla Gatti [53] suggested that species themselves generate higher biodiversity by creating niches for other species and thus increasing the niches available in a given ecosystem, in his Biodiversity-related Niches Differentiation Theory (BNDT). Cazzolla Gatti et al. [54] argued that ecosystems can be viewed as an emergent autocatalytic set in which one group of species enables the existence of other species by creating niches for them. An autocatalytic set is a collection of entities, each of which can be created catalytically by other entities within the set, allowing the set to catalyze the increase of the number of its entities (species). They showed that biodiversity can be considered a system of autocatalytic sets. They argued that such a view explains why so many species can coexist in the same ecosystem.

There is no question that species provide niches for other species, providing sources of food or habitat. All multicellular organisms provide habitat for a diverse microbiome. Trees and other organisms provide habitat for other species. All hosts provide a habitat for their parasites. All prey provide niches for their predators. Provision of niches by species for other species helps explain why life is autocatalytic and provides both evidence for and a mechanism for the ABH.

We only understand a small part of the entire explanation for the strange history of Earth's diversity at the time of this writing.

However, the null hypothesis that organisms do not play a major role in generating higher diversity predicts that after every mass extinction, diversity will return to the same plateau and remain there until the next mass extinction. Diversity should reach the carrying capacity for diversity of the Earth. This could change a small amount over time, but one would not expect it to change very much. Certainly the null hypothesis does not predict diversity to continue exponential growth for 252 million years, until recent times. Earth has not steadily increased in size; there is no mechanism for continued diversity increase with no leveling off to a plateau of diversity in the null hypothesis. And most importantly, what we know about angiosperms, insects, sea life, and other organisms strongly indicates that biology played a major role in the steady increase in biodiversity from the end of the Permian to the present. Therefore, and what is known of its mechanism strongly supports the ABH.

REFERENCES

1. Dasmann, R. F. (June, 1965). *The Destruction of California*. Macmillan Pub. Co.: New York, NY.
2. Mackinnon, J. B. (2013). *The Once and Future World: Nature As It Was, As It Is, As It Could Be*. Houghton, Mifflin, Harcourt: Boston, MA, New York. ISBN 978-0-544-10305-4.
3. Schlichting, P. E., et al. (2019). Efficiency and composition of vertebrate scavengers at the land-water interface in the Chernobyl Exclusion Zone. *Food Webs* 18: e00107. doi: 10.1016/j.fooweb.2018.e00107.
4. Erwin, T. L. (1982). Tropical forests: their richness in Coleoptera and other arthropod species. *Coleopterists Bulletin* 36: 74–5.
5. Gaston, K. J. (1991). The magnitude of global insect species richness. *Conservation Biology* 5: 183–96.
6. May, R. M. (1992). How many species inhabit the Earth? *Scientific American* 10: 18–24.
7. Storks, N. (1993). How many species are there? *Biodiversity and Conservation* 2: 215–32.
8. Mora, C., et al. (23 Aug., 2011). How many species are there on Earth and in the ocean? *PLoS Biology* 9 (8): e1001127. doi:10.1371/journal.pbio.1001127.
9. Barrowclough, G. F., et al. (23 Nov., 2016). How many kinds of birds are there and why does it matter? *PLOS ONE*. doi: 10.1371/journal.pone.0166307.
10. Appeltans, W., et al. (2011). The World Register of Marine Species: an authoritative, open-access web-resource for all marine species. In: *The Future of the 21st Century Ocean: Marine Sciences and European Research Infrastructures. An International Symposium*, 28 June–1 July, 2011, Brest, France.
11. Grassle, J. F. & Maciolek, N. L. (Feb., 1992). Deep-sea species richness: regional and local diversity estimates from quantitative bottom samples. *The American Naturalist* 139 (2): 313–41.
12. Howard, R. (1982). via R. Eyde, pers. comm. to Erwin, in Erwin. *Coleopterists Bulletin* 36: 74–5.

13. Joppa, L., et al. (2010). How many species of flowering plants are there? *Proceedings of the Royal Society B*. doi: 10.1098/rspb.2010.1004.
14. Locey, K. & Lennon, J. (2016). Scaling laws predict global microbial diversity. *PNAS USA*. doi: 10.1073/pnas.1521291113.
15. Hebert, P. N. D., et al. (12 Oct., 2004). Ten species in one: DNA barcoding reveals cryptic species in the neotropical skipper butterfly *Astraptes fulgerator*. *PNAS USA* 101 (41): 14812–7. doi: 10.1073/pnas.0406166101.
16. Lücking, R., et al. (2014). A single macrolichen constitutes hundreds of unrecognized species. *PNAS USA*. doi: 10.1073/pnas.1403517111.
17. Gómez, A., et al. (22 Jan., 2007). Mating trials validate the use of DNA barcoding to reveal cryptic speciation of a marine bryozoan taxon. *Proceedings of the Royal Society B* 274 (1607): 199–207. doi: 10.1098/rspb.2006.3718. ISSN 0962-8452.
18. Brown, D., et al. (2007). Extensive population genetic structure in the giraffe. *BMC Biology* 5 (1): 57. doi: 10.1186/1741-7007-5-57. ISSN 1741-7007.
19. Knowlton, N. (Feb., 2000). Molecular genetic analyses of species boundaries in the sea. *Hydrobiologia* 420: 73–90. doi: 10.1023/A:1003933603879. ISSN 0018-8158.
20. Pfenninger, M. & Schwenk, K. (18 July, 2007). Cryptic animal species are homogeneously distributed among taxa and biogeographical regions. *BMC Evolutionary Biology* 7, Article number 121. doi: 10.1186/1471-2148-7-121.
21. Leigh, E. G. Jr, & Vermeij, G. J. (29 May, 2002). Does natural selection organize ecosystems for the maintenance of high productivity and diversity? *Philosophical Transactions of the Royal Society B* 357 (1421). doi: 10.1098/rstb.2001.0990. Theme issue: The Biosphere as a Complex Adaptive System.
22. Lenton, T. (1998). Gaia and natural selection. *Nature* 394 (6692): 439–47. Bibcode: 1998Natur.394..439L. doi: 10.1038/28792. PMID 9697767.
23. Wagner, P. J., et al. (2006). Abundance distributions imply elevated complexity of post-Paleozoic marine ecosystems. *Science* 314 (5803): 1289–92. Bibcode: 2006Sci...314.1289W. doi: 10.1126/science.1133795. PMID 17124319.
24. Giles, S., et al. (2017). Early members of 'living fossil' lineage imply later origin of modern ray-finned fishes. *Nature* 549: 265–8.
25. Brusatte, S. L., et al. (July, 2010). The origin and early radiation of dinosaurs. *Earth-Science Reviews* 101 (1–2): 68–100. doi: 10.1016/j.earscirev.2010.04.001.
26. Hallam, A. & Wignall, P. B. (1997). *Mass Extinctions and Their Aftermath*. Oxford University Press: Oxford, UK.
27. Lyson, T. R., et al. (22 Nov., 2019). Exceptional continental record of biotic recovery after the Cretaceous–Paleogene mass extinction. *Science* 366 (6468): 977–83. doi: 10.1126/science.aay2268.
28. Smith, F. A., et al. (26 Nov., 2010). The evolution of maximum body size of terrestrial mammals. *Science* 330 (6008): 1216–9. doi: 10.1126/science.1194830.
29. Bininda-Emonds, O. R. P., et al. (29 March, 2007). The delayed rise of present-day mammals. *Nature* 446: 507–12. doi: 10.1038/nature05634.
30. Feng, Y.-J., et al. (18 July, 2017). Phylogenomics reveals rapid, simultaneous diversification of three major clades of Gondwanan frogs at the Cretaceous–Paleogene boundary. *PNAS USA* 114 (29): E5864–70. doi: 10.1073/pnas.1704632114.
31. Brayard, A. (15 Feb., 2017). Unexpected Early Triassic marine ecosystem and the rise of the Modern evolutionary fauna. *Science Advances* 3 (2): e1602159. doi: 10.1126/sciadv.1602159.
32. Yates, A. M. & Warren, A. A. (2000). The phylogeny of the 'higher' temnospondyls (Vertebrata: Choanata) and its implications for the monophyly and origins of the Stereospondyli. *Zoological Journ. of the Linnean Society* 128 (1): 77–121. doi: 10.1111/j.1096–3642.2000.tb00650.x.
33. Benton, M. J. (2005). *When Life Nearly Died: The Greatest Mass Extinction of All Time*. Thames & Hudson: London, UK. ISBN 978-0-500-28573-2.

34. Parrott M. L., et al. (23 Sept., 2019). Eat your heart out: choice and handling of novel toxic prey by predatory water rats. *Australian Mammal*. doi: 10.1071/AM19016.

35. McCook, L. J. (1999). Macroalgae, nutrients and phase shifts on coral reefs: scientific issues and management consequences for the Great Barrier Reef. *Coral Reefs* 18 (4): 357–67.

36. Witze, A. (Nov. 5, 2011). Story one: rain tips balance between forest and savanna: amount of tree cover can shift suddenly and abruptly. *Science News* 5–6. doi: 10.1002/scin.5591801003.

37. Gould, S. J. (1996). *Full House. The Spread of Excellence from Plato to Darwin*. Chapters 12–14, pp. 135–216. Harmony Books, New York, NY.

38. Szewzyk, R., et al. (1994). Thermophilic, anaerobic bacteria isolated from a deep borehole in granite in Sweden. *PNAS USA* 91: 1810–13.

39. McShea, D. W. (1993). Evolutionary change in the morphological complexity of the mammalian vertebral column. *Evolution* 47: 730–40.

40. McShea, D. W. (1994). Mechanisms of large-scale evolutionary trends. *Evolution* 48: 1747–63.

41. Vermeij, G. J. & Grosberg, G. (6 Nov., 2012). Biodiversity in water and on land. *Current Biology* 22 (21): R900–3. doi: 10.1016/j.cub.2012.09.050.

42. May. R. M., et al. (29 Jan., 1994). Biological diversity: differences between land and sea. *Philosophical Transactions of the Royal Society B* 343 (1303). doi: 10.1098/rstb.1994.0014.

43. Vermeij, G. J. & Grosberg, R. K. (2010). The great divergence: when did diversity on land exceed that in the sea? *Integrative and Comparative Biology* 50: 675–82.

44. Sepkoski, J. J., Jr. (1984). A kinetic model of Phanerozoic taxonomic diversity. III. Post-Paleozoic families and mass extinctions. *Paleobiology* 10: 246–67.

45. Foote, M. (2000). Origination and extinction components of diversity: general problems. In Erwin, D.H. & Wing, S. L. (eds.), *Deep Time: Paleobiology's Perspective*, pp. 74–102. Paleobiology 26 (S4), supplement. doi: 10.1017/S0094837300026890. Published online by Cambridge University Press, 26 Feb., 2019.

46. Benton, M. J. (1990). The causes of the diversification of life. In Taylor, P. D. & Larwood, G. P. (eds.), *Major Evolutionary Radiations*, pp. 409–30. Clarendon Press, Oxford, UK.

47. Labandeira, C. C. & Sepkoski, J. J., Jr. (1993). Insect diversity in the fossil record. *Science* 261 (5119): 310–5. Bibcode: 1993Sci...261..310L. doi: 10.1126/science.11536548. PMID 11536548.

48. McGuire, J. A., et al. (5 May, 2014). Molecular phylogenetics and the diversification of hummingbirds. *Current Biology* 24 (9): 1038. doi: 10.1016/j.cub.2014.03.016.

49. Alroy, J., et al. (2008). Phanerozoic trends in global diversity of marine invertebrates. *Science* 321: 97–100.

50. Foote, M. (2010). The geological history of biodiversity. In Bell, M. A., et al. (eds.), *Evolution Since Darwin: The First 150 Years*, pp. 479–510. Sinauer, Sunderland, MA.

51. Bush, A. M. & Bambach, R. K. (2011). Paleoecologic megatrends in marine metazoa. *Annual Review of Earth and Planetary Sciences* 39: 241–69.

52. Kidder, D. L. & Gierlowski-Kordesch, E. H. (April, 2005). Impact of grassland radiation on the nonmarine silica cycle and Miocene diatomite. *PALAIOS* 20 (2): 198–206. doi: 10.2110/palo.2003.p03–108.

53. Cazzolla Gatti, R. (2011). Evolution is a cooperative process: the biodiversity-related niches differentiation theory (BNDT) can explain why. *Theoretical Biology Forum* 104 (1): 35–43.

54. Cazzolla Gatti, R., et al. (2017). Biodiversity is autocatalytic. *Ecological Modelling* 346: 70–6.

3 Natural Selection Is One Mechanism by Which the Autocatalytic Biodiversity Hypothesis Operates

The theory of evolution by natural selection of Darwin and Wallace is one of the mechanisms by which the ABH works. The ability of organisms to adapt to environmental challenges allows them to persist and thus maintains diversity. It also allows species to have macroevolutionary breakthroughs to new adaptive zones, such as the colonization of land by marine vertebrates. This gives access to many new niches and so is followed by adaptive radiation to many new species and forms. Natural selection is also involved in speciation, increasing species number. Adaptation by organisms is a mechanism by which life maintains and increases diversity. The ABH is supported by the fact that speciation is mostly actively driven by organisms, which do not speciate as entirely passive agents of selection.

EXTRAORDINARY ABILITY OF ORGANISMS TO ADAPT BY NATURAL SELECTION

The power of natural selection and life's incredible capacity to respond to selective forces, adapt to extreme environments, and evolve innovative, unusual, and amazing adaptations are phenomenal. This maintains diversity, and increases it when it leads to speciation, and when it leads to macroevolutionary breakthroughs followed by adaptive radiation. Some examples of life's extraordinary ability to adapt, as well as spectacular adaptations, follow.

The archaea *Pyrolobus fumarii* and *Pyrococcus furiosus* live in submarine hydrothermal vents at temperatures higher than boiling water, at 110°C–121°C. Archaea in genus *Pyrococcus* in the Mariana Trench survive in extraordinarily high pressures that would kill people, at 1,100 bars. The cyanobacterium *Synechococcus lividus* lives in freezing ice habitats at −17°C to −20°C. Some bacteria in the genera *Psychrobacter*, *Vibrio*, and *Arthrobacter*, and some archaea in genus *Natronobacterium* live in extremely basic soda lakes with a pH over 11, which would easily damage human skin. Bacterium *Clostridium paradoxum* and some bacteria in genus *Bacillus* thrive in volcanic springs and acid mine drainages with acid strong enough to eat away human flesh, at a pH below 0.1. The Green microalga *Dunaliella salina* grows in highly saline conditions, such as salt evaporation ponds. Bacterium *Deinococcus radiodurans*, some bacteria in genus *Rubrobacter*, and archaeon *Thermococcus gammatolerans* can survive extreme ionizing radiation up to 1,500–6,000 Gy, in the form of cosmic rays, X-rays, or radioactive decay, and UV radiation up to 5,000 J/m^2. Cyanobacterium *Chroococcidiopsis* thrives in extreme conditions of high and low temperatures, ionizing radiation, and high salinity. Other microbes can withstand very high salt concentrations of 0.2 M NaCl, and others can tolerate dissolved heavy metals, including copper, cadmium, arsenic, and zinc, in such high levels that they are toxic to most organisms. The bacterium *Halomonas titanicae* lives at the sea bottom, and sticks to steel surfaces such as ships, and eats them, obtaining iron. Microbial populations that survived in ice for 750,000 years in the Qinghan-Tibetan plateau were subjected to ionizing radiation and kept at 5°F in a laboratory for 2 years, and shown to have fully functional DNA repair mechanisms in freezing conditions. A mix of archaea, bacteria, and fungi in Chile's Atacama Desert survived in several places by going dormant, surviving dessication, and high salinity and UV light. Species of these at various sampling sites grew and thrived when rain came in March 2005 [1]. *Ideonella sakaiensis* is a bacterium that dwells in soil and wastewater, and has evolved the ability to eat a strong, human-made plastic fiber as its main food source [2].

There is a highly diverse subterranean biosphere of prokaryotes that is almost twice the size of Earth's oceans [3], with almost no nutrition, total darkness, extreme heat, and intense pressure. Called the deep biosphere, it has been explored by drilling far down into the Earth, deep in South African mines, and below sediments in the middle of oceanic gyres. Those studying it estimate it to be equivalent in diversity to the Amazon rainforest, and that 70% of prokaryotes exist underground! Underworld biospheres vary immensely depending on geology and geography. Their influence on biogeochemical cycles is unknown.

The deep biosphere contains between 15 and 23 billion tons of microbes, hundreds of times the combined weight of every human on the planet. Their combined size is estimated to be more than 2 billion cubic kilometers, and this could be expanded further in the future. Life was found 5 km underground and well underneath the sea-floor. Researchers found a temperature maximum of 122°C thus far, but believe this record will be broken with further exploration.

A methanogen found 2.5 km below the Earth's surface had been buried for millions of years, possibly as a single organism, without reproducing or dividing, but only repairing and replacing its constituent parts. It likely creates CH_4 in an environment that has almost no energy. Organisms such as this, far beneath the surface, are not metabolically active, but in stasis, using less energy than previously thought possible to sustain life. Life in deep subterranean timescales is different, with some microbes living thousands of years without actively moving. Researchers are attempting to find a lower limit beyond which life cannot exist, but continue to find life as they dig deeper.

Algae, fungi, and amoebae occur about 656 feet beneath the Earth's surface in the US. Fungi exist approximately 1,476 feet below the Earth's surface in Sweden. And our planet's deepest-dwelling metazoa are *Halicephalobus mephisto* and its other parthenogenic nematode relatives, which were found in gold mines in South Africa to be at a minimum of 1,476 feet, and up to 2.2 miles below the surface, where the only food source is prokaryotic, the temperature is extremely high, O_2 levels are very low, and sulfur and other dissolved chemicals are high. They are up to 0.02 inches long.

Thus, most of the Earth's life is under the surface in extreme environments and much of it is metabolically barely active at all, and these extreme environments harbor extraordinarily diverse life.

Permafrost represents 26% of terrestrial soil ecosystems. It is exposed to subzero temperatures and high radiation for geological timescales in a habitat with very low water activity and rates of nutrient transfer. Yet viable bacteria were found in deep permafrost core samples in northeast Siberia that were dormant there for up to about 3 million years [4]. Arctic soils and the underlying permafrost in northern Canada and Alaska have yielded bacteria (*Bacillus* spp., *Azotobacter* spp., sulfate reducers, and some thermophiles), yeasts, fungi, and protozoa (ibid and references therein). High numbers and biodiversity of bacteria exist in permafrost [5]. Some of these may be among the oldest life on Earth. Research suggests that functional microbial ecosystems exist within the permafrost environment and may have important implications on global biogeochemical processes as well as the search for past or extant life in permafrost that is presumably present on Mars and other bodies in our solar system (ibid). Legendre et al. [6] showed the virion *Pithovirus sibericum* had survived 30,000 years frozen in permafrost and could infect an amoeba.

Fruits of *Silene stenophylla* Ledeb. (Caryophyllaceae) excavated in northeastern Siberia from fossil squirrel burrows buried at a depth of 38 m in never-thawed Late Pleistocene permafrost sediments with a temperature of −7°C, about 31,800 years old, produced plants that flowered, fruited, and set viable seeds [7].

Two soil nematode species were revived from permafrost in the Arctic after being frozen there for 30,000–40,000 years [8]. These are the first metazoans surviving long-term cryobiosis in Arctic permafrost.

Subglacial Lake Whillans lies beneath about 800 m of ice in West Antarctica. It has a chemosynthetically driven ecosystem inhabited by a diverse assemblage of bacteria and archaea that can influence Southern Ocean geochemical and biological systems [9].

The deep sea is one of the Earth's most challenging environments. Scientists expected little life there. It is under crushing pressures typically from 200 to 600 atm, but can reach 1,000 atm. The deeper the fish live, the more gelatinous their flesh and more minimal their skeletal structure, as an adaptation to extreme pressure. They have no excess cavities that would collapse under the pressure, such as swim bladders. Temperatures range from 4°C to −1°C. There are fish with antifreeze proteins. It has depths below which no light penetrates that are perpetually completely dark. The sea's mean depth is 4,300 m, so the photic zone is but a tiny fraction of the ocean's total volume. Where there is a little sunlight, many animals have more or less transparent bodies for cryptic coloration. Many have bioluminescence. Ecosystems in the deep sea are almost entirely reliant on sinking organic matter. Only about 1% to 3% of the production from the surface reaches the deep sea, so nutrients are sparse. Oxygen is also low. So animals have slow metabolism, require less oxygen per unit time, and can live long times without food. Many save energy by waiting and ambushing prey, instead of moving and hunting it. In spite of these environmental challenges, the deep sea is remarkably diverse. Though the deep sea is not as diverse as a prairie, it is much more diverse than scientists expected. Actual numbers are not yet known, since new species are being discovered so rapidly. As described in

Chapter 2, Grassle and Maciolek [10] found remarkably high diversity in both species richness and evenness in the deep sea. They found 798 species in 171 families and 14 phyla in 233 30 cm×30 cm samples, and this went up to 1,597 species when more sampling stations were added. They found diversity is maintained by a combination of biogenic microhabitat heterogeneity in a system with few barriers to dispersal, disturbance created by feeding activities of larger animals, and food resources divided into patches of a few square meters or less. That is, diversity is maintained by the organisms.

Some animals have adapted to the deep sea by becoming very large. Whereas "typical" isopods, which are not in the deep sea, are up to 5 cm, giant isopods of the deep sea are generally between 17 and 50 cm. The giant squid (genus *Architeuthis*) gets 13 m for females and 10 m for males from the posterior fins to the tip of the two long tentacles. The Gant Oarfish (*Regalecus glesne*) is the longest bony fish alive, growing up to 11 m.

There are also cases of deep-sea creatures being abnormally tiny; the lantern shark fits in an adult human's palm.

Water temperatures at hydrothermal vents on the ocean floor range from 60°C to 464°C [11]. Pressures are an astonishing 250 atm, and it is pitch black. Yet, they harbor complex ecosystems whose primary producers do not use photosynthesis, but chemosynthesis, using heat from the vents instead of sunlight as an energy source, and using the oxidation of compounds such as hydrogen gas, hydrogen sulfide, or CH_4 to convert one or more carbon-containing molecules, usually CO_2 or CH_4, and nutrients, into more complex organic compounds. These producers include diverse groups of extremophile bacteria and archaea. They grow into thick mats grazed on by copepods, amphipods, and other invertebrates. There is a complex food web with large populations of mussels, cockles, oysters, sea anemones, shrimps, crabs, tube worms, and fish. Higher levels of the food web include gastropods, octopuses, and fish. Notable fish include eelpout, cutthroat eel, ray-finned fish of order Ophidiiformes, and a tonguefish (*Symphurus thermophiles*). Diversity is much higher than the rest of the deep sea, with, for example, over 100 gastropod species found by 1993 [12]. Many vents have high numbers of endemic species, and vent species tend not to be found in other habitat types. There are many mutualistic associations between the chemosynthesizers and multicellular animals (see Chapter 4). Allopatric speciation has produced different species at different vents, and there are cases of convergent evolution between species at different vents.

Unusual adaptations abound. A scaly-foot gastropod, *Crysomallon squamiferum*, in Indian Ocean vents, uses the iron sulfides pyrite and greigite, instead of calcium carbonate, for its dermal sclerites. The ultrahigh pressure may help stabilize iron sulfide for biological purposes, and the armor plating likely serves as a defense against the venomous radula of predatory snails there. The first species to exclusively use light other than sunlight for photosynthesis, a bacterium in family Chlorobiaceae, was found off Mexico's coast, using the faint glow from black smokers for photosynthesis [13].

Eight mya, grasslands spread across the globe. They thrive in the wet season and catch fire in the dry season, spreading forest fires, which kill their tree competitors. Grasses recover quickly from fires, while trees generally do not. Savannas maintain themselves this way. Savannah fires are the most frequent in the world, with two fires per year common in some high-rainfall African savannahs. Some trees have adapted to these high-fire environments by growing underground in both the savannahs of southern Africa and South America. Almost none of these trees originated before the savannahs spread 8 mya, supporting the idea that the selective agent was grasslands. Some have branched networks of stems measuring up to 10 m across underground. Their aboveground shoots are so small they can easily recover from fire. At least 30 different types of trees evolved this strategy independently in Africa, and dozens in the Cerrado, the giant savannah that covers more than a fifth of Brazil. The trees are diverse; 200 underground tree species occur in Africa. This is an example of both creative adaptation in nature, and one group of organisms greatly affecting the evolution of another, causing the unusual adaptation.

Danovaro et al. [14] found an animal in phylum Loricifera, and Yahalomi [15] found an animal, a cnidarian parasite of salmon, both of which live in wholly anoxic conditions, lack mitochondria, and do not carry out aerobic respiration. Tardigrades (phylum Tardigrada) are examples of animals with exceptional adaptations to extreme environments [16]. They are usually about 0.5 mm (0.02 inches) long, have about 1,150 known species, and date to 530 mya, which is during the Cambrian period. They can survive conditions that would kill almost all life. They can live without food or water for over 30 years, desiccating to 3% or less water, then rehydrating and even reproducing. Some can survive extreme cold temperatures close to absolute zero (1 K, −272°C, −458°F), some extreme heat up to 150°C (300°F), some the extremely low pressure of a vacuum, some very high pressures of 6,000 atmospheres (almost six times the water pressure in the deepest ocean trench, the Mariana Trench), and some 1,000 times higher ionizing radiation doses than other animals. They have an uncanny ability to repair DNA from this damage. They could likely survive until Earth is engulfed by the sun in several billion years. They live in the deep sea 4,000 m below sea level, hot springs, tropical rain forests, the top of the Himalayas at 6,000 m, and Antarctica. They could survive global mass extinctions, such as large meteorite strikes, supernovae, and gamma-ray bursts, and are the first animal known to survive in space. They are pioneer species that colonize new environments, providing a food source, and attracting other invertebrates and predators. They thus make some habitats habitable by more species than would be the case without them. Tardigrades are one of the few taxa that can suspend their metabolism to a state in which it is less than 0.01% of normal.

Neotrogla, a genus of four species of barklice, a winged insect that inhabits dry caves in Brazil, has reversed sex organs [17]. Females aggressively seek out males and have a complex penis that penetrates the male's small genital opening and extracts sperm and nutrient-filled seminal fluid from him. If the male were to break away, his abdomen would rip open. It is thought to be an adaptation for the female to obtain nutrients in the low-nutrient cave environments. This is complete sex role reversal. It is surprising because the male and female needed to evolve their genitalia and behaviors in tandem.

A tiny, 2-mm, European water boatman, *Micronecta scholtzi*, a freshwater insect, creates sound up to 99.2 decibels (to a listener a few feet from it), the equivalent of a loud orchestra playing a few feet away, as part of its courtship display [18]. It averages 78.9 decibels at a 1 m distance, comparable to a passing freight train. This is the loudest sound of any animal relative to its body size. It does it by rubbing a ridge on its penis across the ridged surface of its abdomen. Amazingly, they make the sound from an area only about 50 μm across, about the width of a human hair. It is not deafening to humans, because the insect makes the sound from the river bottom, and 99% of the sound is lost when it travels from water to air. Still, it is audible to humans at the riverside. Researchers think runaway sexual selection and lack of predators is why the sound reaches such high amplitudes. Males compete to have access to females, and then produce as loud a sound as possible, scrambling the song of their competitors. In most species, volume is limited because predators select for lower amplitude, which makes the prey harder to detect, but *M. scholtzi* has no known aquatic predators. Insects can use several body parts to produce sound. Some species use their wings, others their legs, head, thorax, abdomen, etc.

The penis of at least one species of damselfly is barbed with spines that collect and eject sperm from the female's vagina of any previous males that she had mated with. In some rodent species, sperm cells have hooks, stick together, and swim cooperatively in clusters.

There are hornets that attack Japanese honeybee hives and eat them. The bees defend themselves by surrounding the hornets in a spherical formation and beating their flight muscles, thus raising their collective temperature beyond what the hornets can withstand.

Aguilarac et al. [19] found Gulf Killifish (*Fundulus grandis*) collected from two Superfund sites on the highly polluted Houston Ship Channel had F1 embryos that were approximately 1,000-fold more resistant to the toxic polychlorinated biphenyl PCB126- and two-to-fivefold more resistant to coal tar-induced cardiovascular teratogenesis relative to embryos from a reference population. The resistance to such high pollution levels resulted from a gene segment acquired from mating with its sister species, Atlantic Killifish (*Fundulus heteroclitus*), which was likely transported in the ballast of ships. Heritable resistance to the toxicity of both polychlorinated biphenyls and polycyclic aromatic hydrocarbons has been documented in several populations of the latter species.

It was thought until recently that the South American Polka Dot Tree Frog (*Hypsiboas punctatus*) was the only amphibian to fluoresce. It glows with an intense blue-green that might play a role in complex courtship and fighting behaviors [20]. Figure 3.1 shows this frog. Recently, Lamb and Davis (2020) [21] showed that biofluorescence is widespread in amphibians, with spectacularly glowing colors and striking variations in fluorescent patterning, and this includes being widespread among

FIGURE 3.1   The Polka Dot Tree Frog fluoresces gorgeously. (From Internet.)

salamander families. It is common in cnidarians, arthropods, and cartilaginous and ray-finned fishes.

The Timber Rattlesnake (*Crotalus horridus*) has heat-sensitive pits that can sense temperature differences of 0.001°C. Tail muscles can shake its rattle almost 90 times per second for hours.

The Laysan Albatross flies over oceans without landing for up to 5 years. It makes the equivalent of three trips to the moon and back in its life.

About 200,000 species of animals have adapted to human habitations. There are 8,000 species of bacteria in house dust. The average home has 100 arthropod species. Showerheads have complex ecosystems with dozens of species of *Mycobacterium*, predatory swimming bacteria, multicellular protists that eat these swimmers, and small worms that devour the protists [22].

ORGANISMS INCREASE VARIABILITY WITHIN POPULATIONS

Biodiversity includes variability within species and within populations. Natural selection can act in conjunction with organisms to decrease variability within species, conflicting with the ABH. Predators selectively find and consume individuals whose color deviates from the prey's substrate, causing stabilizing selection, and they cull slower runners, causing directional selection for speed. Both of these reduce variability in the prey. Variability is likewise reduced during positive frequency-dependent selection, where the more common morph is selected for. And rare alleles, even if adaptive, often disappear from populations because their benefit is not often manifested until their frequency is sufficiently high. However, the rule is that organisms increase within-population variability.

Lewontin and Hubby [23] showed that genetic variability is very high in natural populations, studying *Drosophila*. They found that 39% of the loci they looked at were polymorphic, and between 8% and 15% of the loci in an average fruit fly from the populations they studied were heterozygous. They deemed these results conservative, with the possibility that the figures are higher. The high variability found in this early study of genetic variability surprised researchers at the time. It indicates selection generally acts to maintain and increase variability in populations. If and only if this variability is primarily maintained by organisms is the ABH supported.

I will argue that balancing selection is the primary mechanism by which genetic variability is maintained. It is the active maintenance of multiple alleles in a population, increasing its variability. This is done mainly by heterozygote advantage; negative frequency-dependent selection; and adaptation to environmental heterogeneity.

The dominant paradigm has been that stabilizing and directional selection are the rule, and balancing selection is rare. However, Sellis et al. [24] showed balancing selection is a form of adaption in diploid populations. When a new adaptive mutation first arises in a diploid population, individuals with this variant will typically be heterozygous for it. Using Fisher's [25] model of adaptation, the authors showed that many adaptive variants will gain their advantage from heterozygote advantage.

In heterosis, the heterozygote has a selective advantage over the corresponding homozygotes, maintaining polymorphism. This is often caused by parasites and their vectors. The superior fitness of heterozygotes for sickle cell genes is the result of malaria mosquitoes and parasites. Individuals homozygous for the sickle cell state are the most resistant to malaria, but their red blood cells cannot carry oxygen efficiently. People who are heterozygous are resistant to malaria, and their red blood cells carry oxygen at reduced capacity, but good enough to function. Cystic fibrosis causes viscous mucus to form in the lungs and intestinal tract. The trait is recessive, and heterozygous mice have a higher survival rate from cholera [26]. This heterozygote advantage does not occur in humans [27]. Heterozygotes may also be protected against typhoid [28]. All of these cases of heterosis occur only when the relevant pathogen is present in sufficient numbers.

There is evidence that heterozygosity in humans confers increased resistance to certain viral infections. Significantly more heterozygosity of HLA-DRB1, which plays a central role in the immune system, occurs among those not infected with Hepatitis C virus than those infected [29].

There are too few known cases to draw definitive conclusions, but heterozygote advantage is apparently often biologically driven, mainly by pathogens.

Double-blind studies have shown females prefer the scent of males who are heterozygous at all three major histocompatibility complex loci in humans [30,31]. Penn et al. [32] provided evidence that heterozygosity at these loci results in higher variability and thus a greater ability to combat a wider variety of diseases. It is reasonable to assume that disease organisms have been selective agents for variability in immune systems for all higher organisms.

Frequency-dependent selection is a situation where the fitness of a phenotype or genotype depends on its frequency in the population. The success of a form depends on how common it is. In positive frequency-dependent selection, the fitness of a phenotype or genotype increases with its frequency in the population. Thus, one morph will become fixed and drive the other morph extinct, eliminating the polymorphism and decreasing genetic variability. However, even this can sometimes increase a species total variability by maintaining geographic variation. For example, *Heliconius erato* butterflies in Peru have seven races, each with distinct color patterns. Each race has a corresponding race that is its Müllerian mimic and that it resembles, of another species, *Heliconius melpromeme*. The races that mimic each other live in the same area, and are separated from all other races of both species. Mallet and Joron [33] showed experimentally that butterflies that differ from the locally common color pattern are selected against because the predators have not learned to avoid attacking butterflies with the rare color pattern. So polymorphism is eliminated and genetic variability is decreased in each population. On the other hand, geographic races are maintained in this case by positive frequency-dependent selection, Müllerian mimicry, and predators. The result is higher overall genetic variability of each of the two species. This also has the potential to result in speciation of some of the populations, increasing species richness.

In negative frequency-dependent selection, the rarer a phenotype or genotype is in a population, the higher its fitness, and hence the more it tends to increase. Rare morphs increase and common morphs decrease in dynamic equilibrium. This increases genetic variability by allowing both morphs to persist in the population. It maintains polymorphisms. The following paragraphs support the thesis that it is more common than positive frequency-dependent selection.

Pathogens often cause negative frequency-dependent selection by selectively attacking the most common form of the host. Parasitism of the freshwater

New Zealand Snail (*Potamopyrgus antipodarum*) by the trematode *Microphallus* sp. results in decreasing frequencies of the most common genotypes across several generations. The more common a genotype became in a generation, the more vulnerable to parasitism by *Microphallus* sp. it became [34].

Frequency-dependent selection may explain some of the high degree of polymorphism in the Major Histocompatibility Complex (MHC) [35], although heterozygote advantage and selection for polymorphism to counter challenges of many different pathogens contribute, with the latter likely being the most important factor.

Predators can have a search image, whereby they actively seek and eat the most common morph. Search image is adaptive because it increases the efficiency of the predator at finding the most abundant morph. The common morph can thus become the rare morph, and then the predator switches to eating the other, now-common morph. Frequency of the morphs thus cycles.

In England, the Song Thrush (*Turdus philomelos*) consumes the Grove Snail (*Cepaea nemoralis*), which is famous for having highly polymorphic shell coloration. Clarke [36] showed that this polymorphism is maintained by negative frequency-dependent selection caused by the thrush's search image for the most common morph. Although the situation is more complicated, Clarke's thesis is valid.

The cichlid fish *Perissodus microlepis* in Lake Tanganyika, Africa, feeds by stealthily approaching larger cichlid fish from the rear and biting off mouthfuls of scales from the prey's sides. There are two forms of the small fish. Those that have their mouths twisted to the right (dextral fish) feed on the left side of the prey, and those with their mouths twisted to the left (sinistral) feed on the right side of the prey. The two forms fluctuate around 50% of each type. When the dextral form is more common, the prey is attacked more often on its left side, and so becomes more wary of attacks from this side, and vice versa. So the escape behavior of the prey maintains the polymorphism by negative frequency selection [37].

In many plants, two plants that share the same self-incompatibility alleles cannot mate, a mechanism that increases the probability of outbreeding. Thus, plants with rare alleles of this type have greater mating success, and their alleles increase in frequency.

Common side-blotched lizards have three male morphs and two female morphs. Orange-throated males are the dominant, best fighters who establish large territories with several females. Yellow stripe-throated males do not defend a territory, but stay on the fringes of orange-throated lizard territories, mating with the females on those territories when the orange-throat is not near them, as the territory to defend is large. Blue-throated males are less aggressive and guard only one female; they can defeat the yellow-throated males but not the orange-throats. Orange-throated females lay a great number of small eggs and are very territorial, while yellow-throated females lay fewer, larger eggs, and are more tolerant of each other. Three-way negative frequency-dependent selection results from this, with a mechanism similar to the rock-paper-scissors game, because each morph beats one other morph, but not both. The orange-throat male beats the blue-throated male. If there are many orange-throated males, there are many opportunities for yellow-throated males to sneak in and mate with orange-throated females, defeating the orange-throated males [38,39]. Blue-throated males can beat yellow-throated males, controlling yellow-throated male numbers when they get high.

Interestingly, recent experiments indicate that this pattern of maintaining poly-morphism might indeed be common, even the rule, in populations. In one example, fruit flies were evolved in the laboratory over many generations, having to adapt to new environmental pressures. Although many DNA variants were found to have changed their frequencies in response to natural selection, none of these variants ever reach 100% frequency in the population. It might thus be necessary to rethink the prevailing view of adaptation as a process that predominantly removes genetic variation, seeing it instead as one that could actually play a substantial role in pro-moting such variation. How much this is selected for by organisms as opposed to the physical-chemical environment is not yet known, but the above examples indicate the organisms are generally key players.

Polymorphism can be maintained by environmental heterogeneity. This could be from physical factors, like differences in soil color. But it could be biological factors, from different colors in vegetation. The Pacific Tree Frog (*Pseudacris regilla*) var-ies in color within populations, having green, tan, reddish, gray, brown, cream, and black morphs, with most being a shade of green or brown. They can change color seasonally to better match their environment, lending evidence that the color varia-tion is a result of the varied colors, often of vegetation, that they rest on. This is also supported by the fact that it lives in many habitat types: riparian, woodland, grass-land, chaparral, and pasture land.

Industrial melanism is found in over 70 moth species in Europe and North America. Forest fires blackened trees, resulting in a mosaic of light and dark trees, maintaining polymorphisms of light and dark moth morphs. So trees and fires helped maintain this polymorphism.

Adaptation of populations to different geographic locations causes the formation of races, subspecies, and sometimes even species, which increases biodiversity. The selective forces that select for differences in the various populations are largely bio-logically-caused differences between the habitats of the different populations, such as different vegetation, predators, and parasites. The formation of new populations is often caused by the organisms, by their dispersal.

ORGANISMS DRIVE SPECIATION

This section will attempt to illustrate that organisms are the major drivers of specia-tion in the major speciation models.

Allopactric speciation by vicariance happens when a geographic barrier divides a population into two, isolating the populations. Selection, drift, and lack of gene flow then cause speciation. The process is sometimes actively initiated by the organ-isms themselves. The organisms undergoing allopatric speciation generally are not passive agents that get separated by the appearance of mountains or rivers without playing any role in their separation, as textbooks often assert. They often disperse to new areas and establish new populations with reduced or no gene flow with the par-ent population [40]. This includes birds and insects flying, animals walking, insects and plant seeds being blown by wind or carried by animals, marine larvae being transported in the ocean, and other means of transport. Being transported by wind, animals, or currents is not fully done without the dispersed organism's influence, since organisms have adaptations to hitch rides with these dispersive agents. When a land corridor gradually formed between North and South America, in the late Miocene and Pliocene, taxa such as deer, camelids, horses, and cats spread from North America deep into South America, while possums, armadillos, porcupines, and some other mammal groups went from South to North America. This led to adaptive radiations in novel geographic areas and the evolution of many new species.

Sometimes organisms disperse great distances over unsuitable habitat, often giving rise to novel species, and sometimes resulting in adaptive radiation (ibid). A grassquit from South America colonized the Galapagos Islands, resulting in a radiation into several species on several islands, the Galapogos finches (ibid). This may have involved peripatric speciation. The populations evolve apart from each other by selection and drift until they are different species. The selective agents are partly physical factors, such as climate. But much of the selection resulting in speciation results from differences between the predators, parasites, competitors, commensals, and mutualists between the two environments of the two populations. Differences between the populations in social structure, including mating preferences, also select for speciation. Thus, allopatric speciation by vicariance is mainly driven by biology.

Vila et al. [41] verified a hypothesis of the novelist and lepidopterist by Vladimir Nabokov, showing that Neotropical *Polyommatus* blue butterflies flew from Asia over Beringia to the New World in five separate migrations over the past 11 million years, going as far as Chile. Several species share a common ancestor in Asia, having speciated after arrival. This shows species can actively drive their own speciation by movement to new regions and habitats.

In peripatric speciation, a few founder individuals that are not necessarily a random sample of the parent population found a population on a limited area such as an island and speciate rapidly. They may be blown to a new area by the wind or carried by a sea current, but they often actively walk or fly there. Speciation by dispersal has happened innumerable times when oceanic islands have been colonized from continental populations [40]. This apparently happened with some species of the paradise kingfisher on small New Guinea islands [42]. Celia et al. [43] looked at 56 putative species of the diverse Calcinu hermit crabs and found it has great dispersal abilities and peripateric speciation drove its diversification.

Sympatric speciation is speciation within a single population with no physical barrier to gene flow. One way this could happen is if there were two morphs, and each morph selectively bred with others of its form to the exclusion of the other morph. This is organism-directed. If sympatric speciation is common, it would support the idea that organisms can drive their own speciation, since it relies on the behavior of organisms—different habitat preferences, positive assortative mating, and so on. Also, it shows that species can speciate without the need for physical factors like geographic barriers. This would support the ABH.

If two different subspecies, whether their differences originated sympatrically or not, are in the process of speciation and have reached the point whereby hybrids are less fit, selection will favor those organisms that prefer to mate with members of their own subspecies as an adaptation to prevent them from wasting resources by producing unfit hybrids. This prezygotic isolation will speed up the process of speciation. Coyne and Orr [44] estimated from their data that, for the fruit fly, *Drosophila*, full speciation with complete reproductive isolation takes 1.1–2.7 million years for allopatric species, but only 0.08 to 0.2 million years for sympatric species. They attributed this to prezygotic isolation. This shows that the organisms themselves, through behavior and mate preferences, re-enforced by natural selection, speed up the process of speciation when the process has reached a threshold difference within sympatric populations. It is evidence for the ABH. Nevertheless, since sympatric speciation is not universally accepted and is compatible with the ABH, I will give some examples of it to show its feasibility.

Hiller et al. [45] studied radiations of four Hawaiian arthropod groups and found diversification does not occur through ecological speciation, but largely within a single environment, implying sympatric speciation.

Some claim the great number of species of fish in the cichlid family in the African Great Lakes occurred by sympatric speciation, but it is easy to interpret their speciation as allopatric, because the species are separated and distributed discontinuously in different habitats. However, there are two groups of cichlid fish species found only in two small crater lakes in Cameroon [46]. The DNA in their mitochondria indicates that these fish evolved from one common ancestor in each of the lakes. So speciation occurred within the lakes. The lakes are both essentially uniform habitats with little variation in them, so it is implausible that the fish could have had areas that were spatially isolated from each other with physical barriers to gene flow. There are also two species of Arctic Charr (*Salvelinus*), which is related to the Brook Trout, in a glacial lake in Iceland that appear to have speciated sympatrically [47]. Barluenga et al. [48] presented convincing evidence for sympatric speciation in the Midas cichlid species complex (*Amphilophus* sp.) in a young and small volcanic crater lake in Nicaragua.

The Medium Ground Finch (*Geospiza fortis*) on Santa Cruz Island of the Galapagous Islands is experiencing selection against intermediate beak sizes. Here, different beak phenotypes may result in different bird calls, providing a barrier to gene flow [49].

The constant frequency component of the call of horseshoe bats both determines prey size and functions in aspects of social communication. In the Large-eared Horseshoe Bat (*Rhinolophus philippinensis*), three sympatric size morphs have undergone recent genetic divergence, and this process has occurred in parallel more than once. They echolocate at different harmonics of the same fundamental frequency.

Switching harmonics creates a discontinuity in the bats' perception of available prey that can initiate diversifying selection. Because call frequency in horseshoe bats functions both in locating prey and communication, ecological selection on frequency might lead to assortative mating and ultimately reproductive isolation and sympatric speciation [50].

Sequencing of the genomes of Blue, Humpback, and Gray Whales showed that these whales also speciated sympatrically by exploitation of different niches, and that these species hybridize across species boundaries [51].

Parapatric speciation is speciation between two populations that are contiguous but not overlapping. There is no geographic barrier, so the organisms can move to the adjacent ranges. The habitats differ, so selection drives divergence. It is certainly initiated by active migration to a new area. Many examples of strongly selected genes and phenotypes that differ between populations that interbreed are known [52]. Fence lizards and earless lizards have white morphs that inhabit white sands that differ from gray morphs on darker soils in the White Sands region of New Mexico. They are probably undergoing this form of speciation [53]. These incipient speciations are being driven by life because the lizards likely actively moved to the new habitats, and because predators are the selective agents for the cryptic colorations and thus geographic variations.

The host shift model asserts that speciation occurs by some members of the population actively shifting their food source. It is claimed to occur most readily in phytophagous insects. Here, biology drives speciation by organisms changing behavior, and by the host exerting different selective pressures on two morphs. The Apple Maggot Fly (*Rhagoletis pomonella*) is the most studied case of this [54,55]. The flies in eastern North America originally all fed on hawthorns (*Crataegus*). Then about 150 years ago, this fly first appeared in the northeastern United States as a pest on apples (genus *Malus*), and infestation spread on apples to the west and south. Apples are related to hawthorns. Flies preferring hawthorns had gene exchange reduced with those preferring apples to only 2% because of the difference in food preferences and because of a 3-week difference in mating time between the two. Once some flies were feeding on apples, differences between apples and hawthorns led to selection for differences in the flies feeding on them. Apples ripen and are thus suitable for fly larval development earlier than hawthorns, selecting for earlier development of the flies on apples. So flies on apples emerge from their pupas and mate earlier than those on hawthorns, causing reproductive isolation. The two types of flies have evolved many genetic differences, and the crucial step that started the divergence was a genetically-based switch in food preference by some of the original flies, and this occurred in sympatry. Also, Schwarz et al. [56] showed some tephritid fruit flies speciated sympatrically by shifting hosts and hybridization.

Speciation by hybridization between two separate species to form a new species involves organisms actively promoting speciation by their behavior, so supports the ABH. It was once thought to be rare, but now known to be not too uncommon. It is known in a few animal species, including the Lonicera fly, some fish, a few birds, and some mammals. It appears that several physically divergent but closely related genera of cichlid fishes in Lake Malawi diverged rapidly into a species complex of multiple hybrid species [57]. The Golden-crowned Manakin, a bird, was formed 180,000 years ago by hybridization between Snow-capped and Opal-crowned Manakins [58]. The Clymene Dolphin is a hybrid species [59]. The Red Wolf and Great Lakes Region Wolf are highly admixed hybrids with different proportions of Gray Wolf and Coyote ancestry [60].

Rapid adaptation to global warming and range expansion has occurred in a shark species by hybridizing with another. Hybrids between local Australian Black-tip Sharks and their global counterparts, Common Black-tip Sharks, were found in Australia [61], indicating adaptation of the Australian species to global warming, since it can only live in tropical waters. Hybrids were found 2,000 km down the coast, in cooler seas, showing this is a range expansion of the Australian species. Hybridization enabled a species restricted to the tropics to move into temperate seas. The researchers suspect speciation by hybridization may be occurring in other shark species.

A hybrid species can, rather than being intermediate between its parent species, be beyond the range of either. The sunflowers *Helianthus annuus* and *H. petiolaris* have given rise to three species by hybridization: *H. anomalus, H. paradoxus,* and *H. deserticola* [62]. The hybrid species have characteristics outside the range of either parent species: They grow in drier or saltier habitats, flower later, and one species even has thicker and more succulent leaves and smaller flower heads than either parent species. Likewise, the Liger, an artificial hybrid cross between a male Lion (*Panthera leo*) and a female Tiger (*Panthera tigris*) typically grows larger than either of its parent species. This cross would never occur in nature. So hybrid speciation is a source not only of new species, but species with novel morphological and ecological features. It could even potentially lead to adaptive radiation if an adaptive new form is created.

Species can fuse by hybridization and decrease species number, acting counter to the ABH. Kearns et al. [63] showed Common Ravens (*Corvus corax*) have undergone reverse speciation, with two species, the "California" and "Holarctic", fusing into one in the distant past, without human influence. However, this phenomenon is rare.

Speciation often occurs spontaneously by polyploidy, especially in plants. About 15% of angiosperm and 31% of fern speciation events are accompanied with an increase in ploidy [64]. This is biologically-driven speciation because it is driven by the behavior of genomes. Polyploidy can be an increase of chromosomes from one species, or two species can hybridize, with the resulting offspring having multiple numbers of the parental chromosome number(s).

Thus, although abiotic factors are involved, organisms are major drivers of speciation.

SEXUAL SELECTION INCREASES BIODIVERSITY

Sexual selection is sometimes an important cause of speciation. Females in one population could prefer males that differ from males that females prefer in another population, selecting for males that differ more and more between the two populations until they speciate. This could even possibly happen sympatrically. Researchers have produced mathematical models supporting the hypothesis that divergent sexual selection in different geographic populations of a species results in different male display traits and female preferences [65–67]. It has also been proposed that different sexual characteristics in populations enable individuals to recognize conspecific mates and avoid producing hybrids of low fitness [68].

Comparison of bird groups shows the importance of sexual selection in speciation [69]. Male hummingbirds mate with many females and have elaborate tail feathers, bright colors, and fancy dances that were selected by female choice, with intense competition between males for mates. Males vary greatly from one species to another, while females of different species are much more similar to one another and less colorful. Swifts are birds that diverged from hummingbirds from a common ancestor, so have been around exactly as long as hummingbirds have. So each group has had an equal amount of time to evolve new species. Thus, comparing the number of species gives a comparison of the rates of speciation, unless the extinction rates differ between the two groups. Swift males are monogamous, form pair bonds with females, help raise the young, and do not seek multiple mates, resulting in much weaker sexual selection. They did not evolve bright colors or elaborate feathers, dances, or songs, since there was no significant sexual selection by females among the males. There are 319 species of hummingbirds and 103 species of swifts, indicating sexual selection has led to high speciation rates and diversity in hummingbird (ibid). Similarly, the highly promiscuous males of the various birds of paradise have fantastic plumage, songs, and dances due to sexual selection, in contrast to the non-promiscuous manucode birds, which evolved from a common ancestor with birds of paradise and hence are equal in age to them. There are 33 species of birds of paradise, three species of manucodes (ibid).

Møller and Cuervo [70] have correlated diversity of bird species and subspecies with the evolution of elaborate feather ornaments, such as crests and long tail feathers, by sexual selection. Panhuis et al. [71] showed evidence supporting the thesis that sexual selection was a key factor in speciation in several very diverse groups of species, including Hawaiian *Drosophila*, pheasants, and the great diversity of cichlid fish in some African lakes. In many taxa, including ostracods, flies, ducks, and dolphin, sexual selection sculpts genitals and sperm into an extraordinary diversity of shapes and sizes, often leading to diversification. Genitalia are conspicuously variable, even in closely related taxa that are otherwise morphologically very similar, and this variability is driven by sexual selection [72].

Capuchino seedeaters (birds) rapidly diversified into nine sympatric species via sexual selection [73]. Gene sequencing showed minimal differences in genomes, ranging from 0.03% to 0.3%. All genetic differences were in the same regions, which were likely cis-regulatory regions for melanin and song. Their phenotypes display differences in color and song. Speciation happened with very little genetic change.

The examples above show sexual selection can lead to adaptive radiation. The case of African cichlids illustrates this best. Over 1,500 species of fish in the cichlid family alone dwell in the African Great Lakes around the East African Rift [74]. They contain 10% of the world's fish species. Many are endemic. Lake Malawi alone has more than 700 cichlid species (ibid). Seehausen [75] presented good evidence that much of this speciation happened as a result of diversifying sexual selection, and that sexual selection can increase rates of sympatric speciation and hence diversification.

Sexual selection can play a role in macroevolutionary breakthroughs and the evolution of key innovations. These are essentially always followed by adaptive radiation. As mentioned, it was hypothesized that feathers in dinosaurs developed at least in part for sexual display, and in combination with other selective factors, this led to flying birds.

On the other hand, sexual selection is often associated with a few dominant males keeping harems and mating with the vast majority of females, while most males in the population do not mate. Elephant Seals are an example. This is a case counter to the ABH, where the behavior of organisms greatly decreases genetic variability within populations. Sexual selection's net effect is to increase biodiversity, because the increase in speciation and innovation it causes outweighs the decrease it causes in intraspecific variability.

ORGANISMS CAUSED MANY MACROEVOLUTIONARY TRANSITIONS

The action of the organisms themselves—through their behavior, intraspecific and interspecific competition, predation, mutualism, and other mechanisms—were instrumental causative agents of many of the major evolutionary breakthroughs and key innovations that gave rise to many large taxa. Migration can result in macroevolutionary changes if the new habitat is radically different.

The colonization of land by fungi, plants, invertebrates, and vertebrates were all largely driven by biology.

The first land fungi evolved from sea forms that were washed onto the shore, where they were free of fungivores and intraspecific and interspecific competition. This allowed them to diversify. Some fungi likely came onto land with algae. It is hypothesized that land plants evolved from a semiaquatic green alga that was in an ancient and continuing mutualism with a fungus that was originally aquatic [76]. When fungi and algae were washed by waves and high tides onto land, they had to adapt to a land environment they were not adapted to, but they acquired a new adaptive zone, largely free of herbivores and competitors. The many unfilled niches allowed them to thrive and greatly diversify. A nonbiological factor was that the moon was closer to Earth at this time, so higher tides stranded phytoplankton farther up the shore and for longer times, allowing adaptation to land.

Plants could not have colonized land without their fungal mutualists helping them obtain nutrients and water, and helping them break up the soil. The first land plants evolved from algae and lacked roots, so they needed fungi to obtain nutrients and water. Land plants also needed lichens, which decomposed rock, making soil for plants. The aid from fungi and lichens was necessary for terrestrial plants to evolve into the myriad contemporary forms and support the great diversity of animals we see today.

Invertebrates needed plants and fungi and as sources of food and shelter to colonize land and evolve into the great diversity of terrestrial forms they display today, from insects to arachnids.

The evolution from fish to amphibians was one of the major transitions in the history of evolution of life. It is now clear that this was primarily, although not exclusively, biologically driven. The limbs of at least one transitional form, Acanthostega, were probably adapted for movement through the shallow swamps, dense with vegetation, it inhabited. Vegetation was one factor that selected for limbs preadapted for use on land. Low oxygen content in the water, an abiological factor, would have favored breathing with lungs, preadapting the protoamphibians to land. Land plants and invertebrates that had evolved from the sea would have provided a food source and other amenities on land. The first colonizers of land would have had no predators, nor interspecific or intraspecific competitors, immediately after colonizing land. Whether they crawled onto land and stayed there for gradually increasing time periods or were washed passively by waves and tides is unclear. Selection for laying eggs and larval development in freshwater rather than the sea would have occurred due to the relative paucity of predators and intraspecific and interspecific competitors in freshwater in the beginning stages of the transition.

Each of the four taxa was aided in its colonization of land by at least one other taxon that preceded it or colonized mutualistically with it. All four colonizations were followed by spectacularly large adaptive radiations, producing many new taxa, from species to classes, even two phyla in plants.

Freshwater fish evolved from saltwater ones. They could have swan to freshwater, or been transported there on tides or currents. Intraspecific and interspecific competition and predation could have been selective factors favoring colonization of freshwater, and migration to a radically different environment played a key role, whether active or passive. Invertebrates colonized freshwater before fish did, and were necessary for this transition, providing food. Again, biology provided the selective force and migration resulted in the fish being in a radically different new environment. The adaptations to this new environment—freshwater—created a novel type of fish with a different physiology than that of the marine fish, and was followed by adaptive radiation in the new adaptive zone. Then the great number of species of freshwater fish provided food for numerous invertebrate, bird, mammal, and other species.

The path to modern *Homo sapiens* involved much dispersal and isolation of populations that locally evolved novel characters, then merging and hybridization. Hybridization merged the novel characters and was an essential creative force in the emergence of our species, sometimes igniting innovative evolutionary advances, and possibly creating new species [77]. This model of dispersal, isolation, and local adaptation, and then contact and hybridization with other humans includes populations in Africa, as well as Neanderthals and Denisovans in Europe and Asia (ibid), both of which bred with our ancestors [78,79]. Modern humans likely acquired some advantageous immune system genes from interbreeding with Neanderthals or Denisovans [80].

Organisms can drive their own evolution. Their behavior can cause macroevolutionary breakthroughs to novel adaptive zones and key innovations. This is often accompanied by exaptation. It is always followed by diversification.

The adaptive function of feathers in theropod dinosaurs has been hypothesized to be for flight, for thermoregulation, and for display [81]. The latter represents the organism actively affecting its evolution. Fossil evidence indicates the contour feathers of pre-Archaeopteryx arose for thermoregulation, and such feathers on the forelimbs were secondarily used to capture insect prey as natural nets activated by powerful ventral adductor muscles [82]. This was preadaptation for flapping flight, and exaptation of arms into wings caused by the prey-catching behavior.

Fowler et al. [83] gave evidence that the large claw of Deinonychus was used to grip prey, not rip open its belly, and that flapping of feathers on the arms was for maintaining stability, and this may have been pivotal to the evolution of the flapping stroke. Again, behavior of the animal through the use of its arms and feathers for a function other than flight actively directed a large evolutionary breakthrough to a key innovation, flight. This again is exaptation. Predators were a selective force favoring flight, since flight helped birds escape them. Trees were key selective forces in providing elevation, perches, and elevated food sources, such as nuts and fruits.

Many of the Mollusca, such as bivalves and snails, evolved shells for protection from predation. This led to the evolution of new, highly effective predators, such as fishes with shell-crushing jaws and crustaceans that could either crush or rip shells apart, in the Mesozoic Era (which lasted from 250 to 65 million years ago). These mollusks responded by evolving an amazing diversity of defenses, including thicker shells; spines, ridges, and other growths on the shells; and thicker margins of the shell opening and smaller openings in snails [84]. These mollusk groups evolved shells and then a diversity of species and shell forms due to coevolution with their predators. The predators responded in this coevolution with counter-adaptations to the shells and other defenses in order to be able to subdue and consume the mollusks. This led to diversification of the predators. An example of these adaptations by the predators can be seen in sea stars with powerful muscles, tube feet, and great endurance to outlast and eventually open bivalve shells.

The roles of horizontal gene transfer, including by viruses; mutualism in the evolution of the mitochondrion and chloroplast (and hence of eukaryotes), pollination, and seed dispersal; and viruses in the origin of multicellularity and placental mammals are all examples of major evolutionary transitions driven primarily by life discussed in the appropriate chapters in this book.

All of these transitions were followed by adaptive radiations creating huge numbers of species filling the many novel niches in the new adaptive zones the groups had entered. Then additional new species evolved that took advantage of the new niches created by these species.

REFERENCES

1. Schulze-Makuch, D., et al. (13 March, 2018). Transitory microbial habitat in the hyperarid Atacama Desert. *PNAS USA* 115 (11): 2670–5. doi: 10.1073/pnas.1714341115.
2. Yoshida, S., et al. (11 March, 2016). A bacterium that degrades and assimilates polyethylene terephthalate. *Science* 351 (6278): 1196–9. doi: 10.1126/science.aad6359.
3. Edwards, K. J., et al. (May, 2012). The deep, dark energy biosphere: intraterrestrial life on Earth. *Annual Review of Earth and Planetary Sciences* 40: 551–68. doi: 10.1146/annurev-earth-042711-105500.
4. Shi, T., et al. (1997). Characterization of viable bacteria from Siberian permafrost by 16S rDNA sequencing. *Microbial Ecology* 33: 169–79.
5. Blaire, S., et al. (Aug., 2006). Microbial ecology and biodiversity in permafrost. *Genetics in Medicine* 10 (4): 259–67.
6. Legendre, M., et al. (March 18, 2014). Thirty-thousand-year-old distant relative of giant icosahedral DNA viruses with a pandoravirus morphology. *PNAS USA* 111 (11): 4274–9. doi: 10.1073/pnas.1320670111.
7. Yashina, S., et al. (March 6, 2012). Regeneration of whole fertile plants from 30,000-y-old fruit tissue buried in Siberian permafrost. *PNAS USA* 109 (10) 4008–13. doi: 10.1073/pnas.1118386109.
8. Shatilovich, A. V., et al. (16 July, 2018). Viable nematodes from late Pleistocene permafrost of the Kolyma River Lowland. *Doklady Biological Sciences* 480: 100–2.
9. Christner, B. C., et al. (2014). A microbial ecosystem beneath the West Antarctic ice sheet. *Nature* 512: 310–3.
10. Grassle, J. F. & Maciolek, N. J. (2 Feb., 1992). Deep-sea species richness: regional and local diversity estimates from quantitative bottom samples. *The American Naturalist* 139 (2): 313–41. doi: org/10.1086/285329.
11. Haase, K. M., et al. (2009). Fluid compositions and mineralogy of precipitates from Mid Atlantic Ridge hydrothermal vents at 4°48′S. *PANGAEA*. doi: 10.1594/PANGAEA.727454.
12. Sysoev, A. V. & Kantor, Y. I. (1995). Two new species of *Phymorhynchus* (Gastropoda, Conoidea, Conidae) from the hydrothermal vents. *Ruthenica* 5: 17–26.
13. Beatty, J. T., et al. (2005). An obligately photosynthetic bacterial anaerobe from a deep-sea hydrothermal vent. *PNAS USA* 102 (26): 9306–10. Bibcode: 2005PNAS..102.9306B. doi: 10.1073/pnas.0503674102.
14. Danovaro, R., et al. (2010). The first metazoa living in permanently anoxic conditions. *BMC Biology* 8 (30). doi: 10.1186/1741-7007-8-30.
15. Yahalomi, D. (24 Feb., 2020). A cnidarian parasite of salmon (Myxozoa: *Henneguya*) lacks a mitochondrial genome. *PNAS USA*. doi: 10.1073/pnas.1909907117.

16. Horikawa, D. D. (2012). Survival of Tardigrades in extreme environments: a model animal for astrobiology. In Altenbach, Alexander V., et al. Anoxia. *Cellular Origin, Life in Extreme Habitats and Astrobiology*. 21: 205–17. doi: 10.1007/978-94-007-1896-8_12. ISBN 978-94-007-1895-1.

17. Lienhard, C., et al. (March 2015). Review of Brazilian cave psocids of the families Psyllipsocidae and Prionoglarididae (Psocodea: 'Psocoptera': Trogiomorpha) with a key to the South American species of these families. *Revue Suisse de Zoologie* 122 (1): 121–42.

18. Sueur, J., et al., (2011). So small, so loud: extremely high sound pressure level from a pygmy aquatic insect (Corixidae, Micronectinae). *PLoS ONE* 6 (6): e21089. doi: 10.1371/journal.pone.0021089.

19. Aguilarac, L., et al. (May, 2014). Evolved resistance to PCB- and PAH-induced cardiac teratogenesis, and reduced CYP1A activity in Gulf killifish (*Fundulus grandis*) populations from the Houston Ship Channel, Texas. *Aquatic Toxicology* 150: 210–9. doi: 10.1016/j.aquatox.2014.03.012.

20. Taboada, C., et al. (4 April, 2017). Naturally occurring fluorescence in frogs. *PNAS USA* 114 (14): 3672–7. 201701053. doi: 10.1073/pnas.1701053114.

21. Lamb, J. Y. & Davis, M. P. (27 Feb., 2020). Salamanders and other amphibians are aglow with biofluoresence. *Scientific Reports* 10, Article number 2821.

22. Dunn, R. (Nov., 2018). *Never Home Alone. Basic Books*, Hachette Book Group, Inc.: New York.

23. Lewontin, R. C. & Hubby, J. L. (1966). A molecular approach to the study of GENIC heterozygosity in natural, populations II. Amount of variation and degree of heterozygosity in natural populations of *Drosophila pseudoobscura*. *Genetics* 54: 595–609.

24. Sellis, D., et al. (2011). Heterozygote advantage as a natural consequence of adaptation in diploids. *PNAS USA* 108 (51): 20666–71.

25. Fisher, R. A. (1930). *The Genetical Theory of Natural Selection*. Oxford University Press: Oxford, UK.

26. Gabriel, S. E., et al. (7 Oct., 1994). Cystic fibrosis heterozygote resistance to cholera toxin in the cystic fibrosis mouse model. *Science* 266 (5182): 107–9. doi: 10.1126/science.7524148.

27. Högenauer, C., et al. (Dec., 2000). Active intestinal chloride secretion in human carriers of cystic fibrosis mutations: an evaluation of the hypothesis that heterozygotes have subnormal active intestinal chloride secretion. *American Journal of Human Genetics* 67 (6): 1422–7. doi: 10.1086/316911.

28. Josefson, D. (16 May, 1998). CF gene may protect against typhoid fever. *British Medical Journal* 316 (7143): 1481.

29. Hraber, P., et al. (Dec., 2007). Evidence for human leukocyte antigen heterozygote advantage against hepatitis C virus infection. *Hepatology* 46 (6): 1713–21. doi: 10.1002/hep.21889. PMID 17935228.

30. Rikowski A. & Grammer, K. (May, 1999). Human body odour, symmetry and attractiveness. *Proceedings of the Royal Society B* 266 (1422): 869–74. doi: 10.1098/rspb.1999.0717.

31. Thornhill, R., et al. (March–April, 2013). Major histocompatibility complex genes, symmetry, and body scent attractiveness in men and women. *Behavioral Ecology* 14 (5): 668–78. doi: 10.1093/beheco/arg043.

32. Penn, D. J., et al. (Aug., 2002). MHC heterozygosity confers a selective advantage against multiple-strain infections. *PNAS USA* 376 (17): 11260–4. doi: 10.1073/pnas.162006499. PMC 123244. Freely accessible. PMID 12177415.

33. Mallet, J. & Joron, M. (1999). Evolution of diversity in warning color and mimicry: polymorphisms, shifting balance, and speciation. *Annual Review of Ecology, Evolution, and Systematics* 30: 201–33.

34. Koskella, B. & Lively, C. M. (2009). Evidence for negative frequency-dependent selection during experimental coevolution of a freshwater snail and a sterilizing trematode. *Evolution* 63: 2213–21. doi: 10.1111/j.1558-5646.2009.00711.x.

35. Borghans, J. A., et al. (Feb., 2004). MHC polymorphism under host-pathogen coevolution. *Immunogenetics* 55 (11): 732–9. doi: 10.1007/s00251-003-0630-5. PMID 14722687.

36. Clarke, B. 1962. Balanced polymorphism and the diversity of sympatric species. In Nichols, D. (ed.). *Taxonomy and Geography*, pp. 47–70. Systematics Association: Oxford, UK.

37. Hori, M. (9 April, 1993). Frequency-dependent natural selection in the handedness of scale-eating cichlid fish. *Science* 260 (5105): 216–9. doi: 10.1126/science.260.5105.216.

38. Sinervo, B. & Lively, C. M. (1996). The rock-paper-scissors game and the evolution of alternative male strategies. *Nature* 380 (6571): 240–3. doi: 10.1038/380240a0.

39. Sinervo, B., et al. (2000). Testosterone, endurance, and Darwinian fitness: natural and sexual selection on the physiological bases of alternative male behaviors in Side-Blotched Lizards. *Hormones and Behavior* 38 (4): 222–33. doi: 10.1006/hbeh.2000.1622. PMID 11104640.

40. Futuyma, D. J. & Kirkpatrick, M. (2017). *Evolution*. Fourth Edn. Sinauer Assocs., Inc.: Sunderland, MA.

41. Vila, R. et al. (2 Feb., 2011). Phylogeny and palaeoecology of *Polyommatus* blue butterflies show Beringia was a climate-regulated gateway to the New World. *Proceedings of the Royal Society B* 278 (1719). doi: 10.1098/rspb.2010.2213.

42. Mayr, E. (1954). Change of genetic environment and evolution. In *Evolution as a Process*, Huxley, J., Hardy, A. C., & Ford, E. B. (eds.), pp. 157–180 Allen and Unwin, London, UK.

43. Celia, M., et al. (19 Feb., 2010). Peripatric speciation drives diversisfication and distributional pattern of reef hermit crabs (Decapoda: Diogenidae: *Calcinus*). *Evolution* doi: 10.1111/j.1558-5646.2009.00848.x.

44. Coyne, J. A. & Orr, H. A. (1997). "Patterns of speciation in Drosophila" revisited. *Evolution* 54 (1): 295–303. doi: 10.1111/j.1558-5646.1997.tb02412.x.

45. Hiller, A. E., et al. (June, 2019). Niche conservatism predominates in adaptive radiation: comparing the diversification of Hawaiian arthropods using ecological niche modelling. *Biological Journal of the Linnean Society* 127 (2): 479–92. doi: 10.1093/biolinnean/blz023.

46. Schliewen, U. K. (14 April, 1994). Sympatric speciation suggested by monophyly of crater lake cichlids. *Nature* 368: 629–32.

47. Gíslason, D., et al. (1999). Rapid and coupled phenotypic and genetic divergence in Icelandic Arctic char (*Salvelinus alpinus*). *Canadian Journal of Fisheries and Aquatic Sciences* 56 (12): 2229–34. doi: 10.1139/f99-245. Published on the web: 12 April, 2011.

48. Barluenga, M., et al. (2006). Sympatric speciation in Nicaraguan crater lake cichlid fish. *Nature* 439: 719–23.

49. Huber, S. K, et al. (2007). Reproductive isolation of sympatric morphs in a population of Darwin's finches. *Proceedings of the Royal Society B* 274 (1619): 1709–14. doi: 10.1098/rspb.2007.0224.

50. Kingston, T. & Rossiter, S. J. (2004). Harmonic-hopping in Wallacea's bats. *Nature* 429 (6992): 654–7. doi: 10.1038/nature02487.

51. Árnason, U., et al. (4 April, 2018). Whole-genome sequencing of the blue whale and other rorquals finds signatures for introgressive gene flow. *Science Advances* 4 (4). eaap9873; doi: 10.1126/sciadv.aap9873.

52. Nosil, P. (2012). *Ecological Speciation*. Oxford University Press: Oxford, UK.

53. Rosenblum, E. B. & Harmon, L. J. (2011). "Same but different": replicated ecological speciation at White Sands. *Evolution* 65: 946–60.

54. Bush, G. L. (June, 1969). Sympatric host race formation and speciation in frugivorous flies of the genus *Rhagoletis* (Diptera, Tephritidae). *Evolution* 23 (2): 237–51. doi: 10.1111/j.1558-5646.1969.tb03508.x.

55. Feder, J. L., et al. (2003). Allopatric genetic origins for sympatric host-shifts and race formation in *Rhagoletis*. *PNAS USA* 100: 10314–9.
56. Schwarz, D., et al. (2005). Host shift to an invasive plant triggers rapid animal hybrid speciation. *Nature* 436: 546–9.
57. Genner, M. J. & Turner, G. F. (Dec., 2011). Ancient hybridization and phenotypic novelty within Lake Malawi's Cichlid fish radiation. *Molecular Biology and Evolution* 29 (Published Online): 195–206. doi: 10.1093/molbev/msr183.
58. Barrera-Guzmán, O. A., et al. (2017). Hybrid speciation leads to novel male secondary sexual ornamentation of an Amazonian bird. *PNAS* 201717319. doi: 10.1073/pnas.1717319115.
59. Amaral, A. R., et al. (2014). Hybrid speciation in a marine mammal: the Clymene Dolphin (*Stenella clymene*). *PLoS ONE* 9 (1): e83645. doi: 10.1371/journal.pone.0083645.
60. Vonholdt, B. M., et al. (2016). Whole-genome sequence analysis shows that two endemic species of North American wolf are admixtures of the coyote and gray wolf. *Science Advances* 2 (7): e1501714. Bibcode: 2016SciA....2E1714V. doi: 10.1126/sciadv.1501714.
61. Morgan, J. A. T., et al. (April, 2012). Detection of interspecies hybridisation in Chondrichthyes: hybrids and hybrid offspring between Australian (*Carcharhinus tilstoni*) and common (*C. limbatus*) blacktip shark found in an Australian fishery. *Conservation Genetics* 13 (2): 455–63.
62. Rieseberg, L. H., et al. (2003). Major ecological transitions in wild sunflowers facilitated by hybridization. *Science* 301 (5637): 1211–6. Bibcode: 2003Sci...301.1211R. doi: 10.1126/science.1086949.
63. Kearns, A. M., et al. (2 March, 2018). Genomic evidence of speciation reversal in ravens. *Nature Communication* 9, Article number 906.
64. Wood, T. E., et al. (2009). The frequency of polyploid speciation in vascular plants. *PNAS* 106 (33): 13875–9. Bibcode: 2009PNAS.10613875W. JSTOR 40484335. PMC 2728988 Freely accessible. PMID 19667210. doi: 10.1073/pnas.0811575106.
65. Lande, R. (1981). Models of speciation by sexual selection on polygenic traits. *PNAS USA* 78: 3721–5.
66. Pomiankowski, A. & Iwasa, Y. (1998). Runaway ornament diversity caused by Fisherian sexual selection. *PNAS USA* 95: 5106–11.
67. Turelli, M., et al. (2001). Theory and speciation. *Trends in Ecology & Evolution* 16: 330–43.
68. Futuyma, D. J. (2013). *Evolution*, Third Edn. Sinauer Assocs., Inc.: Sunderland, MA.
69. Mitra, S. H., et al. (1996). Species richness covaries with mating systems in birds. *Auk* 113: 544–51.
70. Møller, A. P. & Cuervo, J. J. (June, 1998). Speciation and feather ornamentation in birds. *Evolution* 52 (3): 859–69. First published: 31 May, 2017. doi: 10.1111/j.1558-5646.1998.tb03710.x.
71. Panhuis, T. M., et al. (2001). Sexual selection and speciation. *Trends in Ecology & Evolution* 16: 364–71.
72. Hosken, D. J. & Stockley, P. (2004). Sexual selection and genital evolution. *Trends in Ecology & Evolution* 19 (2): 87–93. doi: 10.1016/j.tree.2003.11.012.
73. Campagna, L. (24 May, 2017). Repeated divergent selection on pigmentation genes in a rapid finch radiation. *Science Advances* 3 (5). e1602404. doi: 10.1126/sciadv.1602404.
74. Turner, G. F., et al. (2001). How many species of cichlid fishes are there in African lakes? *Molecular Ecology* 10: 793–806.
75. Seehausen, O. (2000). Explosive speciation rates and unusual species richness in haplochromine cichlid fishes: effects of sexual selection. *Advances in Ecology Research* 31: 237–74. doi: 10.1016/S0065-2504(00)31015-7.
76. Pirozynski K. A. & Malloch D. W. (March, 1975). The origin of land plants: a matter of mycotrophism. *Biosystems* 6 (3): 153–64.
77. Ackermann, R., et al. (March, 2006). The hybrid origin of "modern" humans. *Evolutionary Biology* 43 (1): 1–11.

78. Green, R. E., et al. (2010). A draft sequence of the Neanderthal genome. *Science* 328: 710–22.
79. Reich, D., et al. (2010). Genetic history of an archaic hominin group from Denisova cave in Siberia. *Nature* 468: 1053–60.
80. Abi-Rached, L., et al. (2011). The shaping of modern human immune systems by multi-regional admixture with archaic humans. *Science* 333: 89–94.
81. Dimond, C. C., et al. (1 Sept., 2011). Feathers, dinosaurs, and behavioral cues: defining the visual display hypothesis for the adaptive function of feathers in non-avian Theropods. *BIOS* 82 (3): 58–63. doi: 10.1893/011.082.0302.
82. Ostrom, J. H. (1974). Archaeopteryx and the origin of flight. *Quarterly Review of Biology* 49 (1). doi: 10.1086/407902.
83. Fowler, D. W., et al. (Dec., 14, 2011). The predatory ecology of *Deinonychus* and the origin of flapping in birds. *PLOS ONE*. doi: 10.1371/journal.pone.0028964.
84. Vermeij, G. J. (13 June, 2008). Escalation and its role in Jurassic biotic history. *Palaeogeography, Palaeoclimatology, Palaeoecology* 263 (1–2): 3–8. doi: 10.1016/j.palaeo.2008.01.023.

4 Mutualism Is Fundamental and Greatly Increases Diversity

MUTUALISM IS FAVORED BY NATURAL SELECTION, VERY COMMON, AND GREATLY INCREASES DIVERSITY

Symbiosis is defined as a relationship between two species in which both benefit, or as an interspecific relationship in which one species benefits. In the latter definition, if the other species benefits, it is called mutualism; if the other species neither benefits nor is harmed, it is commensalism; and if the other species is harmed, it is parasitism. I will use the term mutualism instead of symbiosis for mutually beneficial interspecific relationships, to avoid the confusion of the multiple meanings of symbiosis.

Mutualism clearly maintains species richness by benefiting both species involved. It also greatly increases species richness. I contend that its importance is underestimated, and that it is of crucial and fundamental importance in organizing and structuring ecosystems, affecting evolution, causing diversification, and increasing species richness. The evolution of mutualistic interactions can result in macroevolution, with spectacular leaps to new forms in novel adaptive zones that then undergo huge adaptive radiations and provide habitat and niches for the diversification of other taxa. Macroevolutionary leaps caused by mutualism occurred at least ten times in the history of life. Mutualism is a major driver of evolution.

My hypothesis that I call the hypothesis of coevolution, mutualism, and diversification (CMD) is as follows. First, two interacting species coevolve a mutualistic relationship. For example, a species of beetle ate pollen on wind-pollinated plants, pollinating them inefficiently. The plant's leaves became modified into petals, and the plant evolved nectar and nectar guides to attract the beetle pollinator and be pollinated more effectively. The beetle evolved to be better at getting the nectar and pollen, and to use the nectar guides. The animal-pollinated flower, a key adaptation, and effective pollinator, with its own key adaptations, coevolved a mutualism.

Second, because of the key adaptations and mutualism, the two species are each in new adaptive zones, with many new, unfilled niches. Thus, both undergo tremendous adaptive radiation. These diversifications occur by coevolution of the two mutualistic taxa. In the pollination example, many new species and even new taxa above the species level evolved in both flower and beetle lineages. Once there were animal-pollinated flowers, other taxa, such as butterflies, bees, birds, primates, and others coevolved with flowers and became their pollinators, with the flowers and their new pollinators both evolving into many new species and taxa above the species level, continuing the tremendous adaptive radiation. Finally, other species and taxa evolve to take advantage of the niches provided by the great number of species created from these diversifications. In the example, predators adapted to eat the new insect and other pollinators; insects and mammals evolved to eat, live on, and use the shade of the new floral species; birds evolved to consume their seeds; and so on. The result was a tremendous increase in biodiversity as a result of mutualism. This model could have been illustrated with seed dispersal, the evolution of the mitochondrion and the chloroplast, plants and their mycorrhizal fungi, and so on. It is a somewhat general rule, but it does not apply to most mutualisms. For example, the wolf-raven mutualism, to be discussed in this chapter, did not lead to macroevolutionary breakthroughs followed by large diversifications.

Population biologists, evolutionary biologists, and ecologists who study species interactions tend to look at predator-prey interactions, host-parasite relationships, and competition because they are easier to model mathematically than mutualism and commensalism. These latter two deserve more attention.

Mutualism is a stable state that evolution tends toward under the right conditions (there must be a fit where both species benefit). A large literature supports this, and following is a sampling of it that is far from comprehensive. Mutualism plays a fundamental role in all ecosystems (see [1]). Doebeli and Knowlton [2] showed that, when increased investments in a different species yields increased returns, mutualism can evolve with ease. Frank [3] argued that mutualism evolves by enhancing the inclusive fitness of all participants, and that cheating in the interaction can be constrained. His model showed that the spread of altruism between species is enhanced by spatial correlations between species in the genetic tendency to give aid to partners. The model shows a tendency for selection to favor mutualism. He posits that mutualisms have transformed whole ecosystems, and that they may be steps in ecosystems that serve their members' common welfare. Leigh [4] developed a general model that showed a tendency for mutualism to evolve. Leigh also pointed out that mutualisms, even brief exchange ones, have transformed entire ecosystems, and that some represent major evolutionary innovations. He stated mutualism evolves most readily between members of different kingdoms, because they can pool complimentary abilities for mutual benefit. Wyatt et al. [5] did a market economy analysis of the plant and mycorrhizal fungal mutualism, finding that individuals are favored to parcel out resources among trading partners in direct relation to the relative amount of resources they receive. Harcombe (2010) demonstrated in the laboratory that mutualism can evolve between two bacterial species, *E. coli* and a *Salmonella* strain, if there is preexisting reciprocation or feedback for cooperation, and reciprocation is preferentially received by cooperative genotypes.

There is a documented example of evolution from parasitism to mutualism. *Wolbachia* are maternally inherited bacteria that cause a reduced egg hatch when uninfected female fruit flies, *Drosophila simulans*, mate with infected males, by causing cytoplasmic incompatibility. Infected females are often less fecund as a result of infection. Theory predicts that *Wolbachia* will evolve toward mutualism with its host because more *Wolbachia* will be produced if it helps its host, and such aid could select for its host to help *Wolbachia*. Weeks et al. [6, 7] found the parasite reduced female fecundity by 15%–20% under laboratory conditions. It then evolved in California populations of *D. simulans* over 20 years with the result that females now show a mean 10% fecundity increase over uninfected females in the laboratory. Their data show smaller but qualitatively similar changes in relative fecundity in nature and that fecundity-increasing *Wolbachia* variants are now polymorphic in nature.

A key question concerning mutualism and the validity of the Autocatalytic Biodiversity Hypothesis (ABH) is how common mutualism is. The quantitative question is: What percentage of interspecific relationships are mutualistic? Although it is hard to quantify, there are some scientists who have made attempts to estimate how common it is. The number is apparently high. For example, associations between plants and fungi, which generally are mutualisms [8], are found in 83% of dicots, 79% of monocots [9], and all gymnosperms [10]. And these relationships between plant roots and mycorrhizal fungi may be as old as the evolution of land plants themselves [11–13]. And mutualisms of plants and their animal pollinators are also highly diverse, common, and ancient [14]. Seed dispersal is a very common mutualism between plants and animals, as this chapter will show. Composite mutualistic organisms include the 2,500 coral species, of which 1,000 are hard coals that build reefs, promoting high diversity, and an estimated 13,500–17,000 species of lichens, extending from the tropics to the polar regions. All metazoa are ecosystems harboring thousands of species of mutualistic and commensal bacteria, archaea, fungi, and other organisms. Taking these microbiomes into account greatly increases the percentage and number of mutualistic and commensal relationships.

Most researchers do not take indirect mutualism into account, but it must be considered to gain a full understanding of mutualism's prevalence. By indirect mutualism, I mean mutualism in which two species benefit each other indirectly through an intermediary species that the two species are both mutualistic with. For example, a mycorrhizal fungus helps a tree obtain nutrients and receives carbohydrate from it, and an earthworm enhances the soil for the tree and receives food from the tree's fallen leaves. The fungus and worm indirectly benefit each other by benefiting the tree, a shared mutualist that benefits both fungus and worm. Any of the tree's pollinators and seed dispersers, as well as some of its microbiome, are additional indirect mutualists of the worm and fungus. Thus, any given species has many mutualists when one considers indirect mutualism. When we consider earthworms improving the soil and so aiding trees, trees mutualistic with their pollinators and seed dispersers, the seed-dispersing birds being preyed on by birds of prey, and so on, we can see that there are long and many-branching webs of indirect mutualisms throughout nature. Thus, there are a spectacular number of mutualisms; and mutualism is very common. These indirect mutualisms maintain and can increase biodiversity.

Any two species that are two trophic levels apart are in an indirect mutualistic relationship. For example, primary producers benefit carnivores that eat herbivores that eat these primary producers. The carnivores benefit the producers by controlling their herbivores.

So, we do not know how common mutualism is in nature, since we do not have a really good estimate of what percentage of all interspecific relationships are mutualisms. Yet, we can say in a general sense that mutualism is economically adaptive, often selected for, often has a tendency to evolve, and is apparently very common.

I will coin the term keystone mutualists, which are mutualistic pairs of species that are both keystone species at least partly because of the mutualism. For example, in Western Australia, the tree, Acorn Banksia (*Banksia prionotes*), is the only source of nectar for several species of birds called honeyeaters that pollinate it during specific times of the year. These birds play an important role in the pollination of numerous other plant species. The loss of this one tree species could cause many honey eater species to decrease or go extinct, making many plant species that depend on honeyeaters die off, making many animal species that rely on these plants to be substantially adversely affected. The examples discussed in detail under headings below, such as plants and their seed dispersers, represent keystone mutualists.

Since I am arguing that mutualism is fundamental in ecosystems and a major driver of diversity, I will now discuss some important mutualisms and how they increase diversity. Most represent keystone mutualists, my hypothesis of CMD, or a combination of the two.

TYPES OF MUTUALISM

GENETIC COEVOLUTIONARY MUTUALISM

In some cases, there is a unique form of mutualism, involving a transfer of DNA. I am coining the term genetic coevolutionary mutualism for this. It can happen when viruses give DNA to their hosts by transduction. If the host receives an adaptive DNA segment or gene, it could undergo a large adaptive change or acquire a key innovation. This could be followed by adaptive radiation into several new species. This in turn would aid the virus by providing it with a greater amount and diversity of habitats. The virus can then undergo adaptive radiation. A species of microbe in the microbiome of a higher organism can transfer DNA to its host. If it gives an adaptive gene or segment of DNA, its host can acquire an adaptive trait or key innovation and diversify, giving the microbe a greater amount and variety of habitats to exploit and diversity to utilize.

PROKARYOTE-TO-PROKARYOTE MUTUALISMS

There are many mutualisms between microbial species, including bacteria-to-bacteria, bacteria-to-archaea, and archaea-to-archaea. Many of these had profound effects on evolution, and many are key in structuring ecosystems. Since over 99% of all life's cellular species are bacteria or archaea, it is highly probable that the mutualisms are numerically astronomical, although only a small fraction of them are known or have been studied. Two of the most important ones will be discussed now.

Consortia of species of methanotrophic archaea and sulfate-reducing bacteria work together to obtain nutrients and energy by anaerobically oxidizing CH_4 [15,16], often forming small aggregates, and sometimes large mats. They are important in atmospheric temperature regulation, since they remove a significant amount of CH_4 from the air. They are globally distributed. How the archaea and bacteria interact and what nutrients they interchange is unknown. No one has been able to isolate them, perhaps because their mutualism is tight and obligatory, though even this is not known. They have a large array of species, with diverse genes for nitrogen metabolism in the consortia. In some marine areas, the process produces hydrogen sulfide, used by commensal filamentous sulfur bacteria and animals, such as clams and tube worms. These animals have mutualistic sulfide-oxidizing bacteria, allowing them to benefit from the hydrogen sulfide. And the consortia aid many other species by producing nutrients for other prokaryotes and some animals.

Microbial mats are multilayered sheets of microbes, mostly bacteria and archaea. They are mainly in water or moist areas on land, but a few are in deserts, and a few are mutualists in animals. Mats of different types can survive at temperatures from $-40°C$ to $+120°C$. They are held together and to their substrates by slimy substances the microbes secrete. Many are made tougher by tangled webs of filaments formed by the microbes. The slime and tangled threads attract other microorganisms that join the mat community. Protozoa, which feed on the bacteria, are among these, as are diatoms, which often seal the submerged mats with thin, parchment-like coverings that protect them. The mats are resistant to antibiotics produced by bacterial and fungal competitors.

Mats date to 3.5 bya, and are the oldest life on the planet represented by good fossil evidence. Since their first appearance up to the present, they have been communities of mutualism, commensalism, competition, and predation. Interestingly, the fossil record of early photosynthesis shows that three distinct bacterial communities lived in vertical layers in a mat along a gradient of oxygen availability, within a few centimeters of each other. Several fossils, all more than 2 billion years in age, demonstrate this. In the top layer, bacteria produced organic carbon and oxygen through photosynthesis. Immediately below them, a middle layer of aerobic bacteria took advantage of the O_2 produced by the top layer to get energy through oxidation of such reduced substances as sulfur, ammonia, iron, manganese, and others, combining it with reduced compounds to produce energy. Typical members of this layer included sulfur-oxidizing bacteria that combined sulfides with O_2 to produce sulfate and energy. Slightly further below and at the very bottom, where there was no O_2, a third community of heterotrophic bacteria lived off nutrients, including dead bacteria, that seeped down from the overlying communities. These anaerobic bacteria used sulfate, nitrate, ferric iron, and other oxidized by-products from the community above them for oxidation of the organic carbon produced by photosynthesis by the bacteria in the top layer, some of which trickled down to them. This layer produced sulfide and other reduced compounds that were used by the middle layer. So the middle and lowest layers were mutualistic, benefiting each other. The bottom layer produced CO_2, used in photosynthesis by the top layer, making the bottom and top layers mutualistic as well. The top and middle layers were commensal, with the middle one benefiting from the top one, but neither helping nor hurting it. Thus, a stable coexistence based on mutualism and commensalism was established early in evolution. There are many mutualisms between soil prokaryotes today.

The mutualisms and commensalisms in the mats were key drivers of many of the major evolutionary transitions of life's history. The first mats depended on the heat and chemicals from hydrothermal vents on the seafloor and in hot springs for energy and nutrients, with no input from the sun. These still exist today. Some microbes evolved anoxygenic photosynthesis, which is done by such organisms as purple nonsulfur and green sulfur bacteria, and does not produce oxygen. This allowed the mats, including nonphotosynthetic species that benefited from the photosynthesizers, to be less dependent on the deep-sea vents and hot springs, and thus live in larger, more remote areas, although they still depended on the diffusion of chemicals from hydrothermal vents. These mats had bacteria that evolved photosystem I (PSI) in one bacterial species, and photosystem II (PSII) in another. These are the two photosystems of oxygenic photosynthesis that together make the light reactions of photosynthesis. The bacteria with PSI and those with PSII were close to each other because they were in the same mat, and this likely helped them to transfer genes and combine the two photosystems, resulting in the two photosystems existing together in one species, possibly with the help of viruses through transduction. This is how the cyanobacteria, the first organisms to carry out fully developed oxygenic photosynthesis, evolved. The key innovation of merging PSI and PSII into one species allowed the cyanobacteria to enter a new adaptive zone, and diversify into the many novel, unoccupied niches that became available. They could also live throughout the ocean, and then in diverse areas on land, fully free from the vents, which allowed yet much more adaptive radiation. This created myriad niches for bacteriophage, heterotrophic prokaryotes that consumed the photosynthetic bacteria, and many other species that benefited from the many new species of bacteria that carried out oxygenic photosynthesis, resulting in the diversification of these beneficiary species. If transduction was involved, this is a case of genetic coevolutionary mutualism, with the phages helping bring the two photosystems together, and benefiting from the radiation of the cyanobacteria into several new host species. Because photosynthetic and other bacteria were near to each other in a mat, the production of O_2 by cyanobacteria selected for bacteria in the mat to evolve enzymes to protect from its toxic effects, and this evolution continued until the bacteria evolved a way to use the O_2 for energy, which is cellular respiration. These new functions then resulted in new mutualistic and commensal relationships involving microbes that photosynthesized and those that respired. The proximity of the respiring bacteria to other bacterial species allowed the ingestion of respiring bacteria and the eventual evolution of the mitochondrion. The proximity of cyanobacteria to other bacteria in the mats allowed the ingestion of the cyanobacteria and subsequent evolution of the chloroplast. Thus, microbial mats and their mutualistic associations were almost certainly instrumental in the evolution of photosynthesis, cellular respiration, and the eukaryotic cell! It is in mats that the increase in atmospheric O_2 began. This would eventually make higher life possible. Mats were key builders and maintainers of the Earth's original and subsequent early ecosystems. The mutualism and commensalism in these mats were crucial in all of these processes.

Mats can form on many types of interfaces, including between water and sediment or rock below it, air and sediment or rock, soil and bedrock, and so on. They also form in certain animals; an example is in the hindguts of some sea urchins. Pathogens in animals form mutualistic mats that are difficult for the host immune system to fight. The species in mats vary according to their environment, but they tend to have ecotypes common to most of them. They have several layers with specific species of microbes dominating and adapted to each layer. As a general rule, the by-products of the microbes of each layer serve as nutrients for groups in other layers. Each mat forms its own food web. In wet environments dependent on solar energy, the top layers are generally dominated by oxygen-making, photosynthetic cyanobacteria, while the lowest layers are generally dominated by anaerobic sulfate-reducing bacteria. They are replete with mutualistic and commensal relationships, a community acting similar to a unified organism.

Every type of metabolism and method of obtaining food is used by mats: oxygenic and nonoxygenic photosynthesis, organic and inorganic respiration, fermentation, herbivory, predation, parasitism, and scavenging.

Many other highly structured mutualistic consortia of various types exist between two types of prokaryotes, which maintain a permanent cell-to-cell contact [17,18]. About 3.5% of prokaryotes are known to be involved in mutualisms with other prokaryotes (ibid). But this has been studied little, so many more such mutualisms may await description. They have not been shown to have the specific morphological adaptations of the mutualisms between prokaryotes and eukaryotes [19]. Common ones, and among the most developed, are phototrophic consortia, which consist of a motile, heterotrophic central bacterium of the beta-Proteobacteria [18] surrounded by up to 69 photoautotrophic green sulfur bacterial cells [17,18,20], called the epibionts, which never occur as free-living cells [21]. They are thus specifically adapted to life in the consortium. Cell division of both the central bacterium and the epibionts is also highly coordinated [22]. The epibionts can sense light, and have a signal exchange affecting the movement of the central bacterium, which only incorporates 2-oxoglutarate in the presence of light and sulfide used by the epibionts [23]. These consortia occur in many stratified lakes around the world (ibid; [22]), where they can be as much as to two-thirds of the total bacterial biomass in the chemocline [24], an example of how common mutualism can be. "*Chlorochromatium aggregatum*" is an example of a highly developed phototrophic consortium [25]. They act like unified organisms, so tight are the mutualisms.

EUKARYOTE-BACTERIA MUTUALISMS

The Mitochondrion

Mitochondria carry out cellular respiration in all eukaryotic cells. The mitochondrion evolved exogenously from the host cell in a process called endosymbiosis [26,27], as it was originally a bacterium capable of respiration that was swallowed but not digested by another cell. The ingested cell started as a parasite, evolved to be a mutualist, and then an organelle, as it lost functions other than respiration. The host cell lost respiratory functions, becoming dependent on the mitochondrion for this. It was mutualistic coevolution. Thus, all eukaryotes are composite organisms, hosting what were once bacteria carrying out their cellular respiration. Imachi et al. [28] found that the sequence of the genome of an archaeon from the deep sea indicated its ancestor may have captured, with branching protrusions, a bacterium that became the mitochondrion, suggesting archaea gave rise to the main part of the eukaryotic cell, the part that includes the cytoplasm.

Mutualistic coevolution of a host and respiring parasite led to a tight mutualism in a composite organism. This mutualism and coevolution led to a key innovation, the mitochondrion and cellular respiration, one of the most important evolutionary transitions in the history of life.

Mitochondria allow the use of O_2 to produce a spectacular amount of energy, resulting in the evolution and persistence of all higher, complex life. Eukaryotes have nearly 200,000 times as much energy per gene as prokaryotes because of mitochondria [29]. Without the mitochondrion, even small multicellular organisms would not be able to produce sufficient energy to survive. So this mutualistic coevolution was necessary for the spectacular diversity of all complex life, and all the interactions and ecosystems that resulted from them. It also allowed a large multitude of additional niches and habitats for prokaryotes and viruses provided by the eukaryotes, allowing for a tremendous increase in prokaryote and viral diversity.

Martin and Koonin proposed that the mutualistic origin of the mitochondrion led to the most important feature defining eukaryotic cells: the evolution of the nucleus, the very essence of the eukaryotic cell [30]. They proposed that mitochondria transferred DNA in the form of group II introns to the DNA of an archaeon host. The new introns spread to many areas in the host DNA, and underwent transitions from group II introns to spliceosomal introns at those positions. Spliceosomal introns lead to the cutting of RNA introns into smaller pieces that are then translated. Splicing of spliceosomes is slow compared to translation of mRNA. This caused strong selection for a membrane to separate the slow splicing of the RNA from the rapid translation of the RNA, so that translation would occur only with intact reading frames. So the nuclear membrane evolved. A cytoplasm that is free of chromosomes, something that does not exist in prokaryotes, was likely a more important innovation than a nucleus with DNA. It allowed many additional structures and processes that distinguish eukaryotes from prokaryotes, such as the cell's cytoskeleton, trafficking of chemicals across membranes, complex pathways for transmitting signals, and so on. So, Martin and Koonin think the evolution of the eukaryotic cell and its many innovations was precipitated by mutualism—the endosymbiotic evolution of the mitochondrion.

Plastids and the Chloroplast

The same endosymbiotic process as in the mitochondrion happened in the evolution of plastids, most notably the chloroplast, the cellular organelle that carries out photosynthesis in oxygenic photosynthesis. Here, a photosynthetic cyanobacterium was ingested by a bacterium of another species, and the same gradual coevolutionary process occurred, with the host and endosymbiont each losing functions until the chloroplast was an organelle.

Although still debated, the morphology, biochemistry, genome, and molecular phylogeny of plastids indicate a single origin of all plastids, in the great diversity of species carrying them [31,32]. There is also is secondary endosymbiosis. Primary endosymbiosis comes from the ingestion of a bacterium by another organism. Secondary endosymbiosis results when the product of primary endosymbiosis is swallowed by yet another eukaryote. This has occurred several times independently, giving rise to a great diversity of photosynthetic organisms, from green algae to red algae to plants. There is even tertiary endosymbiosis, in which the product of secondary endosymbiosis is engulfed by a eukaryote.

The chloroplast is a key innovation that is necessary for the existence of all eukaryotic organisms that perform oxygenic photosynthesis, from unicellular phytoplankton to trees. Plants would not be able to obtain their carbohydrates, and thus would not be able to exist, without chloroplasts. Without chloroplasts, only prokaryotic organisms, like cyanobacteria, would be able to photosynthesize. Without multicellular plants, multicellular animals would only exist as small, simple forms, or not at all. The great diversity of the many types of terrestrial forests, grasslands, and other ecosystems owe their existence to this mutualistic coevolution. Unicellular photosynthetic eukaryotes, such as the phytoplankton of the sea, also need chloroplasts to exist, and they are the primary producers that support all animal life in the ocean, and produce an estimated 70%–85% of the Earth's oxygen. Thus, the endosymbiosis of chloroplasts, making eukaryotic photosynthetic organisms, is responsible for the great diversity of these organisms from unicellular algae to trees. Multitudes of species benefit from photosynthetic eukaryotes, including herbivores, predators that eat these herbivores, pollinators, fruit and nut eaters, fungi, bacteria, archaea, viruses, animals that use plants for shelter, sea life that benefits from phytoplankton, and many others.

Mutualistic coevolution leading to the fusion of different species into composite organisms happened between a number of other taxa, not always involving animals or plants. At least one of the protozoan parasites that cause malaria, *Plasmodium falciparum*, has an unusual organelle on the side of its cell that carries out various metabolic processes. DNA sequencing has shown that it was once a free-living alga. Cyanobacteria of the genus *Nostoc* occur intracellularly in the fungus *Geosiphon* and the angiosperm *Gunnera*, and extracellularly in hornworts (a group of bryophytes), in the water fern *Azolla*, and in cycads. They provide nitrogen to their hosts, and carbohydrate in the case of the fungal hosts, and the hosts provide them with a habitat.

Nitrogen Fixation

Molecular nitrogen in the air, N_2, cannot be used by life. Most nitrogen fixation is the result of a mutualistic relationship between plants and bacteria in their root nodules, which convert N_2 to ammonia, which can be used by most organisms, especially plants. Perakis and Pett-Ridge [33] showed that trees that form mutualistic associations with nitrogen-fixing bacteria take up more rock-derived nutrients than adjacent plants that lack them. Nitrogen fixation is essential to most life. Without it, there would be too little nitrogen useable to life to support much life.

It is essential for almost all species, from microbes to mammals. There are free-living nitrogen-fixing microbes in the soil, and cyanobacteria carry it out in the sea, supporting that ecosystem. It occurs some in the air, mediated by lightning. But it is noteworthy that such a process so fundamental to life as nitrogen fixation is carried out in terrestrial ecosystems primarily by a mutualistic relationship. This relationship also occurs between some termites and fungi [34]. Without this mutualism, life on land would be far less diverse and complex.

Multicellularity from Mutualistic Evolution between Protists and Bacteria

Choanoflagellates are free-living unicellular and colonial flagellate eukaryotes considered to be the sister group and closest relatives of animals. Unicellular ones can be transformed into multicellular organisms by interactions with certain bacteria, which cause their cells to remain together after cell division [35]. The cells form epithelial rosettes that can share an extracellular matrix and bridges between cells. This is likely an adaptation of the bacteria to more easily move from one host to another. Porifera (sponges) have cells called choanocytes that have been shown to have come from choanoflagellates. Also, modern choanoflagellates live in small colonies, and multicellular choanoflagellates and the Porifera are sister groups, both having descended from the same eukaryotic clade, indicating sponges evolved from choanoflagellates [36]. This suggests multicellularity evolved from mutualism between protists and bacteria. And genomic data support the idea that Porifera were the first multicellular animals [37]. However, there is also a hypothesis that viruses were key in the evolution of multicellularity. Both these hypotheses strongly support the ABH.

Bioluminescence

Bioluminescence is the production of light by organisms, causing them to glow. In most species, bioluminescence is produced by chemicals made by the glowing organism itself, but in some, the light is produced by mutualistic bacteria that dwell in their light organs [38]. The habitat where bioluminescence is most important and common by far is the deep sea, where light cannot penetrate. The need for light in this dark environment means that the mutualistic bacteria that provide it and the mutualistic relationship itself are key to structuring the entire deep-sea ecosystem and essential to its high biodiversity. The extent to which this is true is surprising. There are 460 species of marine teleost fish, the main fish family living today, that form mutualistic relationships with bioluminescent bacteria. These fish occur in 21 families and 7 orders. Hundreds of species of invertebrates form such relationships. The deep sea displays spectacular light shows as a result.

The bacteria benefit because their hosts provide them with a mobile habitat, nutrients, and oxygen. The metazoans benefit in various ways, depending on the species. The advantages of bioluminescence include escaping predators through startling them; camouflage through countershading; prey warning predators of their unpalatability; misdirection of predators by such things as certain squid and small crustaceans expelling glowing bacteria; the prey making their predators easier for the latter's predators to see; locating, illuminating, attracting, startling, and confusing prey; warning and deterring competitors; recognition and attraction of and signaling to mates; and navigation in dark, or low-light environments [39]. Bioluminescence may be offensive and defensive in the same organism. It is so important in fish that they have specialized organs called "light organs" to contain their glowing bacteria, and specialized tissues that act as reflectors, lenses, and shutters used to control, direct, and diffuse the light. These organs show that coevolution was important in the development of bioluminescence. It has been proposed, although not proven, that bioluminescent dinoflagellates glow to attract predators of the small animals that eat them. This would be mutualism without the dinoflagellates living in and illuminating their mutualists, two levels up the food web.

There may have been an amazing exaptation in that Rees et al. [40] proposed the systems that evolved to detoxify oxygen when it first appeared as a toxin were likely the foundations for the evolution of many bioluminescent systems.

There are a handful of known genera of bioluminescent bacteria in mutualistic associations with higher organisms, and it is known that there are more unidentified ones. Glowing bacteria are very adaptable, able to leave their hosts and exploit new niches, from eating feces to being pathogenic to marine mammals [41].

Some nematodes that parasitize moth and butterfly caterpillars have glowing bacteria. Some think when the caterpillar dies, the glow might attract scavengers, helping disperse the roundworms.

Some squid have light organs with bioluminescent bacteria. In some, this helps with camouflage by countershading. In most squid, this is done by a light organ without microbes, but in the species *Euprymna scolopes*, the bacteria are an integral component of the animal's light organ [43]. One squid species controls the intensity of light from its bacteria to match the moon's brightness nightly, maximizing camouflage. Squid of the genus *Sepiola* have two bioluminescent bacterial species that coexist, with one favored in warmer temperatures, and the other in colder ones [44]. Squid release their glowing microbes into the sea every day, resulting in areas high in these bacteria [42]. Taking on new bacteria may be dependent on exuding them, and this microecology may be a mechanism to keep selecting more host-compatible strains of mutualistic bacteria. This exudation is also done by fish. In both squid and fish, it has a big effect on the size, density, and distribution of the bacterial populations in large areas of the sea. The squid and fish increase diversity of these bacteria in the open sea by this means.

The Deep Sea Angler Fish (*Bufoceratias wedli*) and Dragonfishes (family Stomiidae), both of which dwell in the very deep ocean, have an appendage on their heads called an esca that has bioluminescent bacteria making a long-lasting glow. They wiggle the glowing esca around to attract prey within striking distance.

The mutualism of bioluminescence came about by coevolution. It has provided advantages, a new adaptive zone, and novel niches to some metazoans, causing their diversification and maintaining that diversity. It increased the diversity of certain bacteria by providing them with habitat, a means of dispersal, and novel habitats to adapt to and hence diversify as their host species diversified.

Eukaryote-Eukaryote Mutualisms (Sometimes Including Prokaryotes)

Lichens

Lichens are composed of a tight mutualism, and are composite organisms, consisting of a species of fungus and a photosynthetic ally called a photobiont. The fungus benefits from carbohydrates produced by photosynthesis by its ally, and from nitrogen fixation from the air if the photobiont is a cyanobacteria. The photobiont benefits because the fungus gives it a habitat, protects it from the environment with its fungal filaments, which also gather moisture and nutrients that are available to both mutualists, and usually anchors the composite to the substrate. Both allies collect water and nutrients through air, rain, and dust, and the fungus retains water, provides a bigger area to gather nutrients, and sometimes provides minerals from the substrate. The photobiont is usually a single-celled green alga of the genus *Trebouxia* [45], which generally cannot grow without the fungus. Another common photobiont is the filamentous green alga, *Trentepohlia*, which can grow independently. About 10% of lichens have cyanobacteria, an example being *Nostoc*, as the main or only photobiont. Lichens with green algae and cyanobacteria restrict their cyanbacteria to special wart-like structures on their surface (ibid). These are called cephalodia, are found in about 3%–4% of lichen species, and likely exploit the nitrogen-fixing ability of cyanobacteria. Lichens with cephalodia can be important contributors of biologically usable nitrogen to the ecosystem. There are only about 100 species of photobionts. About 98% of lichen fungi are in the Ascomycota. Close to half the world's fungi are Ascomycota, and about 40% of these are found only in lichens. Nearly 20% of the estimated 2.2 to 3.8 million fungal species are associated with lichens. Considering the number of mycorrhizal fungi, leaf cutter ant fungi, and other mutualistic fungal species, the percentage of fungal species that are mutualists must be at least 25%, so at least 550,000 species of fungus are in mutualistic relationships. This supports my thesis that mutualism is very common. There are about 13,500–17,000 identified lichen species; there are obviously several times this many in reality. They range from the tropics to the poles.

Lutzoni et al. [46] showed that additions of photobionts have been infrequent during Ascomycota evolution, but that there have been multiple independent losses of the photosynthetic ally. Thus, many major Ascomycota lineages and species of exclusively nonlichen-forming species are derived from lichen-forming ancestors, including important ones to humans, such as *Penicillium* and *Aspergillus*. So lichens increased free-living Ascomycota species richness. And Yuan et al. [47] found lichen-like fossils dating between 551 and 635 million years old, showing the mutualisms evolved well before vascular plants.

At least 52 lichen genera on 6 continents have a third mutualist, a basidiomycete yeast, which evolved with them for 200 million years, making the system at least a 3-way mutualism [48]. The yeast likely produces chemicals that repel predators and microbial competitors, while getting a habitat from the fungus. Basidiomycete yeast may also benefit plants by repelling herbivores; this needs more study.

It is suspected that there is a fourth set of mutualists in lichens: bacterial communities [49,50]. The lichen provides a habitat, protection, and nutrient for the bacteria.

The bacterial contribution could be fighting pathogens and making nutrients more accessible to the fungus and photobiont, but this is not known. It is unlikely that none of the species in the bacterial community help the fungus or photobiont in any way, though there are likely some beneficiary commensal species that do not help either of them.

Thus, the lichen is a holobiont, and a well-balanced self-contained, complex composite organism and miniature ecosystem of several mutualistic species [51,52]. Since there are many lichen species, each with their own fungus and photobiont, the lichen mutualism has evolved independently many times. All of the mutualists and commensals in the symbiotic composite organism would have had to adapt to each other via coevolution; thus, all lichens came about by coevolution.

The composite lichen has quite different properties than any of its component organisms, taking on an entirely new form, with a different morphology, physiology, and biochemistry than either its fungus, photobiont, yeast, or bacteria.

Some lichens are commensal, with the photobiont not benefiting, and some are parasitic, with the photobiont being hurt. In many species, the fungus penetrates the photobiont's cell wall with structures similar to those produced by pathogenic fungi that feed on hosts. However, the vast majority of lichens are mutualisms.

Lichens are ecosystem engineers of tremendous importance in structuring ecosystems, and in the evolution and creation of biodiversity. It is estimated that 7% of Earth's land surface is covered by lichens. The cross-kingdom mutualism has allowed them to live in both a great variety of environments and in very harsh environments, including some of Earth's most extreme ones. They increase biodiversity by living in environments where most life would not otherwise be able to live, including Arctic tundra, hot deserts, cooled lava flows, bare rocks, rocky coasts, exposed soil surfaces, tree bark, as epiphytes on tropical rainforest leaves, and toxic slag heaps. They often alter these harsh environments, allowing other life forms to thrive in them. For example, several kinds of lichens grow on rocks, where they produce acids and other chemicals that break down rocks and other substances. Their root-like organs, called rhizines, penetrate rocks and help break them up. They thus make soil by breaking down rocks. This is key to life, because rocks are too hard and compact to act as a medium for plant growth. Other than purely prokaryotic organisms, only lichens can grow on rocks, and only lichens can break them down. Lichens also convert rocks to very-slow-release fertilizers. Lichens cause rocks to decompose and add minerals to the soil. The lichens themselves add some of the organic components to the soil when they die. Lichens are crucial in creating soil and contributing to soil quality. Lichens prepare the soil for more complex plants, and therefore animals, to take hold, on an ecological timescale. They are often the first stage of ecological succession. They also hold soil together, decreasing erosion by wind and water. On desert sands, they stabilize soil and enrich it with nutrients. Since cyanobacteria in some lichens fix nitrogen, some lichens fertilize the soil with nitrogen in a form available to life. The amount of nitrogen usable by life that they add to ecosystems can be substantial. In the Andes Mountains, above the forests but below the snow, lichens control the amount of water and nutrients in the soil, setting stable foundations for food webs that include unique plants, Andean Condors, and Spectacled Bears. They perform similar functions high in other mountains and near the poles.

They were among the first land-dwelling organisms, suggesting that mutualism was instrumental in the colonization of land. Many scientists believe they set the stage for the colonization of land by plants from the sea. They would have done this by making the soil, adding nitrogen to the soil, preventing erosion, breaking down rocks and adding their minerals, and adding organic compounds when they died.

They also increase diversity significantly in tundra. Greenland and Iceland would have virtually no vegetation if it were not for lichens building the soil. In many tundra areas, lichens are the only winter food source for caribou and moose, which would not exist in these areas without them. Elk also eat them. In some habitats, they are the sole food source for flying squirrels. Lichens are low in protein, but high in carbohydrates.

Also in the arctic tundra, lichens, together with mosses and liverworts, make up the majority of the ground cover, which helps insulate the ground. The insulation allows better health and higher diversity of the lichens, mosses, and liverworts themselves; better growth and higher diversity of the soil biota, including invertebrates; and growth of any small plants that might take root there. The insulation also occurs in the alpine tundra, aiding the growth of plants such as cushion plants.

Lichens absorb large amounts of water, and many species of small animals, from insects and other invertebrates to small mammals, obtain at least some of their water from lichens, especially in winter. Some birds and small mammals, such as flying squirrels, use layers of lichen fragments to build well-insulated nests. Their positive effects on diversity are maximized by the fact that they are very widespread and long-lived. Some lichen species, however, are not obvious benefactors, being parasites on or in leaves.

By building and enriching soil, lichens benefit soil organisms, including thousands of invertebrate species, moles and gophers, soil algae, fungi, bacteria, archaea, plants, and trees, and indirectly all the animals dependent on all of these taxa. The latter includes many mammal, bird, reptile, amphibian, and fungal species. Lichens indirectly help the many species that benefit from caribou and moose, such as wolves, and the species that benefit from the flying squirrels and birds that build their nests with lichens. Thus, they are another example of a mutualism tremendously increasing biodiversity, aiding thousands of species directly and indirectly, and promoting evolution, including in this case the evolution of land plants.

Lichens are networks of tightly interconnected species that can be among the largest organisms in the world, closely associated with trees, rocks, or soil. Coral reefs and higher organisms with their microbiomes are similarly networks. The fundamental unit of life may not be the organism, but the network.

Mixoplankton

The vast majority of protist plankton are mixotrophic in nature, being hybrid organisms that are both predators (feeding through phagocytosis) and primary producers (photosynthesizers). These plankton have recently been recognized as "mixoplankton" [53]. Mixoplankton can be broadly divided into two types—those which have their own photosynthetic apparatus, the constitutive mixoplankton, and those which need to acquire photosynthetic capability from photosynthetic prey, the nonconstitutive mixoplankton (NCM). Foraminiferans and radiolarians are such endosymbiotic mixoplankton. Some of the NCMs are mutualistic communities, with more than one species of photosynthetic plankton growing inside hosts. The following comments apply to all mixoplankton types [54].

Food webs with mixoplankton, bacteria, and phytoplankton (photosynthetic plankton incapable of grazing) are more productive than food webs containing protozooplankton (grazing microscopic plankton incapable of photosynthesis) in lieu of mixoplankton. Such productive food webs with mixoplankton help invertebrates and fish thrive and probably diversify. Although more research is needed, it appears that the fusion of primary productivity with phago-heterotrophy within single organisms has allowed tremendous diversification and increased the overall diversity of oceanic plankton communities and species higher up their food webs.

In coastal waters, plankton blooms can become so dense that they self-limit themselves by shading out the light. In such systems, mixoplankton can outcompete phytoplankton algal blooms by obtaining energy from eating other plankton, including the phytoplankton. In contrast, the purely photosynthetic phytoplankton grow better in spring, when there are more nutrients. Mixoplankton thus help maintain a high standing stock of primary producers in summer, allowing fish and invertebrate populations to thrive (ibid). As a caveat to the above, however, some mixoplankton can form toxic blooms.

There is an area in the central Atlantic Ocean covering thousands of square kilometers with very little nutrients. It was formerly thought that bacteria competed with phytoplankton for the nutrients, and protozooplankton grazing on bacteria returned nutrients to the phytoplankton. But work by Zubkov and Tarran [55] reveals more of a mutualistic relationship, in which bacteria consumed sugars and other nutrients exuded by the mixoplankton. In turn, the mixoplankton ate the bacteria, obtaining more nutrients (phosphate and iron) than they could obtain directly from the seawater. Simulations indicate that such a relationship enhances carbon sequestration, for when mixoplankton interact in a food web with bacteria and phytoplankton, they remove 65 g of carbon per square meter of seawater, compared to 30 g per square meter if the grazers on the bacteria are protozooplankton and the primary producers are phytoplankton [56].

Termites and Their Mutualistic Microbiome

Termites cannot digest the cellulose-containing wood they eat without the help of a complex microbiome in their gut. This mutualism allowed them to exploit this adaptive zone, diversifying into about 3,106 described species, with an estimated few hundred more undescribed. The digestive system of the termite is a complex ecosystem with species from all three domains of life. The eukaryotes are unicellular protists, including the genera *Trichonympha*, *Mixotricha*, *Dinenympha*, and *Euconomympha*. They degrade cellulose and produce hydrogen gas and CO_2. Bacteria in termites include the genus *Treponema*, which produces acetate, used by termites as both an energy and carbon source. *Desulfovibrio* reduces sulfate and transfers hydrogen, acting as a donor of molecular hydrogen. Bacteria of genera

Enterococcus and *Lactococcus* produce lactic acid as the major metabolic end product of carbohydrate fermentation. They tend to inhibit the growth of pathogenic bacteria and contribute to the health of their hosts, helping maintain a beneficial microbiome.

Many of the microbes of the termite gut are commensal or mutualistic with each other. The archaea include *Methanobrevibacter*, which is a mutualist with some of the protists, and produces CH_4. Prokaryotes attach to the cell surfaces of or live within the cytoplasm or nucleus of the protists. *Treponema* are spirochete bacteria that attach to bracket-like structures on the plasma membrane of the protozoan, *Mixotricha*, and contribute to its movement by beating synchronously [57]. *Treponema* benefits by living on and in the *Mixotricha* and accessing the nutrients it produces. It is mutualism. Protists of class Parabasalia, genus *Trychonympha*, have endosymbiotic bacteria that help them break down cellulose. Other protists of this class produce H_2 and CO_2 that is used by their mutualistic methanogen *Methanobrevibacter* (domain archaea) to obtain energy and produce CH_4. The methanogen eliminates H_2, allowing the protists to maintain an optimal pH and carry out decomposition [58]. Additionally, these two groups of microorganisms work together to digest cellulose and enhance its fermentation. Protists produce ATP during fermentation, used by both the many microbes and the termite host. Termites obtain little nitrogen in their diet. *Treponema*, *Spirochaeta*, *Citobacter freundii*, and *Enterobacter agglomerans* fix about 60% of the nitrogen in termites, converting N_2 in the atmosphere to NH_3. The bacterium *Bacteroides termitidis* converts simple compounds into fatty acids, does fermentation, and recycles uric acid waste. This recycling is an extremely important source of nitrogen. The protists in termites need and use the recycled and fixed nitrogen.

The enhancement of biodiversity and productivity due to the mutualisms and commensalisms between termites and their gut microbiome is spectacular. The reputation of termites as pests due to their tendency to eat human dwellings obscures the fact that they are of tremendous positive benefit to ecosystems. None of the myriad species has a net negative impact on natural ecosystems. Less than 185 species are pests to humans. Their ecological benefits follow.

Termites decompose and recycle. They generally feed on fallen trees and stumps, leaf litter, and animal dung. They decompose and recycle huge amounts of organic matter through the ecosystem. Some species also carry out these functions when they build their mounds out of dead plant material, such as dead trees.

They aerate the soil. Groups of species such as drywood termites live in soil, sometimes many feet deep. They build tunnels and carry out other activities underground, breaking down soil particles and increasing soil aeration. This allows plants to grow better, allows important soil organisms to thrive, and regulates Earth's temperature by inhibiting the growth of anaerobic methanogens and hence limiting their production of CH_4.

They add and concentrate organic nutrients to the soil when they cart dead plant material and animal excrement to their underground nests. Upon death, their decomposition adds to soil fertility. The great numbers of them in a colony mean this is not trivial. They improve soil texture and structure with their movements and activities, enhancing soil composition and fertility.

Termites help with water absorption in the soil. If soils are compacted, water could fail to seep down. Their underground tunneling breaks down encrusted and compacted soils, aiding the water absorption necessary to support plants and other soil life.

Pringle et al. [59] found the major players in shaping the African savanna to make it better for life are termites, although the large herbivores, carnivores, scavengers, and others play lesser, but important, roles. The researchers looked at termite mounds 10 m wide and spaced 60–100 m apart, built over centuries in central Kenya's grassland. High densities of animals and plants were found around each mound. Each mound was at the center of a burst of plant growth, and the closer plants were to the colonies, the more quickly they grew. Insect abundance fell off appreciably with greater distance from the mounds, as did the abundance, biomass, and reproductive output of insect-eating predators. Significantly, the centers of biological activity were highly organized in relation to one another and evenly dispersed. Interestingly, at the landscape scale, the evenly spaced distribution of termite mounds produced dramatically greater abundance, biomass, and reproductive output of consumers at different trophic levels than would have been obtained in landscapes with randomly distributed mounds. The even distribution of the mounds over a large area maximizes productivity and biomass over the entire ecosystem, and plays a key role in elevating the services the ecosystem provides. Satellite imagery was even more striking in showing this than the ground observations. The services termites provide that I listed above are likely the reason the mounds enhance the ecosystem's productivity and diversity. Also, termites bring in coarse soil particles to the fine particles around the mounds. These promote water infiltration of the soil, and minimize disruptive shrinkage and swelling of top soil in response to precipitation and drought. And the mounds have high levels of nitrogen and phosphorus as a result of the fertility-enhancing activity of the termites previously discussed.

Termite colonies provide good habitats with the temperature and humidity controlled and kept at favorable levels. They attract many species that utilize them, as well as the predators of these commensals. Kenya Dwarf Geckoes occur in high numbers near the mounds in Kenya. Now I shall discuss the use of termite mounds by animals in Australia only, but the mounds are similarly used by commensals all over the world. Golden-shouldered and Hooded Parrots nest in them, digging into them with their beaks when they are damp and soft. Golden-shouldered Parrots of the Cape York Peninsula use the conical "witches' hat" mounds of *Amitermes scopulus* as nests, while Hooded Parrots of Northern Australia nest on the bulbous mounds of the Spinifex Termite (*Nasutitermes triodiae*). The termites repair the holes created by the birds after nesting is finished. The parrots generally avoid mounds that show signs of recent nesting, giving the termites at least 5 years between nest excavations to repair the nest and and recover. Tree Goannas (*Varanus tristis*) visit the mounds to prey on Golden-shouldered and Hooded Parrot nestlings and eggs. Most Australian kingfishers use the mounds to nest. They sometimes fly at a good speed into the hard mound to create an initial hole, occasionally dying from the impact. They often leave the completed burrow vacant while the termites seal off the inside of the tunnel. This protects their nest from dust and drying air. Most kingfishers also use tree hollows or stream banks.

But the Buff-breasted Paradise Kingfisher uses termite mounds exclusively. The Pilbara dtella (*Gehyra pilbara*), a gecko, lives inside bulbous termite mounds in the day, crawling on their surfaces by night. Pythons of the genus *Liasis* enter in termite mounds, feeding on the geckos there. Female Lace Monitor Lizards (*Varanus varius*) dig nesting holes in termite mounds on the ground and in trees. They lay their eggs there, then leave the termites to patch up the hole, safely sealing the eggs in. Scientists believe they return to the nest at hatching time to help their offspring escape. Several beetle species inhabit mounds, some of which are mutualists, producing a nutritious secretion that the termites eat. The termite workers take care of some beetle larvae as if they were members of the colony. Some beetles are parasites. Some ride on the termites' backs. The larva of one species even expands its abdomen, projecting it over its head, mimicking a termite worker! Various species of silverfish (order Thysanura) and bugs (order Hemiptera) use the mounds as habitat. Earthworms seek refuge in them, but only when rain saturates the soil. The Fat-tailed Antechinus (*Pseudantechinus macdonnellensis*), a small marsupial, lives in bulbous termite mounds of the Spinifex Termite in some areas. When fires rage over the land, quolls, bandicoots, rodents, goannas, and Frilled Lizards use the mounds as protective shelters. (Quolls and bandicoots are marsupials.)

There are mutualistic and commensal relationships between the termite mound dwellers as well. The moth *Trisyntopa scatophaga* lays its eggs in the nests of Golden-shouldered and Hooded Parrots in termite mounds. The larvae live in a silk cocoon at the nest's bottom, feeding on the nestlings' droppings and remains of any young that die. This cleans the nest for the parrots, but the pupae form near the entrance of the nests and sometimes block the chicks from leaving the nest. So this is a case of both mutualism and parasitism. This is not a surprise, since relationships between species can be complex and have combinations of this type. The Green and Golden Bell Frog (*Litoria aurea*) sometimes utilizes the parrots' nesting holes; they make ideal sounding chambers for its call. Blowfly maggots feed on the parrots' droppings and dead (mutualism). After the parrot chicks have left the nest, crickets, spiders (including Redbacks), and more than one species of casemoth larvae inhabit it until the termites patch it up, so there is ecological succession. Redback Spiders are poisonous relatives of the Black Widow, which they resemble, and casemoths are a very diverse group of tiny moths. A Northern Quoll was found sleeping in an abandoned Hooded Parrot nest. In addition to these animals, a number of plant species grow on termite mounds. This is an incomplete summary of some of the animals that use termite nests in Australia. In a similar way, termites provide habitats to many other animals and plants all over the world.

During droughts, rainforest termites show greater activity, resulting in greater soil moisture, partially countering the effects of the drought, and increasing nutrient mixing and seedling survival rates [60].

Endophytes and Plants
Endophytes are endosymbionts, often bacteria or fungi, which use plants as habitats for at least part of their lives without causing disease. They have been found in all plant species studied, and only a small percentage have been characterized. One

leaf of a plant can have many bacterial and fungal endophyte species. Their relationships to their hosts are poorly understood, but many almost surely benefit their hosts by preventing pathogenic organisms from invading them [61]. They do this by direct competition and production of chemicals that inhibit the growth of pathogenic competitors. Some bacterial endophytes increase plant growth. Although fungal endophytes can cause leaves to lose water at higher rates, certain of them help plants survive heat and drought. Many forage and turfgrasses (e.g., *Festuca* spp., *Lolium* spp.) carry fungal endophytes of the genus *Neotyphodium* which may increase their tolerance to drought and resistance to insect and mammalian herbivores [62,63].

The endophytic fungus *Piriformospora indica* (order Sebacinales) is capable of colonizing roots and forming mutualistic relationships with every plant species on Earth, and is not a mycorrhizal fungus. It increases crop yield and plant defense against root pathogens for barley, tomato, maize, and several other food crops [64,65].

Bacterial endophytes, which belong to a broad range of taxa, can become intracellular in root and shoot cells of many plants, losing cell walls while still dividing and metabolizing. Paungfoo-Lonhienne et al. [66] postulated that intracellular bacteria may be a source of organic nutrients or vitamins for plants.

There are several endophytes that inhabit seaweeds and algae.

Endophytes greatly aid plants and helped the tremendous diversification of the plant kingdom. Since this mutualism is ubiquitous in plants, benefiting all plants, it indirectly benefits all species that benefit from plants, including pollinators, seed dispersers, herbivores, soil fauna, mycorrhizal fungi, and so on—an absolutely huge array of species. It also indirectly benefits the beneficiaries of many seaweed and algal species, also a spectacular number of species.

ANIMAL AND PLANT MUTUALISMS, SOMETIMES INVOLVING OTHER TAXA

There are many ways by which plants and animals are mutualistic. These often involve additional species mutualistic with them as well. The main examples of these are discussed below.

Leafcutter Ants, Plants, Fungi, and Bacteria

Leafcutter ants of 47 species [67] in two genera, *Acromyrmex* and *Atta*, farm fungus. They occur only in South and Central America, Mexico, and parts of the southern United States. They are mainly in the rainforest. They cut small pieces of tree leaves, flowers, or grasses, depending on the ant species, and carry them to underground nests. The ants do not eat the leaves, but inoculate them with a species of fungus. They clean the leaves and seeds, removing anything that grows on them other than the fungus, eliminating its competitors and fungal pathogens [68]. If the ants unknowingly bring back leaves toxic to the fungus, it secretes a chemical that warns the ant not to gather any more of that species of leaf. They mulch and fertilize their crop, and their larvae eat the fungus (adults feed on the leaf sap). So they literally farm the fungus, using the vegetation as food and a growth medium for the fungus. Thus, there clearly was coevolution between ant and fungus. The fungus is the larvae's only food. Different species of leafcutter ants use different fungus species, but all fungi used are members of the Lepiotaceae family. The fungus clearly benefits from this, but in addition, occasionally some of the fungus escapes and becomes free-living.

There are several nutrient-supplying bacterial species in a mutualistic relationship with the ants, and a mutualistic or commensal one with the fungus. The bacteria benefit from the ants and leaves by receiving food and a substrate. The bacteria break down sugars from the leaves, making a variety of nutrients that are likely used by the fungus and perhaps directly, certainly indirectly via the fungus, by the ants. Over half the bacterial species are in the family Enterobacteriaceae, whose members include the intestinal bacteria that help animals digest food. The bacteria also transport sugars; make amino acids; and manufacture vitamin B5, which is needed to break down protein, carbohydrates, and fats, and to produce energy from nutrients. All of this benefits both their allies. It is not clear if the bacteria benefit from the fungus. They might get nutrients and/or habitat from it.

Another mutualism, almost surely coevolved, has been found in this system. Rainforest soils and foliage are very poor in nitrogen, a key nutrient that ultimately determines productivity. Bacteria of the genus *Klebsiella* grow on the fungus that leafcutter ants grow on. This bacterium fixes nitrogen from the air at the same rate as nitrogen fixers in the roots of some plants. This nitrogen can be used by the fungus, ants, and likely other bacteria. It's possible that nitrogen-fixing bacteria played a critical role in the evolution of the ant-fungus mutualism. And scientists are studying whether nitrogen-fixing bacteria help break down the ants' cut leaves into a form digestible by their fungi.

There is yet another coevolved mutualism. Leafcutters use antibiotic-producing bacteria of the group Actinobacteria, genus *Pseudonocardia*, that grow in the metapleural glands on the ants [69] to keep their leaf cuttings free of the parasitic fungus *Escovopsis* (phylum Ascomycota, Sac Fungi), which kills the fungus the ants eat. The mutualistic bacteria are nourished by special glandular secretions from the ants. The antibiotics are specifically targeted to attack *Escovopsis* [70].

The term fungal garden is misleading; the ants farm fungal-bacterial communities. This is in fact a community of at least six taxa of mutualists, commensals, and prey (the plants). It is a web of many species of bacteria with the ant, plant, and fungus, and each species except the plant in the system directly or indirectly helps and is helped by every other species in the web, and at least most, but probably all, species coevolved together. For such systems, I coin the term community symbiosis. I have given several examples of community symbiosis in this chapter, including prokaryotic consortia, lichens, and others.

There are about 250 species of ants that farm fungus. Most grow the fungus on dead biomass. Only the 47 species of leafcutter ants grow the fungus on living biomass. Worldwide, there are a total of more than 12,500 described ant species and an estimated total of 22,000. Thus, the mutualism resulted in a relatively small amount of diversification of ant species, although the total species that resulted from it is large, considering all the fungal and bacterial species that coevolved with the ants, assuming these bacterial symbionts are present in all farming ant species, a reasonable assumption. There are also some species of termites in Africa that farm fungus.

The ecological impact and increase in biodiversity that resulted from this coevolved community symbiosis involving several mutualisms in leafcutter ants is tremendous. Leafcutter ants have a much greater ecological footprint than other ants that farm fungus, and are uniquely complex among fungus harvesters. Most ant fungus farmers have only a few thousand ants per colony, with tiny garden plots. Leafcutter colonies have over 8 million individuals, and their farms are huge in area, yielding over a ton of fungus each year. The central mound of the underground nests can reach 98 feet across, and have smaller mounds radiating from it and extending to a diameter of 520 feet, occupying 320–650 square feet. Their impact on the environment is exceedingly large; they are estimated to account for 80% of the numbers of individual animals in the rainforest! Nests of *Atta* spp. occur at up to 6 per hectare in tropical forests, especially in fragmented forests and forest edges. Ants that farm fungus on dead biomass have mutualistic relationships as well, but leafcutters have a tighter mutualism with their fungus, and this may at least partially account for their much greater success than the other fungus farmers. In fact, in all but about 16 species of the non-leafcutter fungal farmers, the fungus can survive without the ant; in leafcutters and the 16 other species, the fungus cannot survive on its own.

Leafcutter ants are among the most conspicuous features of tropical forests. They are the rainforest's dominant herbivores, harvesting parts of 10%–50% of forest plants in their local area [71,72], taking leaves, twigs, bark, flowers, fruits, and seeds (ibid). They have been shown to remove up to 8%–15% of the leaf area of Neotropical forests (ibid; [73]), and up to half the species in their colony's territory annually [71,72]. They harvest 13%–17% of the Cerrado savanna [74]. They are important seed dispersers and predators, since they collect large quantities of seeds and fruits [75]. The tremendous amount of seed collection is exemplified by the fact that the ants collect up to 160,000 seeds of *Miconia* (family Melastomataceae) per day [72]!

The ants are ecosystem engineers that transform ecosystems and habitats, mainly through nest construction and maintenance. They cause community-level changes at spatial scales far beyond their nests, for sapling assemblages differed markedly around nest and foraging areas when compared to control sites [76,77]. They have some impacts that decrease species richness, but overall, their net impact is to greatly increase it. Much research has been and is actively being done on this topic (see [76,78]; and references in both these papers). Nests of genus *Atta* are huge, with up to 8,000 subterranean interconnected chambers reaching as deep as 7–8 m underground [79], with a colossal system of foraging and ventilation tunnels. These augment soil porosity, aeration, water infiltration, and drainage [80]. The ants create big soil mounds of up to 250 m^2 (ibid), habitat for many microbial, plant, and invertebrate species, which support many animals, directly or indirectly. And if the soil is low in nutrients before the ants arrive, their enrichment of the soil aids some plants and the species that benefit from those plants. For nest building and maintenance, they sometimes move more than 20 m^3, or 40 tons, of soil to the surface [81]. The ants mix soil particles, which enhances the soil, and alter soil chemistry and the carbon cycle.

The ants transport soil materials, dead fungus, and other organic detritus into subterranean waste chambers or, in some species like *Atta columbica*, on refuse dumps on the nest surface (ibid; [72]), greatly enriching soil and aiding plant germination (ibid).

Plant roots display more growth at the refuse dumps on the surface. Both waste areas are habitat for a great diversity of soil microbes and litter invertebrates, which in turn support more species of plants and animals (ibid). Heap-working ants in their colony constantly shuffle the waste in the heap, aiding and hastening decomposition. Dead ants are placed at the waste heap's perimeter [82,83]. Some species leave seeds in subterranean refuse chambers, but these do not grow. Seeds may be lost along ant trails [68,84], accomplishing dispersal. Both intra- and interspecific competition are reduced among seedlings originating from seeds defecated by vertebrates and then dispersed by ants (e.g., ibid). The nests with aboveground refuse dumps are nutrient-rich (e.g., [85–87]), and have temperature or moisture regimes that enhance germination.

The carrying of underground soil to the surface inverts soil layers, and creates a new edaphic soil horizon [88], high in minerals, but low in nutrients. They actually create an entirely new layer of top soil, different from the rainforest's typical top-soil, that is habitat for different microbes and invertebrates than standard rainforest topsoil. So they increase species richness by increasing habitat heterogeneity, making a mosaic of habitats. Removal of vegetation causes nest mound soil to also have reduced organic matter (ibid), although refuse dumps tend to have higher concentrations of nitrogen and phosphorus [89]. These "islands of fertility" aid plant recruitment [85], and abet niche partitioning among plant species [85,90,91], thus favoring increased plant diversity. Strangely though, the external refuse piles have a lower diversity of seed species than areas without leafcutters. They are also transient, since nutrients decline precipitously in nests abandoned for over a year.

In grasslands, nests serve as safe sites for the establishment of tree and shrub species, often resulting in islands of trees in steppe or savanna where arboreal vegetation would not normally be favored [85,89,92]. Interestingly, trees and shrubs in these soils develop better defenses against herbivores, such as larger and denser spines. All this shows nest construction and maintenance in grasslands alter soils and nutrient distribution during and after the life of a colony [90]. Again, they increase habitat heterogeneity and thus species richness. The shrubs and trees provide resources for more animal, fungal, and prokaryote species.

Ant nests create clearings with no vegetation, similar to gaps from fallen trees, allowing light to penetrate and new trees to establish themselves after the ants abandon their nests. The ants produce varying soil compositions within these gaps, and the gaps vary greatly in size and hence light availability. Light availability affects which trees and shrubs can grow in a given area [93], as does soil chemistry. Although unproven, it is probable that the trees and shrubs partition different soil types and light niches created by the ants, increasing the number of plant species. By promoting a more varied environment, the ants create more niches and plant diversity (see [72]). And if light penetration is low, the ants help plants by increasing light availability.

They dig up to 8,000 underground chambers as deep as 8 m, interconnected by tunnels, and an extensive system of foraging and ventilation tunnels [79]. These augment the soil's porosity, aeration, water infiltration, and drainage [80]. These effects on the soil and the enrichment the ants provide to it increase plant diversity, and create local areas of exceptional diversity of small soil invertebrates. Nitrogen-fixing bacteria thrive in these areas. In fact, the ants have been shown to create an underground environment with spectacular invertebrate and microbial diversity, and this likely increases plant and animal diversity, since it supplies nutrients and supports food webs with many different prey species, allowing niche partitioning by the predators.

The underground ants are a food source for a number of predators of the eggs, larvae, pupae, and adults. The adults support a number of predators above ground. An example is a phorid fly that lays its eggs on a worker ant's head. The larvae burrow into the ant's body, and eat its insides. A small ant often sits on top of the leaf the worker ant carries to defend it from the fly.

On the negative side, most *Atta* species maintain nest clearings free of plants and debris [76,81,88], reducing soil fertility and plant recruitment. Researchers found the ants created gaps in the forest in around 95% of the colonies studied, ranging in size from 0.04 to 87.9 m^2. Overall, canopy openness and light availability at least doubled in ant nest plots compared with distant understory plots. Though small increments of light availability increase plant recruitment and diversity, these larger increments allow rain to wash the soil away, and the sun to bake and compact it, making it hard to impossible for trees to grow for up to 15 years. Scientists found twice as many tree species and a threefold increase in sapling density far away from the ant nests as near them. All seedlings on the nests died, and the ants also acted as seed predators. In areas where the soil is nutrient-rich prior to their appearance, the ants decrease plant survival. Where an area is disturbed and light and heat penetrate well, they increase desiccation, decreasing plant survival.

In pristine rainforests, the ants are better controlled by their predators, and the plants themselves are less palatable and have better defenses, so the ant colonies are kept at relatively low density, and they increase environmental variability, increasing the number of plant niches and species. In environments altered by humans, which are often on forest edges because of clearing of trees, the ants have less enemies, so there are more colonies per acre, and plants have less antiherbivore defenses, so the negative impacts of the ants on plants is increased, and the ants make the environment more uniform, decreasing the number of plant niches and species. Thus, a natural ecosystem has controls on the ants, and this results in leafcutter ants having a net effect that increases plant, and therefore animal, species richness considerably. The ants also increase soil microbial and invertebrate diversity in natural systems. Human-altered systems cause the ants to have negative impacts on diversity.

In addition to leafcutter ants, Chomicki and Renner [94] showed that the Fijian ant species *Philidris nagasau* gathers *Squamellaria* seeds (plants of family Rubiaceae), and put them inside cracks of trees to farm them. They fertilize them with their feces, harvesting the seeds when the plants are fully grown. The ants live in the plants' domatia, closed structures at the base of plant stems, showing this system coevolved.

Plants, Mycorrhizal Fungi, Rodents, and Helper Bacteria

There is a four-way mutualism between plants, rodents, fungi, and bacteria. Many vascular plants have mutualistic fungi called mycorrhizal fungi associated with their roots. The fungus can colonize within or outside the cells of the roots, but is always closely associated with them. The fungus benefits because the plants give it carbohydrates from photosynthesis, which the fungus cannot obtain on its own. The fungus provides the plant with a much greater root area to absorb more minerals and water. The fungal mycelium, the vegetative part of it, has a smaller diameter than the smallest root hair, so can penetrate soil that roots cannot, increasing the surface area for absorption. The mycelia consist of numerous branching, threadlike hyphae, which are very long and increase the effective length of the plant's roots about tenfold. And the fungi secrete acid that can dissolve or chelate ions. The fungus is in some ways better than the plant at obtaining nutrients. Mycorrhizal fungi are especially helpful to plants in nutrient-poor soils. Unaided roots may be unable to absorb ions that sometimes become immobilized, such as phosphate and iron. The fungus can obtain these for the plant. In fact, plant roots cannot always absorb demineralized phosphate in basic soils. But their fungal allies can access it and feed it to the plant [95]. In some ecosystems, the fungi grab the phosphate directly from leaf litter, and phosphate is recycled without ever going into the soil [96]. The fungi also actually make water from lignin and other chemicals. And fungal habitats are sponges that allow water to infiltrate the soil and stay there a long time, making it available to plants. The fungi break down rocks and make their minerals available to their plant allies. The fungi have extracellular metabolites that provide enzymes, acids, antiviral and antibacterial agents, and messaging molecules to their host plants.

It is generally true that plants with mycorrhizal fungi have enhanced growth rates compared with plants lacking them, and also show significant drought tolerance [97]. The mycelia infuse all terrestrial plant habitats on Earth, hold soils together, hold 30,000 times their mass as soil, prevent erosion, and generate the humus that is a large part of how soil supports life [98]. One cubic inch of soil has over eight miles of mycelia (ibid). Plants with mycorrhizal fungi are often more resistant to drought and diseases [99,100], possibly partly because of their improved water and mineral absorption ability. But the fungi also protect the plants against pathogens [101], including fungi, bacteria, and viruses, even attacking pathogenic fungi. And the fungi protect plants in acidic soils and soils contaminated with high metal concentrations. Plants grown in sterile soils or growth media without mycorrhizal fungi tend to do more poorly than if grown with the fungi. Without these fungi, plant growth can be slowed in early succession or on degraded landscapes [99]. Mycorrhizal plants are often better competitors and better able to tolerate environmental stresses than their nonmycorrhizal counterparts. The mycorrhizal fungus *Laccaria bicolor* lures and kills springtails to obtain nitrogen, some of which it transfers to its host plant. Eastern White Pine with this fungus obtained up to 25% of its nitrogen from springtails [102]! Also, the cell membrane chemistry of fungi is different from that of plants, so each is better at absorbing certain nutrients than the other. This association is extremely important in soil ecology. It appeared 440–460 mya, when the plants were first colonizing land. Plants could not have colonized land without mycorrhizal fungi mutualists. Some mycorrhizae associations are highly conserved since their first appearance in the fossil record.

So tight and important to the plant is this mutualism that roots secrete mucigel, which serves to create a favorable environment for the fungi, showing the two coevolved. It also helps create a favorable environment for nitrogen-fixing bacteria, showing this relationship coevolved. It also has other functions unrelated to mutualism.

The mutualism has allowed a tremendous diversification of both plants and fungi. Mycorrhizal fungi are present in about 80% of terrestrial plant species and 92% of plant families studied, and this is the most prevalent mutualism in the plant kingdom [103,104]. About 90% of both angiosperm and conifer species have mutualistic mycorrhizal fungi with their roots ([105]; for a review, see [104]). All orchids (family Orchidaeae) have mycorrhizal fungi, directly, or indirectly through a host they parasitize, in at least some stage in their life cycle, and the orchid family has about 6%–11% of all species of seed plants [98], with about 26,000 species. There are more than twice as many species of orchids as birds, and four times as many as mammal species. This again shows how common mutualism is.

There are tens of thousands of species of fungi mutualistic with plants [106], out of a total estimated 2.2 to 3.8 million fungus species. A specific example of the hidden diversity below ground that trees depend on is found in the Oregon White Oak (*Quercus garryana*), which forms associations with over 40 species of fungus. It is not atypical.

Amazingly, nutrients, carbon, and water can move between different plants through the mycorrhizal fungi [107], which form a large network of many interconnections, branches, and pathways, sometimes called the "wood wide web". This complex highway is redundant and can repair itself if broken [96]. Carbon has been shown to move between Paper Birch into Douglas Fir trees, thus promoting succession in ecosystems [108], and linking these two trees as mutually benefiting each other. Similarly, nutrients have been shown to move from alder and birch to hemlocks, cedars, and Douglas Firs [96]. Plants can also exchange nitrogen and phosphorous via root fungi. One fungus can connect thousands of trees connected by billions of connections (ibid). In many situations, the mycorrhizae create a fine underground, interconnected mesh that extends well beyond the roots of nearby trees, causing trees that may be far apart to become physically interconnected. This can prove essential for reproduction of trees and the survival of young trees. In forests where little light reaches the ground, such as pine forests, there can be insufficient light for young seedlings to photosynthesize, so they could starve. But the roots of young trees can link with mature trees via fungi, and thereby obtain carbohydrates from them, which could be their parents, or could even be trees of a different species. Thus, the mycorrhizae consist of a huge web of fungus connecting many trees of various species through their roots in a great "superorganism", a mutualistic and commensal, multispecies system of food sharing and communication, involving many species of all four groups of beneficiaries: animals, trees, fungi, and bacteria [109].

A stunning purple orchid called Coleman's Coralroot, which exists in only a few mountain ranges in the southwest United States, receives its carbohydrates from the photosynthesis done by the Arizona White Oak Trees connected to it by mycorrhizal fungi. Fewer than 200 specimens of this endangered plant are known to exist in the wild today. The Phantom Orchid, which cannot carry out photosynthesis, receives carbohydrates from other plants via the mycelia network. Some orchids, called mixotrophs, photosynthesize, but obtain some carbohydrate from trees via fungi. There are many other orchids that are epiphytes that receive their carbohydrates from their host trees which have mycorrhizal fungi; the fungus benefits the orchid in indirect commensalism in these cases.

Recent research has demonstrated that warning chemicals can be sent from the roots of one plant to another underground via the fungi [110]. When the bean plant, *Vicia fava*, is damaged by aphids, it emits volatile chemicals, particularly methyl salicylate, that repel aphids and attract their enemies, such as parasitoid wasps. These chemicals go through the air, and alert other plants of the aphids, causing them to produce the same protective chemicals. But the plants can also send warning chemicals from one plant to another through their roots and mycorrhizal fungi when aphids attack [110]. The plant receiving the warning chemicals can receive them before the aphids attack it, so it is prepared for the aphids in advance. Tomato plants in air-tight plastic bags that prevented chemical communication through the air were shown to send signals via the mycelia network, warning other tomatoes of the early blight disease fungus, *Alternaria solani*. Multiple levels in the food chain are affected, since the chemicals affect the behavior of both herbivores and their predators.

Barto et al. [111] showed that marigolds in the lab can pass negative allelopathic chemicals to each other through their fungi that slow the growth of competing plants and kill nematodes. Lettuce grown in soil where the marigolds that received the allelopathic chemicals had grown weighed 40% less than controls. Achatz et al. [112] showed American Black Walnut Trees passed the allelopathic compound, jugalone, to each other in the wild, and roots of tomato seedlings planted in the soil that received the jugalone weighed on average 36% less than controls.

Spotted Knapweed (*Centaurea maculosa*), Slender Wild Oat (*Avena barbata*), and Soft Brome (*Bromus hordeaceus*) can alter mycorrhizal fungal community composition, favoring fungi they can better connect with. Also, compounds produced by plants to attract bacteria and fungi that benefit them can be detected by insects and worms that consume the plants.

There are bacteria known as mycorrhizal helper bacteria, or MHB. They have been found every time they have been looked for, under many different environmental conditions, and in various plant-mycorrhizal mutualisms; several MHB species have been found in the soil influenced by the fungi of many host plants [113–116]. The evidence is clear that this represents a three-way mutualism, that MHB help both root fungus and plant and are helped by both. Five mechanisms by which these bacteria help the plant and fungus were proposed by Garbaye [117]: (1) facilitating the root-fungal recognition process, (2) enhancement of fungal colonization of the plant, (3) nutritional enhancement of fungal growth, (4) benefiting the surrounding soil, and (5) stimulation of germination of fungal propagules. Certainly the MHB aid both plant and fungus. The majority of bacterial species isolated from the Monterey Pine (*Pinus radiata*) and its mycorrhizal fungus, *Rhizopogon luteolus*, promoted the tree-fungus mutualism [118]. Duponnois and Plenchette [119] concluded that the MHB *Pseudomonas monteilii* strain HR13 significantly promoted mycorrhizal colonization of different acacia species. Mamatha et al. [120] identified a *Bacillus coagulans* strain that was able to increase mycorrhizal fungus levels in fungus-inoculated plants, while Budi et al. [121] reported the discovery of a new *Paenibacillus* strain that improves mycorrhizal formation while acting as a biological control agent against soil-borne fungal diseases. Founoune et al. [115] looked at acacia planted seedlings, and sampled mycorrhizosphere soil, roots, galls induced by root-knot nematodes, and *Rhizobium* (nitrogen-fixing bacerium) nodules after six months of culture. The mutualism between the acacia and ectomycorrhizal fungus was promoted by 14 bacterial strains isolated from the mycorrhizosphere soil, three isolates from the roots, and four from the galls. Shoot biomass of acacia seedlings was stimulated by eight bacterial isolates from soil, six from galls, and seven from roots.

Some bacteria actually work against the plant-fungal mutualism at times, since bacterial isolates repressed the establishment of the mutualism in Douglas Fir (*Pseudotsuga menziesii*) with its mutualist fungus, *Laccaria bicolor* [122]. But as a whole, MHB promote mutualism between plant roots and their fungi. MHB promote the establishment of mutualism by stimulating fungal growth; increasing root-fungus contacts and colonization; and reducing the impact of adverse environmental conditions on the mycorrhizal fungi in the root area [101]. These effects also help the fungi. And as to the third effect, the MHB detoxify poisonous compounds in the soil, such as allelopathic chemicals produced by plant species competing with the plant ally of the fungus (ibid). These compounds are toxic to both the plant and its fungus. MHB also contribute, together with their fungal mutualist, to protection against root pathogens (ibid, and references therein). A significantly higher proportion of the helper bacteria called fluorescent pseudomonads were found to inhibit the growth of seven fungi that attack plant roots when the bacteria were with the mycorrhizal fungus *Laccaria bicolor* with the Douglas Fir root than if the bacteria were in the surrounding soil [122]. MHB produce antibiotics that attack root pathogens. Plants produce chemicals called flavonoids that direct fungal growth towards the fine roots. There is evidence that MHB indirectly facilitate root colonization by mycorrhizal fungi, by inducing the release of the flavonoids [123]. MHB also stimulate the germination of fungal spores for fungal reproduction ([101], and references therein; [124]). MHB stimulate the growth of the plant's central root [125]. They also stimulate the growth of branching (lateral) roots ([126]; reviewed in [117,127,128]), which leads to an increase in area where plant and fungus can interact. MHB also help the fungi-root system obtain soil minerals and organic matter [101]. Few studies have looked at whether the plants, fungi, or both benefit the bacteria, but what has been done indicate both aid the bacteria. In glasshouse conditions, the MHB *Pseudomonas fluorescens* strain BBc6R8 survived significantly better in the soil in the presence of the mycorrhizal strain *L. bicolor* S238N from which it was isolated, whether the roots of the tree mutualist, Douglas Fir, were present or not (Frey-Klett et al, unpublished results, mentioned in [101]). Nonmycorrhizal Douglas Fir roots alone do not improve the survival of strain BBc6R8 [129], suggesting that this bacterial strain benefits directly from the fungus, but not the tree roots. In a like manner, the mycorrhizal fungus *Glomus mosseae* improved the long-term survival of the MHB strain *P. fluorescens* 92rk in the root area of tomato plants (*Lycopersicon esculentum*) [130]. Products the fungi exude serve as nutrients for the bacteria [101]. As far as the plant roots helping the MHB, they help directly by mobilizing nutrients from soil minerals and organic matter (which also aids the fungi), and indirectly by aiding the fungi that help the bacteria. And they provide the bacteria with habitat. Bacterial populations residing in the root-fungus area are a 100 to a 1,000 times larger than those residing in surrounding soils. There is molecular communication between all three players in the system, as well as with other microbes [128].

One mycorhizal fungus, *Suillus tomentosus*, produces specialized structures with its plant host, the Lodgepole Pine (*Pinus contorta* var. latifolia), which host nitrogen-fixing bacteria, which allow the pines to exist in areas of low nitrogen. (Most plants with nitrogen-fixing bacteria have their own root nodules that act as homes for these microbes.)

Mycorrhizal fungi frequently convert the carbon they receive from their plant ally to trehalose, a disaccharide sugar that has been proposed as a carbon sink [131,132]. Trehalose is responsible for the selection of specific bacterial communities associated with the fungus and tree roots in forest nurseries and plantations [129,133,134]. It has been shown to help an MHB strain of *Pseudomonas monteilii* promote the growth of the mycorrhizal fungus *Pisolithus albus* [135]. This bacterium significantly increased fungal growth on a laboratory culture with minimal nutrients with trehalose added, while none of the other seven organic compounds tested produced this effect. This shows coevolution occurred between the fungi and MHB.

MHB often occur inside fungal cells. These bacteria improve the growth of the fungi in the presence of substances from the plant roots [136]. It has been suggested that the presence of bacteria inside the fungus increases the exchange of genes between the bacteria and fungi [137]. This genetic exchange is likely and has good evidence for it ([101], and references therein). This could speed up the evolution of novel traits and functions in the two, in a genetic coevolutionary mutualism. It has also been suggested that some MHB genes make certain antibiotics that are transferred from the helper bacteria to the fungi [138].

The plant-fungus-bacteria complex together perform a crucially important function to the ecosystem. This complex solubilizes minerals in rocks, such as phosphorus and iron, changing them from unobtainable to available to the plants, fungi, and bacteria, and life in general ([101], and references therein). *Weathering and solubilization by this complex of organisms are the only or the main source of all plant nutrients, except nitrogen* (which is made available by nitrogen-fixing bacteria), in natural ecosystems (ibid). This mutualistic complex also sequesters carbon by weathering.

Also, roots provide habitats that promote bacterial diversity (other than MHB) to a spectacular degree. The roots provide bacteria niches that the bacteria partition, so that each of the following aspects of the roots represent different niches and support different species of bacteria: root tips, elongating roots, sites of emergence of lateral roots, older roots, and so on [139]. The root bacteria also differ from those found in the soil not associated with roots. Furthermore, bacterial species vary with plant growth stage [140,141], plant species [142,143], and cultivar within the plant species [144,145]. Bacterial species also vary independently of the plants and their roots, such as from soil types, seasons [144,146], and local climates. Thus, the bacterial species vary in both space and time, and are extremely diverse, largely due to the plants and their roots, but also due to other environmental conditions.

Even more, there are also many species of mutualistic soil bacteria associated with plants, quite apart from those associated with the mycorrhizal fungi. Some solubilize minerals, such as phosphorus; some synthesize plant hormones; some enhance mineral uptake directly; and some protect the plant from pathogens. The latter protect plants from pathogens by competing with the pathogens for a substrate or ecological niche, producing inhibitory and toxic allelochemicals, or inducing the plant to be resistant to the pathogen.

Archaea are also common root inhabitants. Under specific conditions, such as reduced oxygen, they may become abundant and active players in the mycorrhizal system. Poorly understood, they may be ecologically important.

Additionally, a three-way mutualism between a virus, mycorrhizal fungus, and plant has been shown [147]. A tropical grass called panicgrass (genus *Panicum*) from geothermal soils and its fungus (*Curvularia protuberate*) can grow at high soil temperatures in Yellowstone National Park [148]. But this heat tolerance only is present when the plant-fungus system has a virus associated with the fungus. The researchers who found this proposed naming this virus Curvularia thermal tolerance virus (CThTV). This three-way mutualism also occurs if the panic grass is replaced with a tomato plant (*Solanum lycopersicon*) [147]. This finding is not unique, as complex three-way mutualisms with viruses have been found among arthropods, bacteria, and mutualistic bacteriophages [149,150].

This mutualism of plant, fungus, and bacteria has a fourth mutualist: animals. Various species of deer, wild boar, mice, flying squirrels, other tree squirrels, and others (the specific species of each of these groups depends on the ecosystem) eat the fruiting bodies of various fungus species, including truffles in some ecosystems. The fruiting body of a fungus has the spores with which it reproduces; some fruiting bodies are mushrooms. The animal then moves around and defecates the spores of the fungus, dispersing them just like animals disperse plant seeds. Sometimes the animal even buries the spores, placing them closer to plant roots. The fungal spore then grows into the mycelium, growing down to associate with plant roots. Thus, fungi are an important food source for many animals, while the animals disperse their spores. Squirrels may inadvertently disperse fungal spores when they bury nuts, such as acorns, or pine or redwood seeds. All the animals in this mutualism benefit from plants by eating their fruit, seeds, leaves, or bark, and receiving shelter and indirect benefits from the plants. They aid plants by providing them with the mycorrhizal fungi through dispersing them, fertilizing the soil when they defecate, and in some cases dispersing the plants' seeds. The bacteria indirectly benefit from the animals' dispersal of fungi and indirectly benefit the animals by helping the plants and fungi , in a case of indirect mutualism. So, in this four-way mutualistic system of plant (often a tree), animal, fungus, and helper bacteria, each helps the other three and is helped by each of the other three players, either directly or indirectly. The four players all coevolved together. Since each of these four taxa usually consists of more than one and often many species in any given ecosystem, the mutualism actually is an extraordinary web of a great number mutually helping allied species, a community mutualism, a "superorganism".

Many interactions in the rhizosphere are mediated by photoassimilates that are excreted by plant roots. Plant roots release compounds that have evolved to serve mutualistic interactions with soil-dwelling organisms, mainly fungi and bacteria. Plants actually regulate the microbial community around their roots.

These mutualistic signals may have evolved from chemical defenses. They can be (mis)used by specialized pathogens and herbivores [151].

A study found that Deer Mice (*Peromyscus maniculatus*), Harvest Mice (*Reithrodontomys megalotus*), and California Voles (*Microtus californicus*) had twelve species of fungal spores in their pellets, so were dispersing them [152]. The three most common of these fungi were in the Ascomycota. Seedlings growing in the root zones of mature oaks have access to the mycorrhizal network of parent trees, but without small mammals, seedlings germinating outside this root zone may lack mycorrhizal fungi. The same study found evidence that small mammals dispersed fungal spores to the roots of oak saplings up to 72 m from mature oaks. The authors of the study suggested that regeneration of oak woodlands depends on the dispersal of mycorrhizal fungi by small mammals.

Sometimes mycorrhizal fungi become parasitic, taking nutrients produced by the plants during photosynthesis, and giving nothing back, hoarding nitrogen from the plants in times of nitrogen scarcity. Plants have adapted mechanisms to expel the fungus when this happens, although they are not always successful at this, since the fungus has adaptations to stay with the plant. But even when this happens, the fungus is likely usually a net benefit to the plant. Smith and Smith [153] have shown that nitrogen-stingy fungi help transport phosphorous to the plant, even when it at first appears not to be the case; plants with root fungi have higher tolerance to toxins like arsenate; and plants with mycorrhizal fungi tend to outcompete plants without fungi. Occasionally, however, the fungus even acts as a pathogen, not just a parasite.

Some plants not capable of photosynthesis steal carbon from mycorrhizal fungi without giving them anything in return. These plants usually parasitize the system of photosynthetic plants and their mycorrhizal fungal mutualists. A diverse array of plants from many taxa do this, including many orchids. Some do this partially, taking carbon from the mycorrhizal-plant system as a supplement to their photosynthesis. And some orchids exploit fungi that decay wood or litter.

These cases of cheating on both sides are not always mutualism, but do add diversity to the system, since these are cases of species using niches created by other species. In the case of the fungus being pathogenic, the principles discussed in Chapter 11 on the beneficial effects of parasites on their hosts apply.

It is now apparent plants could not have colonized land without the help from fungi. Pirozynski and Malloch [154] hypothesized that terrestrial plants are the product of an ancient and continuing mutualism of a semiaquatic ancestral green alga and an aquatic fungus. The Siluro-Devonian colonization of land by plants, and indeed the very evolution of plants, was possible only through such mutualisms, which allowed plants to obtain water and nutrients (ibid). Since land plants evolved from aquatic green algae, many of the earliest land plants lacked true roots. Instead, they had rhizoids, root-like structures that anchored them to the soil. They could not have obtained the soil nutrients they needed to survive without fungal allies. Of plants living today, plants with poorly developed root systems are much more responsive to mycorrhizal fungi than plants with complex, well-developed, branching root systems [155]. One of the biggest challenges plants faced since their appearance on land has always been extracting water and nutrients from the soil, which mycorrhizal fungi greatly help them to do. Direct fossil evidence supports the need of fungi for plants to colonize land as well. Fossilized plants dating from 400 mya show well-preserved structures that look identical to modern fungal hyphae that penetrate root cells and form highly branched structures [156]. The earliest land plant fossils are 450 million years old, and genetic evidence based on a molecular clock indicates that the mycorrhizal fungal phylum *Glomeromycota* originated between 450 and 360 mya [157].

Fungi also built the soil for plants. *Tortotubus*, a 440 million-year-old fungus, the oldest example of a land-dwelling multicellular organism, recently found in fossilized form, created rich, deep soil, and helped incorporate oxygen and nitrogen into the soil. Its ability to store and transport nutrients through decomposition helped create a nutrient-rich layer of topsoil that was necessary for plant life to colonize land [158].

Plant Communication

When attacked by herbivores, plants emit warning chemicals that warn other plants, even those of other species [159]. Plants receiving the messages produce defensive chemicals before herbivores start eating them. Volatile warning chemicals from lima beans eaten by beetles cause corn seedlings to mount a defense against beet armyworms; tobacco responds to sagebrush chemicals; chili peppers and lima beans respond to cucumber emissions; and willows, poplars, and sugar maples warn each other. This is best viewed as the plant that sends the warning evolved this ability to warn its kin, while the plant that responds to a warning from a different species evolved to pick up the warning, rather than to hypothesize that the plant that emits the warning chemical evolved to warn other species, since it is hard to imagine an advantage of warning different species of danger. It is mutualism in cases in which two species warn each other.

Plants, insects, and fungi have a mutually adaptive communication system. When insects eat plant roots, the plants release volatile chemicals into the air, causing insects above ground to select another food plant, avoiding poisonous defense compounds in the plant and competition with underground insects. Plants also leave specific chemical messages in the soil. Plants that grow where a previous plant left a message can "tell" whether the former one was suffering from leaf-eating or from root-eating insects, and produce the appropriate defensive chemicals. These messages are passed on to insects, strongly influencing their growth and possibly their behavior. Insect communities are influenced by the messages from plants that formerly grew where the insects are present. Scientists grew ragwort plants in a greenhouse and exposed them to leaf-eating caterpillars or root-feeding beetle larvae. They then grew new plants in the same soil and exposed them to insects again. The soil fungi that developed depended on whether the insect had been feeding on roots or leaves. The fungal community then affected the growth and chemistry of the next batch of plants and, as a result, the insects on the plants, both herbivores and their predators. Thus, using the fungal community as an intermediary, the previous plants greatly affected the growth and palatability of the current plants, and the new plants passed down the message in the soil from the past to caterpillars and their enemies (ibid). It is not known how long this message lasts.

Forty out of 48 studies of plant communication confirm that plants detect airborne warning chemicals of emitting plants, and ramp up their production of defensive chemicals or other defense mechanisms in response (ibid).

Predatory invertebrates also cue in on plant volatiles. When beet armyworms consume maize, it releases volatiles that attract parasitoid wasps that lay eggs in the caterpillars' bodies, which the wasp larvae consume (ibid), a process called indirect defense. When spider mites attack lima beans, the plants emit volatile chemical signals that are received by nearby plants of the same species. This causes activation of defense genes in the recipient plants, making them less vulnerable to attack. The genes also cause substances to be synthesized that attract a mite species that is a predator of spider mites (ibid).

Gagliano et al. [160] showed that plants can also communicate with each other via clicking sounds. They used powerful acoustic devices to hear clicking sounds coming from the roots of corn saplings. They placed the young roots in water and played a continuous noise at a similar frequency to the clicks, 200 Hz, and the plants grew towards the source of the sound. The researchers concluded that plants can communicate with each other by making clicking sounds that travel through soil without difficulty. The sounds may warn of threats to the plants receiving them. It is possible that sensitivity to sound and vibrations plays an important role in plants' lives. It remains to be seen whether different plant species can communicate in this way.

As discussed previously in this chapter, mycorrhizal fungi often act as an important channel through which plant communication chemicals can travel, even between different plant species.

Mutualisms Exclusively between Plants and Animals

Plants and animals are in a profound mutualistic chemical relationship. Photosynthesis is the reverse chemical reaction of respiration. Through photosynthesis, plants use water and CO_2 and the energy of sunlight to produce carbohydrates and O_2. Animals eat the carbohydrates and use them as a food source, and breathe the O_2. Through respiration, animals (and plants) use O_2 and carbohydrates to produce CO_2, water, and energy. The energy produced from respiration is necessary for all higher life. Plants use the CO_2 produced by respiration for photosynthesis. This chemical mutualism is essential for animals to survive, and helpful to plants. One might consider it a fortuitous mystery that photosynthesis and respiration are reverse, complimentary chemical reactions.

Plants, phytoplankton, and other photosynthesizers are the basis of the food chain supporting all animals, directly or indirectly. Plants also provide animals with habitat, shelter, shade, and protection from predators. They regulate global temperature and pH through carbon sequestration, regulating levels of CO_2 and CH_4, keeping their atmospheric levels beneficial to life. They regulate local climate and rainfall through evapotranspiration and the provision of shade. They stabilize the soil, give homes to bacteria that make nitrogen available to life, and provide other services to animals. Animals regulate plant populations to the benefit of plants; till, aerate, fertilize, and enhance the soil; pollinate; disperse seeds; and provide other services to plants.

This general mutualistic relationship between the many species in these two kingdoms includes many examples of coevolution that benefited both kingdoms, resulting in a phenomenal increase in biodiversity. Plants increased animal biodiversity enormously in numerous ways, and vice versa. The myriad species that evolved from these coevolved mutualisms created niches that were employed by many additional species from viruses to prokaryotes to protists to fungi to additional species of plants and animals.

Two of the major of the major mutualisms, pollination and seed dispersal, are discussed below.

Plant Pollination and Seed Dispersal by Animals

Pollination and seed dispersal by animals are fundamentally important to angiosperms, and play a crucial role in the maintenance of forest ecosystems throughout the world, while simultaneously benefiting animals. Animals as pollinators and seed dispersers have long been suspected as agents of angiosperm radiation [161–163], and theory indicates these could drive plant diversification by both increasing speciation [164] and lowering extinction [165]. It is now clear that the spectacular evolutionary success of tropical flowering plants over the past 145 million years is attributable to their reliance primarily on animals, rather than the wind, for pollination and seed dispersal. They diversified profusely in the Cretaceous (145–66 mya) and Tertiary (66–2.6 mya) periods, and are still doing so now. This diversification is one of the greatest on land in Earth's history, and has profoundly and fundamentally altered terrestrial ecosystems. The 250,000–400,000 species of angiosperms make up about 90% of all living plant species [166–168]. Essentially all angiosperm species are pollinated and have their seeds dispersed by animals, which is at least two mutualisms per angiosperm species, showing the ubiquity of mutualism. There is good evidence showing insect pollinators greatly promoted early angiosperm diversification [161]. It has been proposed that animal seed dispersal led to the later Tertiary Period angiosperm radiations [169]. There are many causes for angiosperm diversification, and much of the cause of it is not understood, but much of it is a result of pollination and seed dispersal by their animal mutualists. Pollination and seed dispersal diversified animals tremendously as well.

Pollination of Angiosperms by Animals

Pollination is mutualism because the plant benefits by being pollinated, and the pollinator receives a nutritious meal of pollen, nectar, or both. Animals and angiosperms have coevolved in this mutualistic relationship. It is clear that flowers arose as a result of coevolution with animals, because flowers are designed and specifically adapted for animal pollination, and there is no other reasonable explanation for the evolution of the flower and its parts to exist. The only purpose of a flower is to attract animal pollinators. It is also accepted because in the cases where angiosperms species switched back to wind pollination, the flower starts to disappear. Animal pollination is more efficient than wind pollination, and is an efficient agent of cross-pollination that prevents self-fertilization.

The origin of flowers was largely orchestrated by insects. Crepet [161], in reviewing the literature on angiosperm evolution, suggested that there were two very rapid periods of flowering plant evolution. The first was associated with the flower's origin, driven by animals as the selective agents. It is generally thought that coevolution between insects, most likely beetles, and preangiosperms caused the evolution of angiosperms. Although the ancestor of angiosperms is not known, it is thought that beetles ate the flowers of precursors to angiosperms, perhaps structures similar to gymnosperm pollen cones or ovulate cones. They then started pollinating them, herbivore becoming pollinator, and flowers evolved as a result of selection for being more efficiently pollinated. Natural selection favored the modification of leaves into petals and other flower parts, and favored nectar, smells, and colors that reward and attract insect pollinators. This was a huge macroevolutionary step provided by mutualism and coevolution.

The second rapid period of angiosperm evolution Crepet suggested was adaptive radiations of floral forms and pollination mechanisms driven by and associated with the diversification of insect forms. Insects evolved adaptations, including sensory ones, to better obtain nectar and/or pollen. In this coevolution, some of these adaptations resulted in the insects being better pollinators. The flower was a key adaptation that allowed more efficient pollination and allowed flowering plants to enter into a new adaptive zone, resulting in tremendous plant diversification. Angiosperms are the most diverse land plants today, with 416 families, approximately 13,164 known genera, and about 295,383 known species [170]. This major evolutionary transition and diversification happened via coevolution with insects.

Fossil evidence indicates that angiosperms originated at the end of the Jurassic, about 145 mya, and had become the dominant plants, at least in North America, by about 90 mya [171]. Their rapid diversification occurred at end of the Cretaceous and beginning of the Tertiary, when specialized butterflies, moths, and bees appeared. It is believed that mainly beetles at first, and then these latter taxa fueled the tremendous diversification of the angiosperms. Flowering plants became dominant largely as a result of coevolution with flying insects. Later, other pollinators, such as bats, birds, and other insects, radiated tremendously in coevolving with angiosperms, which greatly diversified as a result of these later pollinators. Animal pollinators have increased angiosperms diversity spectacularly, and vice versa, each by acting as a coevolutionary selective force on the other.

Variation in floral preference of conspecific pollinators and variation in conspecific floral types led to multiple speciation events, sometimes with coevolution leading to specialization of flower and pollinator.

The myriad of floral adaptations to advertise to, attract, and meet the needs of animal pollinators is evidence that flowers evolved adaptations to attract animal pollinators, and angiosperm diversity was driven at least in part by coevolution with these pollinators. These are visual, olfactory, and tactile, and have led to a great diversity of flower coloration, shape, fragrance, and so on. They include nectar, which provides a tasty and nutritious meal specific to the tastes and needs of the pollinator, and nectar guides, marks that direct the pollinator to the pollen and nectar [172]. Pollinators have caused plants to evolve a great diversity of species and floral types ([173], and references therein). Bee-pollinated flowers are yellow and blue, with ultraviolet nectar guides, since bees see these colors, but not red. Moths are nocturnal and have a good sense of smell, so the flowers they visit are white or pale to show up in moonlight, often only open at night, and have a strong, sweet perfume that advertises their presence in the dark. Specialized adaptations are particularly prominent in orchids, a very diverse family with about 26,000 species, in 880 genera, which have developed highly specialized pollination systems. It is clear their great diversity is largely from coevolution with pollinators, mainly insects. They even have species that fool male insects into attempting copulation with them by mimicking females of the insect, a deceit done on species of bees, wasps, and flies, as well as mimics that look and smell like rotten flesh that attract fly pollinators. This is not mutualism, but does illustrate flower diversity resulting from coevolution with pollinators. The consensus today is that the tremendous success and diversity of angiosperms is due largely to animal, mainly insect, pollination, but not solely as a result of it, since the ability of flowering plants to grow quickly, obtain nutrients, produce chemical defenses to ward off predators, and tolerate fluctuating climatic conditions are also important factors. Seed dispersal by animals is another important factor.

The fact that the first flowers were plain and likely not attractive to pollinators led some to suggest that angiosperms diversified for other reasons, such as faster growth and being better at obtaining nutrients [174]. This is supported by fossil leaves that indicate angiosperm leaves gained many more veins during the Cretaceous, allowing more water for photosynthesis and growth [175]. Yet simple flowers would have attracted pollinators, because they would be the only flowers in existence, being the first. That could have led to diversification, with complex flowers following from competition for pollinators. Berendse and Scheffer [174] also assert that angiosperms promote soil nutrient release by producing litter that is more easily decomposed, giving them nutrients, in a positive feedback that may have resulted in a runaway process once angiosperms had reached a threshold abundance. While contrary to the idea that pollinators promoted their diversity, this idea still strongly supports the ABH. In fact a combination of factors, with pollination being among the most important, best explains angiosperm success and diversification.

Widely cited figures of the percentage of the Earth's angiosperm species that are animal-pollinated range from 67% to 96% of total species, but these have not been based on concrete data. Baker and Hurd [176] pointed out that the vast majority of flowering plants are insect-pollinated. Ollerton et al. [177] estimated the number and proportion of flowering plants that are pollinated by animals using published and unpublished community-level surveys of plant pollination systems that recorded whether each species present was pollinated by animals or wind, finding a mean of 78% in temperate-zone communities and 94% in tropical ecosystems. By correcting for latitudinal diversity trends, they estimate that worldwide, about 308,000 species, or 87.5% (given *their* estimate of the total number of angiosperm species), are animal-pollinated.

Thorne [166] says that wind-pollinated angiosperms tend to be either self-pollinating, employ vegetative reproduction, or reproduce by apomixis. Thus, this mutualism is very common, and there is a correlation between angiosperm diversity and their being animal-pollinated. Gymnosperms are the closest relatives of angiosperms, and are mainly wind-pollinated. Their 720 species compares to 250,000–400,000 angiosperm species. The difference in pollination methods is one of the several factors explaining this difference. Some gymnosperms and their relatives are animal-pollinated, although not a large percentage of them. Ollerton et al. [177] state that plant-pollinator interactions play a significant role in maintaining the functional integrity of most terrestrial ecosystems. Still, about 12% of angiosperms are wind-pollinated, including the Poaceae, with 780 genera and about 12,000 species, one of the most successful of all plant families, and the Cyperaceae, with 90 genera and 5,500 species. So wind-pollination is a successful adaptation, though much less so than animal pollination.

The immense number of animal pollinators indicates how common mutualism is in animals and how much coevolution with the plants they pollinate has increased animal species richness; this is especially true with insects. It is estimated that there are about 200,000 animal pollinator species [178]. They have also greatly diversified as a result of adapting to flowers. Insect taxa with a great number and percentage of species that pollinate flowers include four insect orders: Lepidoptera, Hymenoptera, Diptera, and Coleoptera. The vast majority of insect species are in these four orders, although not all species in these orders are pollinators. Over half of all eukaryotes are insects, and work by Erwin [179] indicates that a very approximate estimate of total insect species globally might be as high as 20 million. Another insect order that has species that pollinate flowers is Thysanoptera (thrips). There are other insects and invertebrates that pollinate. There are many mammalian pollinators, such as bats, rodents, primates (including lemurs), and the honey possum of Australia, which occupies a niche similar to the hummingbird. About 500 species of flowering plants rely on bat pollination and thus tend to open their flowers at night [180]. Many rainforest plants rely on bats for pollination and/or seed dispersal [181]. Many birds are pollinators. An omnivorous lizard of the genus *Lacerta* that climbs up flowers for nectar on the island of Madeira pollinates. There are about 500,000 insect species associated with flowering plants in some way, including eating them, indicating insects greatly diversified via coevolution with plants by other mechanisms in addition to pollination.

Commonly, angiosperm species have more than one pollinator, and animal species pollinate multiple plant species. This also results in high pollinator and plant diversity. In extreme cases, there are pollinators that service hundreds of plant species, as exemplified by the Western Honeybee (*Apis mellifora*), which visits both wild and cultivated plants [182]. On the other side, the White Mangrove (*Laguncularia racemosa*) is pollinated by at least 65 different insect species [183]. These yield pollination networks that result in many directly and indirectly interacting plant and animal species, supporting and promoting highly diverse communities with the species connected in complex, high-information webs.

Animals also have many adaptations to obtain nectar and/or pollen from the flowers they pollinate. Butterflies have olfactory organs and taste buds on their feet, and a long proboscis to suck nectar. Corbiculate bees, which include honeybees, have body hair that collects pollen, and they use their legs to scrape it into the corbiculae, basket-like structures on each hind leg. These are in fact pollen baskets and serve no other purpose than to store and transport pollen.

There are some unusual variations on the pollination mutualism, resulting in yet more diversity. Some bees use waxes and resins from flowers to build their hives [184]. Neotropical orchids produce terpenoids, volatile chemicals that male euglossine bees collect by scraping their brush-like foreleg tips all over the flowers and transferring it to storage sacs on the back legs. The accumulated fragrances induce the males to form leks, are released by the males at their display sites in the forest, and are thought to act as a perfume that attracts females [185]. The bees' preference of volatiles changes throughout the year seasonally as the species of orchids producing them bloom and die. Tephritid fruit flies also obtain plant compounds for use in sexual behavior. A group of Central American orchids called bucket orchids are equipped with a small bucket structure behind the flower. Oil from the flower drips into the bucket, and attracts bees with its odor. When the male bee smells the oil, he goes to the flower to collect an oily substance that he uses it as a perfume to attract females. But he often falls into the bucket. The only way to escape is through a tube full of pollen. He takes this pollen to the next flower and pollinates it. There are many species of bees and bucket orchids, many specialized to their own sole species of pollinator or flower. Numerous orchid species have nearly microscopic blooms that release a mildew-like odor that attracts small pollinating flies. Yucca moths lay their eggs in the Yucca flowers they pollinate, and their caterpillars eat some of the seeds [186].

Clarke et al. [187] showed that the buff-tailed bumblebee (*Bombus terrestris*) detects changes in the charge petunia flowers emit that signal information important to the bee, including how much pollen they have, if another bee has visited them recently, and what type of flower they are. This advertising is always truthful. It is adaptive to the flower, because it encourages the bee to visit if it signals it has pollen. If it does not have pollen, it would not benefit from a bee visit, and it keeps the bee from losing interest in future visits. Bees build up a positive charge in flight. Flowers are negatively charged. As a bee approaches a charged flower, a small electric force builds up that can transmit this important information. Floral electric fields vary in structure and pattern, and bumblebees can discriminate these. The electrical signals work in conjunction with the flower's other signals, like color and scent.

Sunflowers use internal circadian clocks, acting on growth hormones, to follow the sun across the sky during the day, keeping warm [188]. Warm sunflowers attract five times as many pollinating insects, mainly bees, as those that are not warm.

Figs are not fruits, but a collection of tiny flowers inside a fleshy compartment. The fig tree is pollinated only by the fig wasp. The "fruit" has a hole in it which only the fig wasp can widen just enough for her to lay her eggs inside. The larvae develop inside the "fruit"—i.e., inside a flower field—getting covered with pollen the whole time, while they feed on the fleshy outside part. When they reach the adult stage, they fly off to mate. Then the females find other fig "fruit" to lay their eggs in. In the process of doing that, they will pollinate the fig tree, since the pollen from when they were larvae in the fig flowers is still stuck on their bodies. There are almost 700 species of fig plants. In the majority of cases, each species is pollinated by only one or a few wasp species [189]. So a tremendous diversity of fig and wasp species was created by coevolution, specialization, and speciation of several fig species and their mutualistic pollinating wasp species.

A study indicates that, at the family level, insects did not evolve or diversify primarily as a result of coevolution with the flowers they pollinate, for Labandeira and Sepkoski [190] found the fossil record shows the greatest expansion and diversification of insects at the family level began 120 million years before the advent of angiosperms, which was 145 mya. Thus, insect radiation occurred about 245 mya, soon after the Permian extinction. And 85% of insect mouth part types evolved not to exploit angiosperms, but were largely in place nearly 100 million years before flowers appeared. Insect diversification at the family level actually slowed when angiosperms became dominant. Sepkoski said that angiosperms may have evolved to take advantage of insect types already in existence, not vice versa, as many had supposed, leading credence to the idea that insects were a major causative agent in the evolution and radiation of angiosperms, as discussed above. They did, however, conclude that coevolution with plants sparked the great diversification of insects about 245 mya. Instead of angiosperms, they think it was gymnosperms that coevolved and interacted with insects, and were a major factor in their spectacular radiation at that time. This would obviously still support the ABH, since it is diversification of one taxon as a result of its interaction with another. Angiosperms had a tremendous positive impact on the diversity of both the Coleoptera suborder Phytophaga and butterflies that consumed plants, albeit by herbivore-host coevolution, not pollination.

However, at the species level, it appears insects diversified greatly due to coevolution with angiosperms, and at least partly because they pollinate them. They are the class with the most pollinators and the most total species. The four insect orders that have the highest percentage of species that are pollinators also have the most total number of species. This includes the Coleoptera, with many pollinators, and about 400,000 species [191], having more species than any other order of any multicellular organism, constituting almost 25% of all known animals species, and about 40% of all described insect species, and with new species constantly being found. Hymenoptera have many pollinating species, and 150,000 recognized species, and many more to be discovered. Lepidoptera seem to have the highest percentage of pollinating species of any order; they have a total of 180,000 described species in 126 families [192] and 46 superfamilies [193], making up about 10% of all species of organism (ibid). Diptera have 120,000 described and an estimated 1,000,000 species. Mosquitoes are nectar-feeding pollinators, and have about 3,500 described species and many not yet discovered. Some flies pollinate. The correlation between diversity and percentage of pollinators is present, though not as strong for Diptera as for the other three orders.

Many taxa have benefited or are dependent on the great diversity of flowering plants and pollinators that resulted from this mutualism. Nonpollinating taxa that benefit from the success of angiosperms include soil biota, such as earthworms, pseudoscorpions, ants, earwigs, springtails, sow bugs, and other invertebrates, numbering in at least the tens of thousands of species; and soil fungi, algae, protists, bacteria, archaea, and viruses, collectively likely totaling millions of species. Other beneficiaries include the many insects, rodents, ungulates, and others that feed on fruits, nuts, bark, and leaves of trees, shrubs, and bushes, or use them for shade, shelter, hiding places, or homes. They include birds that nest in trees or eat seeds, tropical fruit-eating bats, and primates that live in and/or feed off trees. Indirectly, predators of these beneficiaries of angiosperms benefit from flowering plants. These include frogs, salamanders, turtles, lizards, predatory birds, insect-eating bats, predatory cats and canines, and many more. These predators also benefit directly from forests or plants that provide them with habitat. Tremendous numbers of species of various kinds benefit from angiosperms in rainforests, cloud forests, boreal forests, mangrove forests, prairies, grasslands, and deserts.

The pollinators benefit countless species as well. Insects are food for many species of bats, birds, lizards, and other insects. These in turn are food for snakes, birds of prey, and many other species. Some insect pollinators control plant populations, preventing plants that are better competitors from driving their competitors locally extinct, increasing plant diversity. Some pollinators are also seed dispersers, not necessarily of the plant species they pollinate. Examples are some ants and some fruit bats. There are many cases of chains of indirect benefits emanating from pollination. For example, moles eat pollinating insect larvae underground, and aerate the soil with their tunnels, benefiting many species. So the mutualism of pollination directly and indirectly benefits many species of virus, prokaryote, fungus, plant, and animal, greatly increasing and maintaining species diversity of all these groups.

An example of the importance of pollination and mutualism involving humans is found in *Apis mellifera*, the Western Honeybee. Humans benefit the bee by providing boxes for their hives in great quantities, greatly increasing their numbers, at least until our pesticides and other destructive environmental practices started causing the collapse of their colonies. Their contribution to our well-being through pollinating our crops is stunning. They pollinate 80% of our crops worldwide, and 70 out of our top 100 food crops, which supply about 90% of the world's nutrition. Bees pollinate the best, most healthful of our foods, which are fruits, nuts, and vegetables. Worldwide, they pollinate food valued at $350 billion per year. Civilization would possibly not survive a total bee collapse.

To comprehend the importance of the pollination mutualism, imagine that there were not any angiosperms due to loss of pollinators, or vice versa. Then all species directly or indirectly dependent on either of these mutualists would decrease in number or disappear, resulting in crashes of ecosystems and an incomprehensible loss of biodiversity. The loss of the lepidoptera alone would be a catastrophe for predators of the caterpillars, such as birds, wasps, and so on; for predators of the adults, such as birds and bats; and for the plants they pollinate. Tens of thousands of species would go extinct.

Seed Dispersal by Animals

Many kinds of animals disperse plant seeds (see [194], and references therein). Animals are key dispersers of seeds in tropical forests, which are notably dependent on fruit-eating birds and mammals for dispersal of their seeds.

Seed dispersal by animals provides plants with a reliable dispersal agent, helping the plant species in at least seven ways. It prevents the offspring from being shaded out by its parent. It reduces the competition between offspring, and between the parent plant and offspring, for nutrients, water, and sunlight, which they would experience if they germinated near the parent plant. It increases the probability of cross-fertilization, preventing inbreeding and keeping the variability of the population high. It increases the probability that seeds and seedlings will escape such

natural enemies as invertebrate and vertebrate seed-eaters, herbivores, and pathogens. Widespread dispersal by mobile animals such as birds and bats allows plants
to colonize new habitats. Animals often selectively drop seeds in areas favorable to
survival of the plant. For example, the Latin American tree species, *Ocotea endresi-
ana*, is dispersed by several bird species, including the Three-wattled Bellbird. Males
of this species perch on dead trees to attract females, with the result that they often
defecate beneath dead trees. This is favorable to the seedlings because there are
higher light levels and less pathogenic fungi. Finally, seeds surrounded by fleshy
fruits that go through animal digestive tracts often have greater germination percent-
ages and speeds, and germinate at different times, decreasing seedling competition.
All these factors increase seedling survival rates. Although the first five advantages
of seed dispersal I listed are also conferred by wind dispersal, the last two are unique
to animal dispersal, and animals can be more effective than wind at some or all of
the first five. The widely accepted Janzen-Connell hypothesis states that seedlings
have a better chance of survival if they are dispersed far from their parent and where
few plants of their species grow, because there are more herbivores and pathogens
attacking them if they are near others of their species [195,196]. Sometimes animal
seed dispersal distances are limited, likely reducing gene flow between populations,
increasing plant speciation and hence diversity. The animal benefits by getting a
tasty, nutritious, energy-rich meal.

Seed dispersal resulted from coevolution, for plants evolved tasty and nutritious
fruits that attract animals, giving them an incentive to eat them, while animals
evolved better ways of finding and obtaining the fruits. Animals often disperse seeds
far from the parent plants as a result of eating the fruit, moving away, and then spit-
ting out or defecating the seeds. Mutualistic fruit dispersal is used by most tree species. An estimated 51%–98% of canopy and subcanopy trees in Neotropical forests
and 46%–80% in Paleotropical forests are vertebrate-dispersed [197]. This is mainly
by mammals and birds, with bats and primates being key taxa in mammals.

Tropical forests are dependent on fruit-eating birds and mammals for dispersal
of their seeds, and could not exist without them. Fruit-eating bats are important pollinators and seed dispersers, especially in the tropics. The vast majority of fruit bats
disperse plant seeds by eating the fruit and carrying it in their body on the wing
before defecating it at a distant site. At least 300 plant species in the Old World are
known to rely on fruit bats (family Pteropodidae) for their propagation, from pollination, seed dispersal, or both. Bats are fruit dispersers of major importance in tropical
forests, essential to keeping the system diverse, healthy, and functioning. Since they
defecate while flying between their day roosts and feeding areas, they are among
the most effective of the tropical seed dispersers. Species of flying foxes, large fruit
bats of the Old World, are among the best examples of this, for they carry seeds several miles in their nocturnal foraging flights. Two tropical, fruit-eating bat families,
the Old World flying foxes (Pteropodidae), and the New World leaf or spear-nosed
bats (Phyllostomidae), each contain over 100 species that disperse seeds from hundreds of species of tropical trees and shrubs. One study found bat and wind dispersal are the most effective ways for fast-growing trees to colonize new habitat [198].
This could result in the creation of new species. If either dispersal method ceased to
exist, plant diversity would be greatly reduced, resulting in much fewer animal species

as well. It is estimated that each square meter of rainforest floor receives between 12 and 80 bat-dispersed seeds annually [197]. And the seeds of some species, such as the Cecropia tree, have higher germination rates after they have passed through the intestines of bats. Flying foxes, especially of the genus *Pteropus*, promote plant speciation, because they can carry seeds many miles—sometimes between islands in the southeast Pacific—in their nightly foraging trips [199].

Let us consider the importance of just one bat species in the rainforest. In but one night, a single Seba's Short-tailed Fruit Bat may eat up to 60,000 seeds, with an average in the tens of thousands. Even with a very conservative estimate that each bat eats 1,000 seeds each night, just one colony of 400 bats would disperse 146 million seeds a year. Each cave roost in the Costa Rican tropical dry forest at Santa Rosa contains up to 400 of these bats. Most seeds land where they cannot germinate, but if but one-tenth of 1% germinated, 146,000 new seedlings would result from this one colony.

There are also mammalian dispersers in the Carnivora, Rodentia, Proboscoidea, Perissodactyla, and Artiodactyla. Elephants play an important role. Primates are important seed dispersers, eating the fruit of trees and defecating the seeds a distance from the place of ingestion.

Birds are important seed dispersers, although a small percentage of species, such as some species of parrots, are destructive feeders, only dispersing seeds that go through their gut abnormally. Most commonly, bird dispersers eat fruit, like berries, and defecate the seeds far from the parent plant. Fruits have evolved many specialized enticements for birds. Some even provide high levels of antioxidants that protect birds from damaging free radicals that would be produced by the heavy oxygen consumption of birds in flight. Chili peppers evolved their hot burn to deter mammals that would chew and destroy their seeds from eating them. Birds are not sensitive to chili peppers, and are their dispersers. One species of cassowary disperses the seeds of 37 species of trees in Australia's Daintree Rainforest in north Queensland. The three extant species of cassowary live in New Guinea, nearby islands, and northeastern Australia, and disperse the fruit seeds of several hundred species of plants and trees from at least 26 families, sometimes over a distance of one kilometer.

In *The Origin of Species*, Darwin [200] explains that he had the leg of a Red-legged Partridge that had been preserved for three years with a six-ounce mud ball stuck to it. He broke up and watered the ball to test his own theories on seed dispersal. Eighty-two plants sprouted from five different species! Fruit bats and birds in tropical rainforests play such an important role in seed dispersal that this ecosystem would collapse if either were removed.

Less known are some reptiles that transport seeds, including turtles, lizards, and alligators. The Galapagos Tortoise is important in dispersing some cactus fruit [201]. The local variety of tomato there can germinate only after passing through such a tortoise, but not through other animals [202]. The Eastern Box Turtle (*Terrapene c. carolina*) has been found with mature seeds within it of, among others, *Rubus*, *Fragaria*, *Prunus*, and *Vitis*, and has a well-developed sense of smell and ability to see orange-red [203]. In southern Europe and USA, wild strawberries are eaten by turtles. On the Canary Islands, an omnivorous lizard of the genus *Lacerta* eats and disperses fruits of the rubiaceous shrub *Placama pendula* in its feces [204].

There are numerous other examples. Seed dispersal is considered an important force shaping the ecology and evolution of vertebrate and tree populations.

In the Amazon River system, many species of fish eat the fruit that drops into the river. Some fish species migrate in the rainy season to the upper reaches of streams in the Amazon River system, where fruits grow more densely. Here, the fruits start to ripen in the rainy season when the water is high and the forest flooded, so fish have the best access to them. The fish and trees thus coevolved for fruit dispersal by fish. Worldwide, more than 100 different fish species have been found with viable seeds in their guts.

There are some snail species that disperse seeds. Dung beetles in deserts contribute to germination by burying feces with seeds, thus supplying them with fertilizer. Seeds have been found in earthworms, which mostly defecate them underground; Darwin recognized that they can act as dispersers. Some fungi and mosses have their spores transported from one dung heap to another by flies.

Ants also disperse seeds of shrubs and understory herbs [205], and are by far the most important seed dispersers among the invertebrates. Dispersal by ants, called myrmecochory, is done by many ant species, and seed-dispersing ants are of fundamental ecological importance as a result. Adaptations to ant dispersal have evolved independently at least 100 times in 55 Angiosperm families [206,207], and are present in 77 plant families. They are estimated to be present in at least 11,000 Angiosperm species, but probably up to 23,000 species, which is 9% of flowering plant species [205]. Myrmecochory occurs in every major biome on every continent except Antarctica [207]. In areas where it is most common, such as in northern temperate forests of eastern North America, it is found in up to 30%–40% of understory herbs [205]. Seeds of myrmecochorous plants have a lipid-rich attachment called the elaiosome, which attracts ants. They feed the elaiosome to their larvae, discarding the otherwise intact seed in an underground chamber. Elaiosomes evolved independently several times from a wide variety of plant tissues [205], showing strong selective pressure favoring them.

Besides increasing distance from the parent plant, the plant benefits because ant dispersal tends to be directed to nutrient-rich or protected sites that enhance germination and establishment of seedlings. Ants carry the seeds to their underground nest and discard them there or carry them out [208]. Myrmecocory coevolved. Plants have evolved mechanisms to manipulate the ants' behavior and hence seed fate [209], directing the seeds to desirable areas. For example, some plant species make rounder, smoother seeds that are difficult for the ants to remove from the nest. The nest is ideal for seed germination, as it is full of plant nutrients, such as nitrate and phosphorous. This is especially important in areas with low-nutrient soils. Ant dispersal decreases ingestion of seeds by seed predators when ants sequester the seeds in their underground nests. This is vital where seed-eaters are common. In certain forests, seed-eaters eat about 60% of all dispersed seeds within a few days, eventually consuming all seeds not dispersed by ants [210,211]. However, this relationship is complicated by the fact that seed predators that have a taste for elaiosomes can increase seed predation [212].

A study was done to test the extent to which ant dispersal affects plant species diversity [206]. Researchers looked at 101 angiosperm lineages in 241 genera from all continents but Antarctica, and found that ant-dispersed plant groups had an average of twice as many species as related groups that were not ant-dispersed. This is because dispersal of seeds by ants isolates plants, increasing speciation rates, and it also benefits the plants, decreasing extinction rates. Contrasts in species diversity between sister groups demonstrated that diversification rates were higher in myrmecochorous lineages in most biogeographic regions. This study showed that seed dispersal by ants is an important worldwide driver of plant species diversity. The study demonstrates the best example of an effect of mutualism on large-scale diversification known thus far. Myrmecochory is a globally important driver of both plant and ant diversity.

Some seed predators, which include many rodents, such as squirrels, and some birds, such as jays, also disperse seeds, hiding them in caches. They forget where they hid some of the seeds, which then can germinate. For this reason, several plant species have evolved to encourage seed predators, not discourage them. I propose a hypothesis stating that seed predators coevolved with plants to become seed dispersers, with the plants evolving fruits or nuts that attracted seed predators.

Seeds are sometimes dispersed secondarily, as when dung beetles disperse seeds from clumps of feces in the process of collecting dung to feed their larvae.

To get an idea of how common plant seed dispersal by animals is, Restrepo et al. [213] did a study on the percentage of genera and species in the four families of New World mistletoe that are animal-dispersed. Mistletoes represent the largest assemblage of hemiparasitic woody angiosperms. The study found a total of only one genus and twelve species of Mistletoe had their seeds dispersed by the wind, while 23 genera and 650 species had their seeds dispersed by animals via fruit ingestion. In addition, two genera and 25 species had their seeds dispersed by a combination of ballistic dispersal and the possible help of animals, including having the sticky seeds get attached to vertebrate feathers and fur. This means animals potentially aided in dispersal in 25 genera and 675 species, out of a total of 26 genera and 687 species studied. Thus, animals likely help with dispersal of mistletoe seeds in the New World in 96% of the genera and 98% of the species. We do not know if this is typical; more studies need to be done to determine the extent to which plants in general depend on animals for seed dispersal. But as a study of all the New World genera of the largest assemblage of hemiparasitic woody flowering plants, it certainly lends evidence to the hypothesis that a large majority of flowering plants are animal-dispersed. Also note that all plants that bear fruit or nuts are animal-dispersed, and this is a gigantic number.

The mutualism of animal dispersal of seeds stabilizes communities, allowing spectacular species diversity. Taking the above estimate by Stoner and Henry of 51%–98% of canopy and subcanopy trees in Neotropical forests as being animal-dispersed, these forests would have a maximum of only 2%–41% of their current tree species if it were not for their seed dispersers. Since other trees, animals, and fungi are dependent on the trees that would not be present, the percentage of trees would be much lower than this, and the number of animal and fungal species would be much less. There would be a similar loss of prokaryote and virus diversity. The soil would not be as healthy and capable of supporting diversity without the leaf fall from a high diversity of trees. If only the bats were removed, the rainforest ecosystem would collapse. Conversely, if all the plant species whose seeds are animal-dispersed were removed, many animal species would die, adversely affecting their predators and prey, and the plant species they pollinate. The system would have a small fraction of its original diversity. These arguments apply to all forest types.

There is little doubt that animal dispersal of seeds has increased plant diversity a great deal. Plants have evolved fruits of various colors, tastes, and nutrients to attract seed dispersers. This caused diversification of plants. Animals often carry the seeds long distances to new, isolated areas, allowing geographic speciation of the plant species. Animals sometimes even deposit the seeds on islands where the plant species did not previously occur, which can result in founder effect speciation of the plant. And animals have adapted to the plants they disperse, resulting in diversification of the animals. They have evolved teeth, digestive systems, smell, taste, preferences, seed burial, and other morphologies and behaviors in response to plant seeds and fruits. This has resulted in adaptive radiation. It has been suggested that primate color vision evolved largely to detect fruits against backgrounds of leaves, while fruits evolved colors to be seen, in a coevolutionary process, and there is good evidence for this ([214], and references therein). The mutualistic coevolution of seed dispersal has greatly increased the diversity of both dispersers and angiosperms. Incidentally, Isbell [215] gave evidence for an alternative hypothesis that posits that this color vision developed when early primates underwent antagonistic coevolution with venomous snakes, developing better vision for color, detail, and movement, and the ability to see in three dimensions, helping them see snakes. This phase of the evolution of primate vision was not from fruit-eating, but further supports the ABH.

Bello et al. [216] found hunting decreased hardwood tree diversity and abundance in 31 tropical Atlantic forests in Brazil, with larger trees being replaced by smaller species, resulting in a 10%–15% loss in carbon sequestration. This is significant, since tropical forests are responsible for about 40% of carbon sequestration worldwide. The main reason for the tree loss is the decrease in seed dispersers. Hunting is an inadvertent experiment that shows the importance of animal seed dispersers.

Plants are composite organisms of eukaryotes and former prokaryotes that coevolved to be their chloroplasts and mitochondria. They are part of a large coevolved, mutualistic, superorganism-like, very tight interacting network of endophytes; mycorrhizal fungi, helper bacteria, and rodents that inoculate them with these fungi; pollinators; seed dispersers; other plant species they communicate with; mutualistic soil organisms; predators of their herbivores; epiphytes; and even their herbivores. Their key roles in carbon sequestration, climate regulation, albedo regulation, stabilization of the water cycle, O_2 production, soil retention, river formation, and other such services to the ecosystem that in turn supports them are further manifestations of this coevolved, mutualistic, interacting network.

CORAL REEFS

Coral is a composite of two species living in an intimate mutualism, very much as if they were one organism. The coral polyps are in the phylum Cnidaria. Their mutualists are photosynthetic unicellular dinoflagellates called zooxanthellae that live in their tissues, providing 90% of the organic nutrients that nourish the polyp, including amino acids that build proteins, and supplying glucose and glycerol for energy. They also supply up to 90% of the polyp's energy requirements [218,219]. Without zooxanthellae, coral would grow too slowly to form a reef. The polyps secrete hard exoskeletons of calcium carbonate that support and protect the algae and themselves, and form the coral reef, with an average of one million algal cells for every cubic centimeter of coral. Polyps also provide a constant supply of the CO_2 the algae need for photosynthesis. Also, polyps obtain nitrogen from zooplankton, and share some of it with the algae, which need it. Coral absorb nutrients directly from water, including nitrogen and phosphorus. There are several types of corals; most reefs are built by stony corals (order Scleractinia). The coral reef is an ecosystem of incredible biodiversity, having the highest number of species of any aquatic ecosystem, marine or freshwater, and higher species diversity than any terrestrial ecosystem, with the possible exception of tropical rainforests. In terms of number of phyla, coral reefs are more diverse than rainforests, since some reef-dwelling phyla, such as Echinodermata, do not occur on land. There are a huge number of coral species, each polyp being associated with its specific alga species, and each alga with a unique polyp species. The hard, carbonate reef provides a three-dimensional habitat for a tremendous diversity of animals. Although coral reefs cover less than 0.1% of the world's oceans, about half the area of France, they support over 25% of all marine species [220]. Over 4,000 species of fish, often very colorful, inhabit coral reefs. The amazing diversity includes great numbers of species of seabirds, worms, crustaceans, mollusks (including cephalopods), cnidarians, sponges, tunicates, echinoderms (including sea stars, brittle stars, sea cucumbers, and sea urchins); a few species of sea snakes and sea turtles; and even visits by whales and dolphins. Inside coral are a great diversity of crabs, shrimps, snails, and other invertebrates. Niche partitioning is the rule on reefs. Each coral species provides niches for—and has associated with it—many species of fish and invertebrates, many of which are not associated with the other coral species on the same reef. Partitioning happens temporally too, with the same corals used at different times by different species. Reefs are among the world's most productive ecosystems, with primary productivity typically producing 5–10 g of carbon per square meter per day [221,222]. Reefs can produce up to 35 tons of fish per square kilometer per year [223]. Organisms can cover every square inch of reef, and these organisms often provide complex and varied habitats for additional species. Reef biomass is correlated with species diversity. Since most coral reefs are under 10,000 years old, the diversity was generated exceptionally rapidly, with surprisingly fast speciation rates.

Vermeij et al. [224] showed coral larvae detect coral habitat from long distances by being attracted to reef sounds, mainly from fish and crustaceans. This benefits coral, which then aid the fish and crustaceans by providing habitat. The sound of the habitat enhances the habitat by recruiting coral, and helps its keystone species.

Before present-day coral reefs, there were other reefs. From a few thousand years after hard skeletons were developed by marine organisms to today, there were reefs. They always created a three-dimensional environment, protection, and food for animals, plants, and prokaryotes, greatly increasing biodiversity. Coral reefs flourished at their greatest extent in the Middle Cambrian (513 to 501 mya), Devonian (416 to 359 mya), and Carboniferous (359 to 299 mya), due to the extinct coral order Rugosa, and in the Late Cretaceous (100 to 65 mya) and the entire Neogene (23 mya–present), with the stony corals dominating. Not all reefs were made by corals. In the early Cambrian (542 to 513 mya), calcareous algae and small animals with a conical shape likely related to sponges called archaeocyathids formed reefs. In the late Cretaceous (100 to 65 Ma), extensive reefs were created by rudists, a group of bivalves that had some forms that possessed a lower valve that was a roughly inverted spike-like cone attached to the seafloor or neighboring rudists, and an upper valve that was much tinier and flat, and acted as a lid. Not all rudists formed reefs, but the conical ones did. They ranged in size from a few centimeters to well over a meter in length. Rudist reefs sometimes reached heights of hundreds of meters and often ran for hundreds of miles on continental shelves. They once fringed the North American coast from the Gulf of Mexico to the present-day Canadian Maritime Provinces, creating three-dimensional habitat for a high diversity of species. These noncoral reefs may have lacked mutualistic algae, and we do not know how much mutualism they harbored. Nor do we know their diversity in detail. So we cannot compare their diversity with today's coral reefs to test how much mutualism might be responsible for coral reef diversity.

There is a spectacular amount of mutualism and commensalism on coral reefs. Some of these directly involve coral and hence the reef structure. Many species of sponges are mutualists with coral, and necessary for reef survival. Coral provides them a habitat; they live in crevices in the reef. Some reef nutrients exist in a form coral cannot use. Some species of sponges filter this, converting it to small particles that can be utilized by coral [225], as well as by some algae. Yet these and other sponges could overrun the reef if not controlled. This is prevented because nudibranchs, sea anemones, and sea turtles, mainly Hawksbill Turtles, eat and control sponges.

Seaweed is controlled and kept at a level beneficial to the reef by its herbivores, including certain species of sea urchins and sea slugs. Foliose macroalgae would inhibit and stunt coral growth and even greatly reduce coral and reverse its development, as well as negatively impact the reef's resilience, and other algae would overgrow and suffocate the reef, but for herbivory of them by certain species of parrotfish and some sea urchins, such as the Long-spined Sea Urchin (*Diademed antillarum*), which lives in holes in the reef. These herbivores ensure the algae is maintained at a more or less optimal level for reef health and diversity. Without them, the reef system would be destroyed, so they are keystone species.

The algae in the coral make a chemical compound that is transported to the polyps, which modify it into an effective, natural sunscreen capable of blocking the sun's harsh ultraviolet rays in the tropics, protecting polyp and algae from them. Amazingly, coral-eating fish incorporate and benefit from this sunscreen protection, so it is clearly passed up the food chain.

Some cyanobacteria fix nitrogen, providing usable nitrogen for reef organisms, and benefit from the nutrients provided by reef organisms. Cyanobacteria are a phylum. Hence, they consist of many species in mutualistic relationships with many reef species. Many cyanobacteria are keystone species because supplying nitrogen in a form usable to life is very important to sea life.

Many invertebrates use the coral substrate as habitat, living in existing crevices, as is done by polychaete worms and crustaceans, or boring into the coral, as is done by some sponges, bivalve mollusks, and sipunculid and Christmas tree worms. These are mainly commensal, but, as will be discussed, some aid the coral.

Cryptobenthics make up about half of the coral reef fish species, and are a crucial component of healthy reefs. Most are shorter than two inches and hide in crevices. Their larvae feed in richer offshore waters, a few hundred meters from the reef, then swim back to the reefs when they mature, transferring nutrient to the reef when they are eaten, otherwise die, or defecate. They are thus in a mutualistic relationship with the reef system, bringing it nutrient and getting a habitat from it. Their limited movement means populations of the same species become easily isolated from one another, so they speciate easily. They are also responsible for much of the reef diversity, both because of their own high diversity and because they support much of the reef's food web. Essentially every reef predator eats them; 70% of them are eaten every week. They comprise 60% of the fish biomass that is eaten on a reef. They reproduce very rapidly, replacing the ones that are eaten, keeping the food supply of their predators abundant. They are short-lived r-strategists; entire generations turn over in weeks; they typically have seven generations per year. They are in 17 families, and have common names such as blennies, dottybacks, clingfishes, and dragonets. They spend most of their lives as juveniles. Their young make up 70% of juvenile reef fish.

Pocilloporid corals in Mo'orea, French Polynesia, provide a home to several species of coral guard-crabs, genus *Trapezia*. This coral is a favored food of the Crown of Thorns Sea Star, which could devastate the reef, but for the protection by the crab, which defends its home by pinching, shaking, and nipping the sea star's tube feet. The defense is effective. Intermediate-sized and small crabs of this genus defend the coral against predators of their size classes. Crabs do not attack predators that do not match their size class. Small crabs (*T. serenei* and *T. punctimanus*) defend against sea snails (*Drupella cornus*), which large crabs ignore, and medium-sized crabs (including *T. bidentata* and slightly larger *T. serenei*) protect from cushion stars, *Culcita novaeguineae*. Large crabs defend against the Crown of Thorns. The crabs also protected smaller coral species on the host coral, and this greatly aided recovery from massive Crown of Thorns predation when this species had a population explosion due to human effects. The coral offer shelter and produce fatty deposits in their tips that the crabs eat [226], showing coevolution in this mutualism.

An intermediate amount of predation by the Crown of Thorns increases coral species number, since it prevents the competition of coral species from running its full course and sending the species that are poorer competitors locally extinct. But too much predation decreases diversity. Crab defense ensures optimum predation levels for coral species richness, which in turn causes greater species numbers of the many species that benefit from the coral.

The coral *Oculina arbuscula* can be overgrown by seaweeds like Dictyota and Sargassum, and hence outcompeted in temperate regions, where herbivores often fail to control seaweeds. It is thus unable to grow in these areas. These seaweeds have chemical defenses against herbivores, such as fish, sea urchins, and other crabs. But if the coral harbors its mutualistic partner, the crab *Mithrix forceps*, which removes seaweeds and invertebrates growing on or near the coral, it persists on reefs in temperate North Carolina. This crab eats all local seaweeds in laboratory choice assays and is not deterred by chemical defenses. In reduced light, where the seaweed cannot grow, the crab does not benefit the coral [227].

Many examples of mutualism and commensalism on the reef do not directly involve the coral. Cleaner fish of the family Labridae feed on ectoparasites on larger fish. About 28 species of clownfish (family Pomacentridae, subfamily Amphiprioninae) of the Indian and Pacific Oceans, and the Great Barrier Reef and Red Sea, are immune to sea anemone's stingers, and get protection from their predators by staying among the anemone's tentacles. The clownfish are generally highly host-specific, and the sea anemone genera *Heteractis* and *Stichodactyla*, and the species *Entacmaea quadricolor*, are frequent hosts. They also use the anemone as a safe nest site. The fish also receive food left over from the anemone's meals and occasional dead tentacles, though they eat other food too. The anemone is thought to get protection from polyp-eating fish, such as butterfly fish, which the clown fish chases away, and from parasites, as well as nutrient from the feces of the clown fish. There are algae that are mutualistic with the anemone, which provide carbohydrates from their photosynthesis that help anemone tissue growth and regeneration. The nitrogen excreted by the fish increases the amount of algae the anemone can incorporate in its tissues. Movement of the fish enhances water circulation around the anemone. The resulting increased aeration of the water benefits the metabolism of the anemone, and both fish and anemone respiration. The increased metabolism of the anemone results in it increasing its size, which allows the fish to get better protection from it. The bright color of the fish may lure the anemone's predators—small fish—to it, providing it with food, though this is not certain.

Groupers, fish of the family Serranidae, are farmers that cultivate seaweed by removing creatures feeding on it, such as sea urchins. They remove inedible seaweeds, influencing the abundance of edible seaweed species. This is both mutualism and herbivory.

The trematode worm, *Podocotyloides stenometra*, has a complex life cycle involving three hosts: a snail, coral, and a fish. The trematode first infects the snail, and then it infects coral by chewing its way into coral polyps. In the coral, the encysted flatworm lives off the energy reserves of the coral, severely depleting the energy resources of the coral and causing "coral zits", or round, pink pimples. Certain fish feed preferentially on the infected coral colonies. This regulates the worm population, preserving the coral and the diverse ecosystem it supports. The fish is both a mutualist to and predator of the coral.

The coral themselves benefit from phytoplankton, seaweed, and coralline algae, especially small ones called turf algae, all three of which pass nutrients to corals and use the coral as habitat [228]. Coralline algae are important in reef construction.

They provide calcareous material to the reef structure, help cement the reef together, and strengthen it by depositing limestone in sheets over the reef surface. Coralline algae also construct algal ridges, carbonate frameworks that prevent oceanic waves from striking the reef and adjacent coastlines, protecting these habitats from erosion. Significantly, they do this in those parts of the reef subjected to the greatest forces by wave action, such as the reef area facing the open ocean. Algal ridges require high and persistent wave action to form.

Coralline algae are also important primary producers on the reef. Many species of them produce chemicals which promote the settlement of the larvae of some herbivorous invertebrates, particularly abalone, and appear to enhance metamorphosis and survival of larvae through the critical settlement period. This is mutualism, because the adult herbivores consume epiphytes that grow on the corallines and prevent light from getting to them. In an intimate mutualism/herbivore relationship, the most common Indo-Pacific coralline, *Hydrolithon onkodes*, is grazed and cleaned at night by the chiton, *Cryptoplax larvaeformis*, which lives in burrows it makes in the alga. The grazing and burrowing produces a strange growth form in which the algae make almost vertical, irregularly curved lamellae.

Some coralline algae not associated with reefs develop into thick crusts which provide microhabitat for many invertebrates. For example, off eastern Canada, juvenile sea urchins, chitons, and limpets suffer nearly 100% mortality due to fish predation unless they are protected by knobby and undercut coralline algae. This is probably an important factor affecting the distribution and grazing effects of herbivores within some marine communities. Also in eastern Canada, herbivores associated with corallines can promote diversity of the dominant seaweeds by generating patchiness in the survival of their young stages, preventing better competitors from driving other species locally extinct.

Additionally, other organisms on the coral deposit calcium carbonate as their own skeletal structure beneath and around themselves, contributing to the skeletal structure of the reef [229]. Since coral skeletons are also calcium carbonate, this action helps the coral, pushing the coral head's top upward and outward. This is mutualism, since the coral give these organisms habitat.

Minor contributors of calcium carbonate include sponges, soft corals (a different order than hard corals, Alcyonacea), fire corals (not true corals, in class Hydrozoa), and many other invertebrate species. The contribution of each of these groups is small, but collectively their contribution can sometimes be substantial, helping the coral grow. The coral provides all of these calcium carbonate donors with a habitat. Also, grazing fish such as parrotfish, sea urchins, and other species break coral down into fragments that settle into spaces in the reef structure, creating habitat for other species, and, when the fragments are further broken down, sand. This sand often becomes sandy bottoms in associated reef lagoons, where it acts as habitat for several species.

Commensalism and mutualism also occur with species that do not live in the water. In time, coral dies and breaks down and forms coral and limestone islands and atolls, which are habitat for marine birds, such as herons, gannets, pelicans, and boobies, and reptiles, including monitor lizards, semiaquatic snakes like *Laticauda colubrine*, and the Saltwater Crocodile. Midway Atoll has nearly three million sea-birds in 17 species, including two-thirds (1.5 million) of the global population of Laysan Albatross, and one-third of the global population of Black-footed Albatross. The defecation of the animals provides nutrients to the water, benefiting the coral and other sea life, so most of these relationships are mutualistic. In fact, when introduced island rats kill seabirds, reef diversity decreases.

Finally, a phenomenon called Darwin's paradox states that coral reefs have very high diversity and productivity in spite of being poor in nutrients. The question is why? One major reason is that coral reefs are extremely efficient recycling systems. This greatly reduces the nutrient inputs required. They recycle a much greater percentage of their nutrients than any marine system, including the open ocean, so less nutrients are needed to support the ecosystem. The reef is a tight-knit system of interacting organisms that use nutrients very efficiently, very little nutrient is wasted, and the food web passes nutrients up trophic levels with little loss. This is all related to the high degree of coevolved mutualism and commensalism on the reef. Reefs are coevolved systems that evolved a high efficiency of nutrient cycling. Like rainforests, the high diversity aids efficient cycling of nutrients through the food webs. Seaweed and coralline algae, especially turf algae, transfer nutrients to corals. Sponges excrete nutrients that corals use. Efficient filter feeders, coral consume about 60% of the phytoplankton that drifts by in the Red sea. Corals also absorb nutrients directly from the water, including inorganic nitrogen and phos-phorus. The key point is that high biodiversity causes efficient use of resources and nutrient cycling.

A boundary layer of still water surrounds a submerged object, acting as a barrier. Waves breaking on the extremely rough edges of corals disrupt the boundary layer. The roughness of coral surfaces cuts through the barrier in turbulent water, allowing coral access to nutrients.

Organisms also provide nutrients, so the solution to the paradox is not entirely mechanisms that increase diversity in spite of low nutrients. Seabirds deposit nutri-ents from afar when they defecate or die in reefs. Their nutrient input increases reef species richness. Reef fish and invertebrates add nutrients through their defecation, shells, and death. Cryptobenthics bring nutrients to the reef from richer offshore waters. Cyanobacteria provide biologically usable nitrogen via nitrogen fixation. Seagrass meadows and mangrove forests provide nutrients to the reef through dead plants and animals. They act as nurseries for the larvae and juveniles of many reef fish. Mangroves also provide nurseries for the larvae of coral and other reef invertebrates.

All of this tremendously high diversity and these intricate symbiotic relationships are possible only because of the mutualism between the polyps and algae.

Müllerian Mimicry

Müllerian mimicry is mutualism, in which two or more unpalatable species rein-force their protection from predation by mimicking one another's warning col-oration, appearance, sounds, or smells. Although it maintains and substantially increases biodiversity, it did not cause the extraordinary amount of adaptive radi-ation and increase in diversity that all of the previously discussed mutualisms did. It is common in butterflies. Unpalatable *Heliconius* butterflies in Central and South America, *H. melpomene* and *H. erato*, have geographic races that differ from other races of their own species, but resemble races of other species that are sympatric with them. Batesian mimicry is commensalism in which a palatable species gains protection from predation by mimicking the warning coloration, appearance, sounds, or smells of an unpalatable species. In Peru, these species are involved in a mimicry ring with other *Heliconius* species, three other genera, and a moth. This ring includes Müllerian, Batesian, and mildly poisonous, quasi-Batesian mimics [230,231].

Some species are both Batesian models and Müllerian mimics. For example, the Plain and Common Tiger Butterflies are Batesian mimicry models as well as Müllerian mimics of each other. The Common Crow Butterfly, a model in a Batesian system, is involved in India in a Müllerian mimicry suite with other members of its genus, *Euploea*, of which it is the most common member. Butterfly species that are Müllerian mimics maximize the advantages of this mimicry. They tend to live in the same area, fly at the same height, and prefer the same forest type. Batesian mimicry, being commensalism, is discussed in Chapter 5, which is on commensalism.

The butterfly family Nymphalidae shows a great deal of Müllerian mimicry. It includes the subfamily Ithomiinae, a large radiation of exclusively New World butterflies, occurring in humid forests from sea level to 3,000 m, from southwestern United States to Argentina. The subfamily contains approximately 370 species in 45–50 genera and ten tribes; all are unpalatable, and all are involved in extensive Müllerian mimicry rings, with one another as well as many butterfly species not in their subfamily. The Nymphalidae also have the subfamily Heliconiinae, which has the tribe Acraeini, a diverse group of unpalatable tropical butterflies. The center of diversity is in Africa, where there are hundreds of species involved in mimicry complexes. Müllerian mimicry is also seen in bumblebees; large complexes in the velvet ant genus *Dasymutilla*; poison arrow frogs; coral snakes; birds of genus *Pitohui*; and likely the mammalian families Mustelidae, Viverridae, and Herpestidae. Thus, Müllerian mimicry has generated and maintained a great number of species, increasing diversity in several taxa.

The Monarch Butterfly (*Danaus plexippus*) caterpillar eats various species of milkweed plants (*Asclepias* spp.), which have evolved potent poisons to fend off these larvae. The caterpillars have evolved the ability to store the poison away from their vital organs, so that they are not harmed by it. The butterfly retains the poison in a similar way, safe from the toxic effects. The poison makes birds sick when they attempt to eat the larva or butterfly. They spit the prey out after tasting it. It could kill them if they ate it. The Monarch has thus turned the plant's defense against it into a defensive advantage of its own. Birds learn to recognize and avoid the contrasting orange and black pattern of the Monarch Butterfly after one trial. The toxic Viceroy Butterfly (*Limentis archippus*) overlaps the Monarch's range and resembles it, in a case of Müllerian mimicry. The adoption of chemicals plants evolved as defenses against them by insect herbivores is not uncommon. The previously discussed *Heliconius* butterflies pick up their toxins from the passion flowers their caterpillars eat, toxins the plants evolved to defend against these caterpillars. Co-opting toxins that their prey evolved to defend against them is not limited to invertebrates.

The Tiger Keelback Snake (*Rhabdophis tigrinus*) is a venomous colubrid snake found in East and Southeast Asia that stores toxins from the toads it eats, and exudes these from glands in its neck as defense when attacked by its predators.

There is coevolution between the plant and its insect herbivore, with the plant evolving greater toxicity and the insect resistance, eating the part of the leaf with less toxin, storing the toxin away from its vital organs, and so on. One would expect the insects to evolve faster than both its predators and the plant, because of the insect's relatively short generation time. Yet the system is extraordinarily stable.

ADDITIONAL EXAMPLES OF MUTUALISM IN NATURE

There are many examples of mutualism in nature that did not cause the tremendous adaptive radiation and increase in biodiversity that the above examples did. Still, they maintained or increased diversity. The following list of them is far from comprehensive, but attempts to illustrate how common and important the relationship is.

Syntrophy is a relationship whereby one species lives off the products of another species. Many times microbes are at least one member of this association. Though usually commensal, it is often mutualistic. There is a large community of microbes that break down leaf litter, releasing nutrients for plants, clearly mutualism. Many mutualistic and commensal relationships are syntrophic.

Methanospirillum hungatei, a methanogenic archaean, is of considerable environmental significance because of its unique ability to form mutualistic relationships with syntrophic bacteria to break down organic matter and produce CH_4 gas. It consumes hydrogen and formate. In mutualistic, syntrophic cooperation with microbes that produce hydrogen or formate, *M. hungatei* degrade some wastes, such as phtalate isomer-containing wastewaters. Microbes growing poorly on amino acid substrates can have their rate of growth dramatically increased by *Methanospirillum* metabolizing the hydrogen waste produced during amino acid breakdown, preventing a toxic build-up.

Hydrothermal vents at the bottom of the deep sea exist where no light penetrates. So photosynthesis cannot occur. Thus, heat and chemical synthesis by bacteria and archaea provide the energy to drive the system. Some archaea can live at temperatures of 235°F. The bacteria and archaea use hydrogen sulfide and CH_4, both poisonous to oxygen-breathing animals, to produce the energy that is the primary production that runs the system. Strange, about 6-foot-long, white tube worms, with gills that look like red plumes, called Pompeii Worms (*Alvinella pompejana*), with no mouths or stomachs as adults, live in tubes on the sides of scalding black-smoker chimneys. It is believed that the worms stay near the sulfur-rich vent fluids to provide nutrients for chemosynthetic bacteria that live in them. The bodies of the tube worms serve as habitat for the bacteria. They have mouths when very young, when they take in the sulfur-eating bacteria. Their mouths disappear as they grow. They absorb O_2 from seawater and hydrogen sulfide from vent fluids, which feed the bacteria inside them. Their bacteria feed them with chemicals made by chemosynthesis. Similar tube worms with similar mutualisms with internal microbes are at other deep-sea vents. Worms at the vents off the Galapagos Islands called Thick-plumed Rift Worms (*Riftia pachyptila*) are an example. The gills of giant clams and mussels that live at deep-sea vents are also full of mutualistic bacteria, allowing them to grow very rapidly and very big. Siboglinid tube worms, such as *Tevnia jerichonana* and *Riftia pachyptila*, which often form an important part of the hydrothermal vent community, have no mouth or digestive tract. Some species of this group are in cold seeps. They have about 285 billion bacteria per ounce of tissue, which they are entirely reliant on and supply a habitat to. They have red plumes containing hemoglobin, which combine with hydrogen sulfide or CH_4 and transfer it to the bacteria living in the worm. The bacteria nourish the worm by oxidizing either the hydrogen sulfide or CH_4 and feeding it carbon compounds. Also, the bacteria are necessary for embryogenesis and molting in the worms.

There are also crustaceans and mollusks that form similar mutualisms with prokaryotes. The bacteria at deep-sea vents were thought to use only two sources of energy: hydrogen sulfide, used by sulfur-oxidizing mutualists, and CH_4, used by methane-oxidizing mutualists. But now researchers have even found mussels that have a mutualism with a bacterial species that uses hydrogen as a fuel source. The mussels are the first species known to be fueled by the hydrogen that belches from the deep-sea vents. Since the discovery of hydrothermal vents in 1977, scientists have found more than 500 organisms that had never been seen before, all potentially mutualistic with microbes. The prokaryotes in these systems tend to transfer the energy that runs these systems directly to mutualistic metazoan hosts that they live in.

Cold seeps on the seafloor are often slightly hotter than the surrounding water, and have similarities to hydrothermal vents. Their communities are often dominated by bivalves with mutualistic relationships with chemosynthetic bacteria, an example being a vesicomyid clam that provides a habitat to sulfur oxidizing bacteria that provision it with nutrients. Both the hydrothermal vents and cold seeps are systems powered by mutualism. Cold seeps are diverse and vital ecosystems, part of the foundation of the ocean food web.

The flatworm, *Paracatenula*, which has neither mouth, anus, nor gut, harbors the chemosynthetic bacterium, *Riegeria*, which feeds its host with lipids and proteins, and probably sugars, fatty acids, and vitamins, made from combining CO_2 and hydrogen sulfide. The worm has a specialized organ called a trophosome where the bacteria live. The bacteria maintain the primary energy reserves in the mutualism as well as providing nutrition. *Riegeria* can make up half of the worms' biomass. The mutualism allows the worm to live in nutrient-poor environments. It is found worldwide in warm temperate to tropical subtidal sediments. Five species of *Paracatenula* have been described; several more have been morphologically and molecularly identified, but not described.

Some species of Nudibranchs, which are commonly called sea slugs, have mutualistic relationships with phytoplankton [232]. One group, the sacoglossan Nudibranchs, maintain live chloroplasts and other plastids from the algae they consume, employing the carbohydrates the algae produce from photosynthesis for their own energy. Another group of Nudibranchs keeps the entire phytoplankton alive in their tissues for the same purpose, giving it a home. As in coral, the mutualistic algae are dinoflagellates. There are also dinoflagellates that are mutualistic with clams, photosynthesizing for them, while receiving a habitat in which to live in return.

The Yellow-spotted Salamander (*Ambystoma maculatum*) has a mutualism with an alga [233,234]. It has jelly-coated eggs that prevent them from drying out and protect them from predators, but inhibit the O_2 diffusion needed for development. The green alga, *Oophila amblystomatis,* photosynthesizes and produces oxygen in the jelly. The alga benefits by metabolizing the nitrogenous waste produced by the embryos and using the CO_2 produced by the respiration of the embryos. The alga is not found anywhere in nature except in the salamander's eggs.

Certain species of flat worms, arrow worms, some moon snails, three or four species of the venomous blue-ringed octopus, pufferfish, the Western Newt, toads, and perhaps other species produce a very potent poison called tetrodotoxin, sometimes in combination with other poisons. It is used for defense, or, in the octopuses, primarily to kill prey and secondarily for defense. The toxin is produced by mutualistic bacteria that the animals harbor of the genera *Actinomyces, Aeromonas, Alteromonas, Bacillus, Pseudomonas*, and *Vibrio* [235].

Fossils in South Africa from 3.22 bya indicate microbial communities colonized trapped gas bubbles, which may have been by-products of microbial metabolism [236]. At this point, there was no ozone layer to protect them from solar UV radiation during low tide if they were at the seashore. Inside the bubbles, bacteria would have been protected from the UV. The researchers think the microbes glued sand and cells together, forming slimy, carpet-like biofilms, under water. Dissolved iron would have screened out UV. The microbes in the biofilm would have been of several species working together for mutual benefit. This exemplifies mutualism and adaptive ecosystem engineering.

De Fouw et al. [237] found a mutualism whereby seagrass off West Africa accumulates floating bits of dead leaves and debris that are broken down by microbes. The microbes release sulfides that can poison the seagrass. Clams of the genus *Loripes* and other lucinid clams benefit from O_2 from the seagrass roots. The sulfides nourish mutualistic bacteria in the clams, and these bacteria release sugars that feed the clams. The bacteria convert the sulfides to harmless sulfates, aiding the seagrass. Almost all four groups—seagrass, clams, sulfide-producing microbes, and clam-dwelling bacteria—are mutualistic with each of the other four, directly or indirectly. There are more than one species of each group except the seagrass, so this is much more than a four-way mutualism; it is a set of mutualisms involving a multitude of species. Not all seagrass meadows have lucinid clams.

Intertidal seagrass meadows at Banc d'Arguin in Mauritania had 1,500–2,000 clams per square meter in the top 19 cm of mud. Drought from anthropogenic climate change caused exposure of the plants, which then failed to oxygenate the clams and their bacteria. A positive feedback loop occurred. The clams and their bacteria died. So the seagrass was killed by the sulfides released by the microbes that decomposed the leaves and debris. It is the breakdown of the mutualism that caused the cascade of death, for only the areas where the clams survived had healthy seagrass.

Various ant species in Australia obtain sugar water from the glands of caterpillars of Lycaenid butterflies, the family which includes the blues. The caterpillars make a sound that signals the ants that the nutrient is coming. It is thought the caterpillar secretes appeasement chemicals that cause the ant not to view it as prey. The ants protect the caterpillar from insect predators by fighting them off, and from avian predators because they often cover the caterpillar, and are distasteful to birds. There are also Lycaenid larvae that prey on ants, that are fed by regurgitations of ants, that prey on or feed on the secretions of bugs (order Hemiptera), and are preyed on by ants. Some ants that prey on some Lycaenid caterpillars also attend and eat the secretions of other Lycaenid caterpillars. So various relationships have evolved besides mutualism; it is complex.

Sea anemones of the genus *Calliactus* often grow on the backs of hermit crabs, getting transportation and eating scraps from the crab's meals. And the crab uses its pincers to drive away the anemone's predators, such as sea stars and fireworms. The crab gets protection from predators, such as octopuses, which the anemone fends off with its barbed, stinging tentacles. Also the crab is hidden under the anemone. Crabs actively recruit these passengers, which co-operate in the recruitment. The crab pokes an anemone with its pincers, and this causes it to release its grip from its substrate, which might be a rock or tide pool bottom. The crab holds it in place as it attaches to the crab's shell. This clearly occurred via coevolution.

The Hawaiian Shrimp Goby looks out for danger for either of two species of snapping shrimp (*Alpheus rapax* and *A. rapacida*), which are almost blind. The shrimp constructs and maintains burrows in the seabed, while the fish stands guard. The fish wiggles its tail against the shrimp's antennae or into the burrow entrance to warn the shrimp if a predator approaches. During construction, shrimps leave the burrow to deposit excavated sand. Shrimps use their long antennae to maintain constant contact with the goby. Sometimes the goby even hovers above the shrimp, allowing it to take its load further from the burrow's entrance. In return, the goby uses the shrimp's burrow as a home, sleeping in it at night, and bolting into it to escape predators.

A wide variety of fishes, including wrasse, cichlids, catfish, pipefish, and gobies, are called cleaner fish, because they obtain food by eating ectoparasites and dead skin on larger fish of many species. Cleaner fish advertise with bright colors, often including a bright blue stripe along their body's length. There are cleaning stations where larger fish line up to be cleaned, and open their mouths to allow cleaner fish access to their mouth parasites. The larger fish do not attempt to eat the cleaner fish. The Banded Butterflyfish (*Chaetodon striatus*) cleans the Green Sea Turtle, *Chelonia mydas*. There are cleaner shrimps and other animals that have the same mutualistic relationship with large fish. Cleaners sometimes cheat, consuming mucous or tissue, adding parasitism to the mutualism. Mimics of cleaners that do not eat parasites imitate the behavior and coloration of cleaner fish, fooling and biting off tissue of larger fish, then swimming away.

Thirteen bryozoan species grow as colonies on certain hermit crab species off New Zealand, making a helicospiral-tubular extension of the crab's commensal gastropod-shell home. The bryozoan colony receives a substrate on which to live, and grows in step with the crab, with tube development appearing to be controlled by crab morphology and activity. This gives the crab a continuously growing home. Hermit crabs occupying bryozoan tubes very rarely have to endure the danger of predation when they move into new shells when they outgrow their shells or fight other hermit crabs for new shells, which are a limited resource [238].

American Badgers and Coyotes work as a team in hunting, and each benefits from the association. Badgers dig up burrows of rodents such as prairie dogs and ground squirrels to catch and eat them. Coyotes stand nearby, and often catch and eat the rodent when it runs out of an alternate exit to escape the badger. Coyotes catch an estimated one-third more rodents when teaming up with badgers than when hunting alone. Some biologists think the badger does not benefit from this relationship. But it almost certainly does, since the Coyote sometimes scares the rodent back into its burrow, where the badger can dig quickly and capture it. And the presence of the Coyote causes the rodent to tend to stay underground, where the badger has a better chance of catching it. The badger is a better digger and has a better sense of smell to locate the prey with its nose, while the Coyote is faster and has better vision. And the Coyote specializes on the part of the hunt that is above the ground, the badger on the part that is underground. The two species engage in play behavior together, which may reinforce their hunting bond. But they are not friends. They are competing for the meal, even as mutualists. The one who catches the prey does not share it with the other. The teamwork also saves energy, for they spend less of their valuable energy working as a team in pursuing elusive, fast prey.

Common Ravens (*Corvus corax*) associate with Gray Wolves (*Canis lupus*) in winter in Alaska and Canada as a mutualistic foraging strategy [239]. They circle above prey, such as Caribou (*Rangifer tarandus*) and other species, signaling the prey's location to wolves, which have learned to look for circling ravens. Ravens stand sentinel for wolves while wolves eat, are more nervous and alert than wolves at kill sites, give them added vision and hearing, and alert wolves to danger, including animals that might steal their food. Ravens grab parts of the carcass as wolves are eating, and finish the carcass after the wolves are done. They also lead wolves to carcasses too tough for the ravens to break into and eat without the wolves opening them up first. Wolves howl before they go on a hunt, and ravens fly toward the wolves when they hear this signal. Wolves may respond to certain raven vocalizations and behavior that indicate prey is nearby. This mutualism is re-enforced by behavior that resulted from eons of natural selection. Ravens peck at wolves' behinds, pull their tales, and fly just out of reach, causing the wolves to chase them, in a game that reinforces their bond. Wolf cubs seem to enjoy chasing ravens. They play for 10 minutes and more. If the wolf tires of the teasing game and stops chasing, ravens cry out in annoyance and try to get the wolves to play again, often succeeding. These two intelligent species have a bond, and appear to "like" each other. The relationship is also competition, for once wolves kill the prey, ravens scavenge and steal up to a third of it from the wolves. Studies show the once commonly held belief that wolves are in packs to hunt more efficiently is wrong, because meat consumed per wolf decreases as the pack size increases in the absence of scavengers, since pack animals have to share meat [240]. But if scavengers are present, meat per wolf increases with pack size, because packs better defend the kill from scavengers, the main one of which is the raven, at least in the Pacific Northwest. Thus, forming packs and sociality seem to have evolved in wolves as a result of selection pressure from ravens, in an example of one species influencing another's evolution. Only with packs and social behavior could wolves have become preadapted to domestication into dogs. Hence without ravens, humans might very well not have dogs! Wolves have also adapted to eat very fast and to be able to eat a great deal—up to 20 pounds of meat at one sitting, to avoid theft by ravens. Human evolution would also have been quite different had we not evolved mutualistically with wolves, aiding each other in hunting. Humans shared food with wolves and eventually domesticated them as dogs. Remarkably, when wolves were reintroduced to Yellowstone National Park after having been absent from there since the 1940s, the ravens and wolves seemed to "remember" their relationship, and it was quickly re-established.

In Everglades National Park and nearby, American Alligators stay under nests of birds, such as herons, egrets, ibises, and storks. This keeps away predators of chicks, such as raccoons and opossums, which eat all eggs or chicks in the nest, if they can get to them. The alligators eat only chicks that fall out of the nest. Sometimes the parents push chicks that they cannot feed and that would starve, feeding the alligators, and losing less chicks than they would to raccoons or opossums. Birds often produce more chicks than they can raise. Alligators under the nests average 6 pounds heavier than those that live about a half-mile away [241].

Birds called oxpeckers (*Buphagus africanus* and *B. erythrorhynchus*) in Africa feed on the backs of various species of antelope, zebra, elephants, hippopotamuses, monkeys, and other large animals, eating their parasites, such as ticks. The bird obtains a meal; the mammal has its parasite load reduced. The birds also produce a hissing scream when startled, and hence warn the mammals they are perched on of approaching predators. However, the birds are also vampires, sucking blood out of open wounds on their hosts created by the ticks. The mammals tolerate them because they help more than they hurt. But this shows the complexity of interspecific relationships, and how one species can be both mutualist and parasite.

Nitrogen is a limiting nutrient in most terrestrial ecosystems, and had been thought to be supplied only by free-living bacteria and bacteria in plant root nodules. But Nardi et al. [242] have found evidence that mutualistic nitrogen-fixing microbes of diverse forms are widespread in the hindguts of arthropods, with nitrogen-fixing rates as high as 10–40 kg/ha/year being possible.

Soil archaea and bacteria and plants of several species are in a mutualistic nutrient cycle. Plants supply leaves that break down into detritus consumed by primary fermenting bacteria, which produce alcohols. The alcohols are eaten by secondary fermenting bacteria, which produce hydrogen, CO_2, and acetate. The alcohols are also eaten by sulfate-reducing bacteria, which produce CO_2 and acetate. Acetotrophic methanogens consume the acetate and hydrogenotrophic methanogens consume the hydrogen. Both of these last two are archaea, and both release CH_4. Finally, there are often methanotrophs that convert the CH_4 back to CO_2. Plants use the CO_2 and hydrogen. Thus, this is a nutrient cycle with many mutualists.

The pitcher plant, *Darlingtonia californica*, catches many insects for its nitrogen. It provides a home for bacteria in its fluid that lower the surface tension, causing insects to sink instead of float [243]. Bacteria also help pitcher plants digest their prey (ibid).

Ingestion of the soil bacterium *Mycobacterium vaccae* by mice can decrease anxiety and improve learning [244]. The mouse provides the bacterium with a habitat.

Viruses of fungi that are plant pathogens reduce the virulence of the fungus to plants. This has been demonstrated experimentally in several pathogenic fungi. The best-known example of this is the hypovirus that attenuates the virulence of the chestnut blight fungus, *Cryphonectria parasitica*. In this plant-virus mutualism, the benefits are indirect. The plant provides a habitat for the fungus that the virus lives off of, and the virus controls the fungus that attacks the plant.

Some coronaviruses are harmless to certain species of horseshoe bats that carry them. The viruses can spread to and act as pathogens to bat species that are competitors with and to predators of the bats that carry the virus (Villarreal, 2020). This is mutualism between the horseshoe bats and their viruses.

Acacia trees of various species (family Fabaceae) are mutualistic with various ant species, most notably in Mexico, Central and South America, and Africa. An example is the Bullhorn Acacia (*Vachellia cornigera*) and the ant *Pseudomyrmex ferruginea*. There are many species of acacia and several species of ants that share this mutualism, and each acacia species tends to ally with but one ant species. Some acacia species provide small, reinforced structures that the ants hollow out and use for nests. Acacias feed ants with extrafloral nectaries, small stores of sugar-rich nectar on their stems, and nutritious structures called beltian bodies on their leaf tips. The ants guard their acacia home, attacking any animal that may try to feed on it. This is often insects, but may include large herbivores; for example, in Africa, ant guards of the genus *Crematogaster* will attack a mammal, such as a giraffe, if they nibble on the plants, swarming its face, biting and stinging, until the giraffe gets so irritated that it leaves. Acacias have evolved mechanisms to prevent the ants from attacking their pollinators. The extrafloral nectaries and beltian bodies are bribes located away from the flowers, removing any incentive ants would have to go to the flowers [245]. This is important, because the ants would attack pollinators. Acacia flowers produce volatile compounds that are repellent to ants, causing them to stay away from the flowers (ibid). The flowers release these chemicals particularly when they are producing large quantities of pollen. Acacia species that have the closest relationships with their ants—those that provide homes for their ants—produce the chemicals that are most effective at keeping the ants away from their flowers. The repellent chemicals are specific to ants. In some cases, the same chemicals even attract bee pollinators. In fact, researchers suspect that the repellents mimic signaling pheromones the ants use to communicate danger, telling other ants to retreat. Thus, not only did a mutualistic relationship evolve, but a potential mechanism to ruin it was overcome by an adaptation of the acacia. Mutualistic bacteria colonizing the ants inhibit the growth of pathogenic bacteria on the plants' leaves and increase plant health [246]. The increase in ant, acacia, and bacterial diversity as a result of this mutualism is very high.

In addition to leafcutter ants, a number of other taxa farm. About 330 species of termites in the subfamily Macrotermitinae farm fungus. The fungus-farming African genus *Macrotermes* builds jagged mounds several meters high that are used commensally as homes by rodents, mongooses, lizards, snakes, insects, and many other animals. Termites and ants have evolved sophisticated ways to tend their fungus gardens, involving weeding and pesticides.

Bark and ambrosia beetles (Coleoptera, Curculionidae, weevils) drill holes in dead trees, and then plant fungus that they farm and eat. There are 7,000 species of these beetles and the behavior evolved 11 times in them. Trees likely contributed to the evolution of this strategy, since they evolved the production of toxins that defend them against attack by hungry beetles. In South American rainforests, many trees have sawdust falling off from these drilled holes with fungus gardens. Every group of bark and ambrosia beetles has its own unique collection of fungi. So the mutualism aided the evolution of about 7,000 species of beetles and about the same number of fungal species. They carry the fungi in specialized pockets on their bodies, in their mouths, on their backs, and in their armpits. In natural situations, the beetles exploit only dead trees, so they not only do not harm trees, but help the ecosystem by recycling them, returning their nutrients to the soil and ecosystem. But humans are introducing them to areas they do not normally live, and changing the climate and environment, so they have started to attack live trees and have become a problem. Life has natural controls that keep the beetles using dead trees in the absence of human intervention. These could be tree defenses against the beetles, regulation of climate, and limits on the natural ranges of the beetle species. But we do not fully understand what the controls are.

The slime mold, *Dictyostelium discoideum*, farms bacteria that it provides a habitat for [247]. It spends part of its life as a single-celled amoeba that eats bacteria that grow on decomposing leaves on forest floors. When there is little food, hundreds of thousands of the amoebas come together and fuse into a multicellular organism. This crawls to a new area as a slug, then forms what's called a fruiting body supported by a stalk. The fruiting body bursts to disperse some spores that formed from the amoebas. Sometimes this process happens without any crawling. Some strains keep bacteria alive in their fruiting bodies, farming them. When given antibiotics that kill their bacteria, the farming strains ate some reintroduced bacteria, and kept some without digesting them. Nonfarming controls simply consumed bacteria and digested them, and did not do as well in some soils as the farmers. Since slime mold ancestors are among the earliest eukaryotic life to colonize land, slime molds are likely the oldest terrestrial farmers. They are also the simplest life forms to farm.

The damselfish, *Stegastes nigricans*, weeds out competitors of the red alga, *Polysiphonia*, in the Indo-Pacific. It defends a territory, and excludes herbivores that eat the alga, such as sea urchins and herbivorous fish. The alga is an inferior competitor and would be extirpated by its competitors and predators if it were not for the fish. This mutualism occurs in many species of the algal genus and 18 species of damselfish, in Thailand, Borneo, the Okinawa Islands, the Great Barrier Reef, the Maldives, Mauritius, Kenya, and Egypt. Neither the fish nor the alga could survive without the other. The fish increases primary producer diversity by maintaining a genus that could not otherwise survive. It is interesting that it helps the poorest algal competitor as a mechanism of increasing diversity via ecosystem engineering and farming. The fish is both a mutualist and herbivore of the alga. Not only do the algae benefit from the damselfish's activities, but numerous small bacteria, protists, and animals utilize the lush algal fronds as habitat. They in turn add nutrition to the fish's diet, in a many-species mutualism [248].

The Marsh Periwinkle (*Littoraria irrorata*), a snail of the salt marshes of the east coast of North America, chews the grass *Spartina* to prepare it for cultivation and weaken its defenses, and then defecates on the chewed blades to fertilize them [249]. This allows a fungus to grow on the grass. The snails eat this fungus, not the grass. This is a farming mutualism between snail and fungus. It shapes about 3,200 km of marshy coastline.

Experiments indicate Saltmarsh Cordgrass (*Spartina alterniflora*) would be locally eradicated by this periwinkle, but for the fiddler crab, which digs burrows for its home. This aerates the marsh soil, allowing the cordgrass to grow better, and survive in the face of periwinkle predation [250]. The cordgrass and crab are in a mutualistic relationship, since the cordgrass provides habitat for the crab. Cordgrass and other salt marsh plants stabilize the marsh by trapping and binding sediments for use by themselves and other plants as substrate to grow on. Saltmarshes also are a nursery for fish and invertebrates, deliver nutrients to coastal waters, filter toxins out of the water, and provide food and habitat for both terrestrial and aquatic vertebrates and invertebrates. Thus, the burrowing of the crab indirectly benefits many species, and its mutualism with the cordgrass stabilizes a diverse ecosystem, greatly increasing its biodiversity.

Some ants and bees ranch, using aphids for their honeydew, which they obtain by stroking their backs with their antennae. The ants move the aphids to areas on the plants with the best sap, for the production of higher-quality honey dew. They protect the aphids from predators. When it rains they may move them to sheltered places, even sometimes into their own nests. But the ants sometimes clip the wings off aphids to stop them from flying away, and apply chemicals on their feet that prevent their wings from developing. Thus, this is mainly mutualism, but partly parasitism.

This is but a small percentage of the many mutualisms in nature. They illustrate that it is common and important in increasing biodiversity.

MUTUALISM, COMMENSALISM, AND PARASITISM ARE OFTEN HARD TO DISTINGUISH, AND ARE COMPLEX

Mutualism, commensalism, and parasitism can be very complex and hard to interpret, and there are cases of two species with two or more of these relationships at once, and/or where it is not clear if the relationship is mutualism or commensalism, or even mutualism or parasitism. For example, some think the clown fish contributes nothing to the sea anemone, and the relationship is commensal. Leaf hoppers in Africa produce a drop of honey dew when certain geckos bob their heads near them. They also signal the gecko with the waving of their abdomen when they are about to produce the nectar. It is not known what the insect gets in return, but it is thought to be protection from predators by the gecko's presence. The insect's behavior indicates that this is mutualism, but it could be commensalism. And some mycorrhizal fungi are parasites. Also, it is best to think of mutualism as a continuum ranging from loose associations of two species to two species that will not survive unless they are always together, such as the alga and fungus that combine to form each lichen "species".

The Great Spotted Cuckoo (*Clamator glandarius*) lays its eggs in the nests of Carrion Crows (*Corvus corone*) in Spain, reducing the average number of crows chicks fledged from 2.6 in nests without cuckoo chicks to 2.1 with them when predation is not a major problem [251]. The cuckoo chick releases a caustic, stinking slime that smells worse than feces when a predator comes to the nest. Predators avoid this. This secondarily protects the crow chicks. In the vicinity of predators, nests with at least one cuckoo have a 76% chance of fledging at least one crow, while those without have only a 54% chance. This is a rare case of an animal that is both a parasite and a mutualist.

Toxoplasma gondii is a parasite that can infect any warm-blooded host, but can only reproduce in cat intestines. When it infects mice, it causes them to lose their aversion to cat urine, making it more likely a cat can catch and eat the mice [252]. This is adaptive for the parasite, helping it obtain a host. The cat and parasite are another example of both mutualism and parasitism.

Male Splendid Fairy-wrens, Australian birds, sing a special song when they hear the call of Butcherbirds, one of their predators [253]. This gets the female to pay closer attention to his call, increasing mating success. They are socially monogamous, forming male-female pairs that last their entire lives, but sexually promiscuous, mating predominantly with birds outside of their home pair. Here a predator is also acting as a mutualist, at least to some of the male prey birds.

Cattle Egrets benefit from eating invertebrates stirred up by grazing mammals, but do not generally help the mammals. But they sometimes eat parasites off their bodies.

Barnacles have grown on scallops for at least 15 million years, getting a substrate to live on. They compete with the scallop for food, having similar diets, and add weight to it. But they protect the scallop from marine snails that drill holes in their shells to eat them. So the relationship could be competitive, parasitic, commensal, mutualistic, or any combination of these, depending on the specific case.

Epiphytes are in every major plant group, and benefit from getting a habitat higher and out of the shade, with better access to sunlight. They usually do not hurt the host tree and are commensal, but may be so abundant they weigh it down and even make it fall, being parasitic. Some, like the strangler fig, drop roots down from the canopy and grow around and kill the host tree. They also provide habitat for invertebrates, tadpoles, frogs, and other animals, which do not help them usually (commensalism). But do these animals sometimes provide the epiphyte with nutrients if they defecate or die in them (mutualism)?

Lovas-Kiss (2020) [254] showed a small percentage of carp eggs eaten by Mallard Ducks can survive passage through the guts of the ducks. They can potentially be dispersed to new bodies of water this way. This is a combination of predation and mutualism. (Predation is a form of mutualism, but usually not because it aids in dispersal). Life is thus diverse not only in species richness, but diversity and complexity of interspecific relationships, consistent with the ABH.

WHEN ONE CONSIDERS INDIRECT MUTUALISM, MUTUALISM IS VERY COMMON

If one considers indirect mutualisms, mutualism is much more common than one would imagine. For example, the larvae of the mussel genus *Mytellus* settle preferentially on filamentous algae such as the seaweed, *Endocladia muricata*. Limpets graze heavily on *Endocladia*, and one would expect this to decrease mussel settlement and growth. But the sea star, *Pisaster ochraceus*, devours limpets. This allows more growth of *Endocladia*, and hence mussel colonization. This helps the sea star, since its main prey is the mussel. So we have two mutualistic relationships. The sea star indirectly helps the seaweed, which indirectly helps it. The sea star indirectly helps the mussel, and preys on it. Thus, the mussel helps it directly. Any two species separated by two trophic levels are mutualists. For example, aspen helps wolves by feeding their prey, deer. Wolves help aspen by controlling the deer that eat aspen. Indirect benefits of ecosystem engineers to other species and indirect aid these engineers receive from their beneficiary species are common. Earthworms help numerous plant species by improving the soil. These plants help many herbivorous species. Some of these herbivores are eaten by predators that help earthworms by also eating their predators. This is a network of indirect mutualisms. Similar networks exist in all ecosystems with ecosystem engineers. And as we shall see in Chapter 11, all ecosystems have ecosystem engineers.

CHEATING IN "MUTUALISM"

Mutualism has cases of one species cheating, and sometimes the other species evolving mechanisms to counter it. In some cases, one species receives no benefit, in which case it is not mutualism. In the case of ant seed dispersal, for example, some ant species that eat plant seeds or eliaosomes do not disperse them, and are hence only seed predators, not mutualists. Some plants have nonremovable eliaosomes or fake the presence of a nonexistent reward with chemical cues. Ants sometimes can discriminate between real, removable eliaosomes and those of cheaters, preferentially taking the real ones.

Several species of plants, especially orchids, have evolved deceptive mechanisms to lure insect pollinators ([172], and references therein). One method is to attract scavengers by mimicking rotting meat, both in looks and odor. Here, the flower is often an orchid, and the target pollinators are generally flies. Some orchids mimic female insects, often bees, causing the male to attempt to copulate, fertilizing the flower.

Some mycorrhizal fungi can receive carbon from their hosts in high phosphorous environments where they do not give much nutrient to the host. Plants vary in their ability to expel these cheaters [255]. Nonphotosynthetic plant species sometimes steal carbon compounds from photosynthetic plant species through the fungal network. There are other instances of cheating in pseudomutualistic relationships. Even in cases of cheating, one species benefits and the other is not extirpated, so diversity is maintained.

INTERACTIONS IN NATURE INVOLVING MUTUALISM ARE OFTEN EXTREMELY COMPLEX, WITH HIGH DIVERSITY MAINTAINING THEIR STABILITY

The complexity of interactions among organisms can be astounding. For example, some dinoflagellates form cysts, and stay dormant, sometimes up to years, during unfavorable conditions. They emerge from the cysts and become swimming dinoflagellates when good conditions return. In good conditions, the cysts can bloom into huge populations overnight. In the Chesapeake Bay, blooms of the dinoflagellate *Akashiwo sanguinea* develop quickly, and face competition from cryptonomads, which have a high growth rate, and can shade out the *A. sanguinea*. When this occurs, the *A. sanguinea* can shift to a predatory mode, and eat the cryptonomads.

Various species of *Amoebophrya* are parasitic dinoflagellates that attack *A. san-guinea*. With increasing population density of *A. sanguinea*, the risk of attack by *Amoebophyra* goes up, and this can lead to an abrupt, large decline of *A. sanguinea*. Viruses, fungi, and parasites can attack *Amoebophyra*. There is competition and predation in this system. The species that attack *Amoebophyra* help *A. Sanguinea*, which helps them, being host to their prey. So there is also mutualism in this complex system. Diversity and complexity keep the system stable. Negative feedback also maintains the stability of the system. When the population of any species increases, it is regulated by its predators and parasites.

Some parasites are sometimes mutualists with their hosts [256]. An example is found in Latin American rainforests, where a species of oropendula is parasitized by cuckoos, raising the cuckoo's young in its nest. Bot flies parasitize the oropen-dula. Cuckoos eat the bot flies. Parasitoid wasps parasitize the bot flies. If the bot fly is present and the wasp is not, the oropendula tolerates the cuckoo, because it benefits from the cuckoo controlling the fly. But if the fly is not present or is parasitized by the wasp, the oropendula does not greatly benefit from the cuckoo, so it expels its young from the nest if it detects them. This is an elegant, complex system showing the complexity of interspecific interactions and how species aid each other.

There are very diverse ecosystems called pleustons floating as islands on the open sea, with many species of bacteria, archaea, animals, and plants. Most abundant are Blue Buttons (*Porpita porpita*, colonies of hydroids), By-the-wind Sailors (*Velella velella*, hydroids), and deep-purple snails in dense patches. More than 1,000 of these pleustons were collected in 20 min. They also may contain Portuguese Man-of-Wars, nudibranchs, sea anemones, barnacles, copepods, gastropods, color-changing crabs, protists, marine fungi, specialized bacteria, and even insects. Pleustons can differ from each other in species composition. There are also freshwater ones. The species help each other mutualistically by providing a shared floating habitat. There is also every other interspecific interaction on the pleustons.

ECOSYSTEMS HELP OTHER ECOSYSTEMS THAT ARE SOMETIMES FAR AWAY FROM THEM; THIS IS OFTEN CARRIED OUT BY LIFE

Ecosystems often provide services such as nutrients or protection to other ecosys-tems, in "commensalism between ecosystems". Sometimes two ecosystems aid each other in "mutualism between ecosystems". I discuss both types together here, rather than putting the commensalism ones in the commensalism chapter.

Nutrients are carried from the photic zone to the deep sea by marine snow, which comes from organisms, mainly in the form of sinking feces, parts of organisms, and dead organisms. Bacteriophage in deep-sea sediments liberate nutrients, many of which come from near the surface, by lysing prokaryotic cells. Some of these nutri-ents are used by deep-sea organisms. A great portion of these nutrients are trans-ported to the near-surface waters by animals, contributing to photic zone ecosystems.

Many species of squid migrate between 500 m deep, some going twice that deep, in the day, to within 150 m of the surface to feed at night, transferring nutrients from deep to shallow waters if eaten near the surface, and in the other direction if consumed in the depths. Most transfers are from deep to shallow. Cephalopods have migrated in this manner for at least 150 million years.

Shrimp that normally dwell in the middle of the ocean's water column hundreds of meters down flee to the seafloor when waters become turbulent from storms above the sea. There they are eaten by squat lobsters, transferring energy from the mid-sea to deep sea, as storms bring the two ecosystems together [257].

Coral reefs on the one hand and mangrove and seagrass ecosystems on the other display "mutualism between ecosystems". Fringing reefs physically protect mangroves and seagrass from waves and currents, and produce sediment in which the mangroves and seagrass can take root. Coral reefs can reduce wave energy by up to 95%. Reefs protect land ecosystems from erosion, especially wetlands and islands, notably the islands between India and Sri Lanka. Mangroves and seagrass protect coral reefs from influxes of fresh water, which harm these saltwater ecosystems, and from large influxes of silt and pollutants. They both supply reefs with nutrients, and provide food and shelter as nurseries for the larvae and juveniles many reef fish, with the subadults and adults migrating from these ecosytems to reefs. This is true in the Caribbean [258] and the Indo-Pacific. Mangroves also provide nurseries for the larvae of coral and other reef invertebrates.

In the Gulf of Mannar near the southern tip of India, coral reef and seagrass are interrelated and support each other. The seagrasses bordering the coral reefs play an important role in limiting sedimentation on the coral reefs and help in protection. The coral protect the mangroves from destructive wave action and erosion of islands near them. Lamb et al. [259] found that there was a twofold reduction in disease levels in coral reefs adjacent to seagrass meadows compared to those not adjacent to them, in Indonesia. Seagrasses are also great at removing excess nitrogen and phosphorous from coastal waters, helping themselves, mangroves, and coral. The coral reefs help in controlling erosion of the islands, preventing destruction of the seagrass. The systems supply each other with nutrients. Many fish species use both coral and one of the other two habitats for shelter, food, and breeding. During the windy, turbulent season, the water is highly turbid and dense seagrass beds are disturbed by strong waves. The nearby reefs shelter them [260].

Forests exchange nutrients with rivers and the sea. A study that used fish carcasses as bait at the site of the Chernobyl nuclear accident showed that nutrient flows freely between aquatic and terrestrial ecosystems [261]. The dead fish were quickly eaten by land animals. This was not due simply to their placement out of the water, since animals routinely catch fish and transfer the nutrients to land. In North America, Grizzly Bears and salmon transfer nutrients from the sea and rivers to the forest. Salmon, rich in nitrogen, sulfur, carbon, and phosphorus, swim up rivers, sometimes for hundreds of miles. The bears capture the salmon to eat them, carrying them onto dry land, defecating and dispersing partially eaten fish. It has been estimated that the bears leave up to half of the salmon they harvest on the forest floor. This benefits a huge number of species of land plants (including trees), which benefit great numbers of animals. Bald Eagles and Osprey transfer vast amounts of nutrients from freshwater rivers and lakes to land through fish consumption and defecation.

Sea birds such as sea gulls, cormorants, boobies, Anhingas, and others eat large quantities of ocean fish, often diving several meters deep to catch them. They transfer many nutrients to land when they defecate there. Often islands they nest on are covered in guano. This often results in forests. The forests that were once on Easter Island were largely fertilized by sea birds.

Nutrients are also transferred by life from land to sea. Water washing from island forests to the sea carries a tremendous amount of nitrogen as a result of sea bird guano. This nitrogen nourishes the sea's phytoplankton. It is released slowly, so does not cause eutrophication. Giant Manta Rays prefer to feed in waters close to native forests, and not palm forests, which birds tend to avoid as nest sites. So nutrients are returned to the sea from islands because of the birds, forest, and rain.

Another way nutrients go from land to sea is sharks eat songbirds. Songbirds migrate across the Gulf of Mexico, and are forced to land on water in bad weather. Tiger Sharks consume them from below; they are easy prey. Large numbers of birds are eaten by sharks in migration every year [262]. Thus, large amounts of nutrients are constantly transferred between terrestrial and aquatic ecosystems, the two systems supporting each other. This movement of nutrients from sea to land and back enhances productivity and diversity far more than if the nutrients remained locked up in one ecosystem or the other, or in one group of organisms, such as fish.

Stumps of Cypress trees over 2 m in diameter, spanning at least 0.5 square miles, about 8 m below the surface in the Gulf of Mexico, that are about 52,000 years old, have formed a reef, utilized by fish, crustaceans, sea anemones, and other life burrowing between the roots of dislodged stumps. Dead trees from a terrestrial ecosystem, perhaps carried to sea by a storm, are providing habitat for diverse sea life.

The trees of the Amazon rainforest form clouds above them by evapotranspiration. These clouds move, forming a "river" in the sky. It is the Earth's largest river. It flows over the rainforest to the Andes, which send it south. It flows 2,000 miles south, well beyond the Amazon, dropping rain as it travels. It transforms what would otherwise be desert to fertile plains. It supplies much of the water to the Pantanal, the largest and most diverse wetland on Earth and the largest concentration of easily seen wildlife in South America. By the same mechanism, the rainforests of Borneo and the Congo basin also create rivers in the sky that travel beyond these rainforests and provide water to other ecosystems. All three rainforests increase diversity of ecosystems well beyond their boundaries by providing them with water. They affect weather in places far from them.

Atmospheric mineral dust plays a vital role in Earth's climate and biogeochemical cycles [263]. The largest source of atmospheric mineral dust on Earth is the Bodélé Depression, a dried-up lakeshore in the Sahara (ibid; [264]). This depression is made up of the shells of dead diatoms, which can reach as much as a half a mile deep on the ocean floor. Winds blow these fossil diatoms from the Sahara across the Atlantic Ocean to the Amazon, Caribbean islands, and Everglades. They fertilize these areas with micronutrients, such as iron and phosphorous, increasing primary productivity in them as well as the equatorial Atlantic Ocean. The increased productivity increases carbon sequestration. Up to 6.5 Tg of iron and 0.12 Tg of phosphorous are exported from the Bodélé Depression every year (ibid), showing the depression is a significant annual nutrient supplier to these ecosystems. The entire Amazon basin is fertilized annually by 27 million tons of diatom shell dust by this method (ibid). There is significant phosphorous input to the Luquillo Mountains in Puerto Rico by this mechanisms as well [265]. The diatoms are acting as commensal keystone species well after their death.

Ecosystems often interact without biology as the mediator. Dust blown from the Gobi Desert provides nutrients, mainly phosphorous, to Sierra Nevada ecosystems in California. This accounts for the ability of otherwise phosphorous-poor soil to support giant sequoia trees. Some 18%–45% of dust comes from Asia, and a smaller amount comes from California's Central Valley [266]. Erosion would deplete the nutrients of many Sierra Nevada forests if not for these inputs.

The Great Plains of North America are rich in minerals supplied by erosion from the Rocky Mountains, allowing the growth of grasses and hence the diverse prairie ecosystems there. The Ganges River washes iron and other nutrients from the Himalayas to the Ganges delta in Bangladesh, greatly enhancing diversity in that region. Such cryptic transport of minerals creates fertile soil that allows plants and their ecosystems to thrive and is the basis of food webs all over the planet. The Earth's systems are profoundly interconnected. Organisms are not isolated from processes that occur far away in time and space. The land, sea, air, and life are interconnected, constantly feed back on one another, have coevolved with each other, and work to increase diversity and aid life. There are many other cases of both geological and other nonliving factors and life working together with a resultant increase in diversity.

CONCLUSION: EMINENT IMPORTANCE OF MUTUALISM

Mutualism is ubiquitous in nature, occurring between species at all levels of complexity, from biomolecules to mammals. DNA, RNA, and protein coevolved mutualistically during the chemical evolution of life. Every eukaryote has a diverse microbiome with thousands of microbial species, a great many of them mutualistic with the host, and many mutualistic with other microbial species in the microbiome. Higher plants could not have succeeded on land without their fungal mutualists. It is safe to say that, even not counting the microbiome, every species is in a mutualistic relationship with at least one other species. The many examples in this chapter showed it is very common and influential in promoting diversity.

REFERENCES

1. Thompson, J. N. (2005). *The Geographic Mosaic of Coevolution*. University of Chicago Press. ISBN 9780226797625.
2. Doebeli, M. & Knowlton, N. (July, 1998). The evolution of interspecific mutualisms. *PNAS USA* 95 (15): 8676–80.
3. Frank, S. A. (1994). Genetics of mutualism: the evolution of altruism between species. *Journal of Theoretical Biology* 170: 393–400.
4. Leigh Jr, E. G. (2010). The evolution of mutualism. *Journal of Evolutionary Biology* 23 (12): 2507–28. doi: 10.1111/j.1420-9101.2010.02114.x.
5. Wyatt, G. A. K., et al. (Sept., 2014). A biological market analysis of the plant-Mycorrhizal symbiosis. *Evolution* 68 (9): 2603–18. doi: 10.1111/evo.12466.
6. Harcombe, W. (6 July, 2010). Novel cooperation experimentally evolved between species. *Evolution* 64 (7): 2166–72. doi: 10.1111/j.1558-5646.2010.00959.6.

7. Weeks, A. R., et al. (17 April, 2007). From parasite to mutualist: rapid evolution of *Wolbachia* in natural populations of *Drosophila. PLOS Biology.* doi: 10.1371/journal. pbio.0050114.
8. Lewis, D. H. (1973). Concepts in fungal nutrition and the origin of biotrophy. *Biological Reviews* 48: 261–78.
9. Trappe, J. M. (1987). Phylogenetic and ecologic aspects of mycotrophy in the angiosperms from an evolutionary standpoint, In Safir, G.R. (eds.), *Ecophysiology of VA Mycorrhizal Plants.* CRC Press: Boca Raton, FL.
10. Newman, E. I. & Reddell, P. (1987). The distribution of mycorrhizas among families of vascular plants. *New Phytologist* 106: 745–51.
11. Pirozynski K. A. & Malloch D. W. (March, 1975). The origin of land plants: a matter of mycotrophism. *Biosystems* 6(3): 153–64.
12. Pirozynski, K. A. (1981). Interactions between fungi and plants through the ages. *Canadian Journal of Botany* 59 (10): 1824–7.
13. Stubblefield, S. R., et al. (Dec., 1987). Vesicular-Arbuscular mycorrhizae from the Triassic of Antarctica. *American Journal of Botany* 74 (12): 1904–11. Stable URL: http://www.jstor.org/stable/2443974.
14. Crepet, W. L. (1983). Chapter 3, The role of insect pollination in the evolution of the angiosperms, In Real, L., (ed.), *Pollination Biology.* Academic Press, Inc. (Harcourt, Brace Jovanovic): San Diego, CA.
15. Milucka, J., et al. (2012). Zero-valent sulphur is a key intermediate in marine methane oxidation. *Nature* 491: 541–6.
16. Knittel, K. & Boetius, A. (2009). Anaerobic oxidation of methane: progress with an unknown process. *Annual Review of Microbiology* 63: 311–34.
17. Overmann, J. (2001). Phototrophic consortia. A tight cooperation between nonrelated eubacteria, In Seckbach, J. (ed.), *Symbiosis: Mechanisms and Model Systems,* pp. 239–55. Kluwer: Dordrecht, The Netherlands.
18. Overmann, J., & Schubert, K. (2002). Phototrophic consortia: model systems for symbiotic interrelations between prokaryotes. *Archives of Microbiology* 177201–208.
19. Huber, H., et al. (2002). A new phylum of Archaea represented by a nanosized hyperthermophilic symbiont. *Nature* 417: 63–67.
20. Fröstl, J. & Overmann, J. (2000). Phylogenetic affiliation of the bacteria that constitute phototrophic consortia. *Archives of Microbiology* 174: 50–58.
21. Glaeser, J., & Overmann, J. (2004). Biogeography, evolution, and diversity of epibionts in phototrophic consortia. *Applied and Environmental Microbiology* 70 (8): 4821–30. doi: 10.1128/AEM.70.8.4821-4830.2004.
22. Overmann, J., et al. (1998). The ecological niche of the consortium "*Pelochromatium roseum*". *Archives of Microbiology* 169: 120–128.
23. Glaeser, J. & Overmann, J. (6 Oct., 2003). The significance of organic carbon compounds for in situ metabolism and chemotaxis of phototrophic consortia. *Environmental Microbiology.* doi: 10.1046/j.1462-2920.2003.00516.xC.
24. Gasol, J. M., et al. (1995). Mass development of *Daphnia pulex* in a sulphide-rich pond (Lake Cisó). *Archiv für Hydrobiologie* 132: 279–96.
25. Wanner, G., et al. (2008). Ultrastructural characterization of the prokaryotic symbiosis in "*Chlorochromatium aggregatum*". PMCID: PMC2394997. doi: 10.1128/JB.00027-08.
26. Margulis, L. & Sagan, D. (1986). *Origins of Sex. Three Billion Years of Genetic Recombination.* Yale University Press: New Haven, CT. pp. 69–71, 87. ISBN: 0 300 03340 0.
27. Martin, W. F. & Müller, M. (2007). *Origin of Mitochondria and Hydrogenosomes.* Springer Verlag: Heidelberg, Germany.
28. Imachi, H., et al. (8 Aug., 2019). Isolation of an archaeon at the prokaryote-eukaryote interface. bioRXiv. doi: 10.1101/726976.

29. Lane, N. & Martin, W. (2010). The energetics of genome complexity. *Nature* 467: 929–34. doi: 10.1038/nature09486.

30. Martin, W. & Koonin, E. V. (2006). Introns and the origin of nucleus-cytosol compartmentalization. *Nature* 440: 41–5.

31. McFadden, G. I. & van Dooren, G. G. (July, 2004). Evolution: red algal genome affirms a common origin of all plastids. *Current Biology* 14 (13): R514–16. doi: 10.1016/j.cub.2004.06.041. PMID 15242632.

32. Gould, S. B., et al. (2008). Plastid evolution. *Annual Review of Plant Biology* 59 (1): 491–517. doi: 10.1146/annurev.arplant.59.032607.092915. PMID 18315522.

33. Perakis, S. S. & Pett-Ridge, J. C. (2019). Nitrogen-fixing red alder trees tap rock-derived nutrients. *PNAS USA* 116: 5009–14.

34. Sapountzis, P. (2016). Potential for nitrogen fixation in the fungus-growing termite symbiosis. *Frontiers in Microbiology* 7: 1993. doi: 10.3389/fmicb.2016.01993. PMC 5156715. PMID 28018322.

35. Dayel, M. J., et al. (2011). Cell differentiation and morphogenesis in the colony-forming choanoflagellate *Salpingoeca rosetta*. *Developmental Biology* 357: 73–82.

36. King, N. (2014). The unicellular ancestry of animal development. *Developmental Cell* 7(3): 313–25.

37. Pisani, D., et al. (2015). Genomic data do not support comb jellies as the sister group to all other animals. *PNAS* 112 (50) 15402–7. doi: 10.1073/pnas.1518127112.

38. Hastings, J. W. (1978). Bacterial bioluminescence: an overview. *Methods in Enzymology* 57: 125–35.

39. Haddock, S. H. D., et al. (2010). Bioluminescence in the sea. *Annual Review of Marine Science* 2: 443–93. doi: 10.1146/annurev-marine-120308-081028.

40. Rees, J.-F., et al. (1998). The origins of marine bioluminescence: turning oxygen defense mechanisms into deep-sea communication tools. *Journal of Experimental Biology* 201 (8): 1211–21.

41. Ruby, E. G. & Lee, K. H. (1998). The *Vibrio fischeri–Euprymna scolopes* light organ association: current ecological paradigms. *Applied Environmental Microbiology* 64: 805–12.

42. Lee, K. H. & Ruby E. G. (1994). Effect of the squid host on the abundance and distribution of symbiotic *Vibrio fischeri* in nature. *Applied Environmental Microbiology* 60: 1565–71.

43. Tong, D., et al. (2009). Evidence for light perception in a bioluminescent organ. *PNAS USA* 106 (24): 9836–41. Bibcode: 2009PNAS.106.9836T. doi: 10.1073/pnas.0904571106. PMC 2700988. PMID 19509343.

44. Fidopiastis, P. M., et al. (1998). A new niche for *Vibrio logei*, the predominant light organ symbiont of squids in the genus *Sepiola*. *Journal of Bacteriology* 180: 59–64.

45. Brodo, I. M. & Sharnoff, S. D. (2001). *Lichens of North America*. Yale University Press: New Haven, CT. ISBN 978-0300082494.

46. Lutzoni, F., et al. (21 June, 2001). Major fungal lineages are derived from lichen symbiotic ancestors. *Nature* 411: 937–40. doi: 10.1038/35082053.

47. Yuan, X., et al. (May, 2005). Lichen-like symbiosis 600 million years ago. *Science* 308 (5724): 1017–20. doi: 10.1126/science.1111347.

48. Spribille, T., et al. (29 July, 2016). Basidiomycete yeasts in the cortex of ascomycete macrolichens. *Science* 353 (6298): 488–92. doi: 10.1126/science.aaf8287.

49. Grube, M., et al. (2009). Species-specific structural and functional diversity of bacterial communities in lichen symbiosis. *The ISME Journal* 3: 1105–15.

50. Barreno, E., et al. (2008). Non photosynthetic bacteria associated to cortical structures on *Ramalina* and *Usnea thalli* from Mexico. Asilomar: Pacific Grove, CA. Abstracts IAL 6-ABLS Joint Meeting, p. 5.

51. Casano, L. M., et al. (March, 2011). Two *Trebouxia* algae with different physiological performances are ever-present in lichen thalli of *Ramalina farinacea*. Coexistence versus competition? *Environmental Microbiology* 13 (3): 806–18.
52. Honegger, R. (1991). Fungal evolution: symbiosis and morphogenesis, In Margulis, L., & Fester, R. (eds.), *Symbiosis as a Source of Evolutionary Innovation*, pp. 319–40. The MIT Press: Cambridge, MA.
53. Flynn, K. J., et al. (July, 2019). Mixotrophic protists and a new paradigm for marine ecology: where does plankton research go now? *Journal of Plankton Research* 41 (4): 375–91. doi: 10.1093/plankt/fbz026.
54. Mitra, A. (April, 2018). The perfect beast. *Scientific America* 318 (4): 36–43.
55. Zubkov, M. V. & Tarran, G. A. (11 Sept., 2008). High bacterivory by the smallest phytoplankton in the North Atlantic Ocean. *Nature* 455: 224–6.
56. Mitra, A., et al. (25 April, 2016). Defining planktonic protist functional groups on mechanisms for energy and nutrient acquisition: incorporation of diverse mixotrophic strategies. *Protist* 167 (2): 106–20. doi: 10.1016/j.protis.2016.01.003.
57. Abe, T., et al. (2000). *Termites: Evolution, Sociality, Symbioses, Ecology*. Kluwer Academic Publishers: Dordrecht, Netherlands; pp. 209–27 and 307–32.
58. Ohkuma, M. (Nov., 2001). Symbiosis within the gut microbial community of termites. *RIKEN Review* 41: 69–72.
59. Pringle, R. M., et al. (25 May, 2010). Spatial pattern enhances ecosystem functioning in an African savanna. *PLoS Biology*. doi: 10.1371/journal.pbio.1000377.
60. Davies, A. B., et al. (May, 2015). Seasonal activity patterns of African savanna termites vary across a rainfall gradient. *Insectes Sociaux* 62 (2): 157–65.
61. Faeth, S. H. (2002). Are endophytic fungi defensive plant mutualists? *Oikos* 98: 25–36.
62. Clay, K. (1988). Fungal endophytes of grasses: a defensive mutualism between plants and fungi. *Ecology* 69: 10–16.
63. Cheplick, G. P. & Faeth, S. H. (2009). *Ecology and Evolution of the Grass-Endophyte Symbiosis*. Oxford University Press: Oxford, UK.
64. Varma, A., et al. (June, 1999). *Piriformospora indica*, a cultivable plant-growth-promoting root endophyte. *Applied and Environmental Microbiology* 65 (6): 2741–4. PMC 91405. PMID 10347070.
65. Kogel, K.-H., et al. (2005). The endophytic fungus *Piriformospora indica* reprograms barley to salt-stress tolerance, disease resistance, and higher yield. *PNAS USA* 102 (38): 13386–91. doi: 10.1073/pnas.0504423102.
66. Paungfoo-Lonhienne, C., et al. (1 Jan., 2013). Bruijn, F. J. de, ed. *Rhizophagy—A New Dimension of Plant–Microbe Interactions*. John Wiley & Sons, Inc., Hoboken, NJ; pp. 1199–207. doi: 10.1002/9781118297674.ch115. ISBN 9781118297674.
67. Speight, M. R., et al. (1999). *Ecology of Insects*. Blackwell Science, Oxford, UK; p. 156. ISBN 0-86542-745-3.
68. Leal, I. R. & Oliveira, P. S. (1998). Interaction between fungus-growing ants (Attini), fruits, and seeds in cerrado vegetation in southeast Brazil. *Biotropica* 30: 170–8.
69. Zhang, M. M., et al. (2007). Symbiont recognition of mutualistic bacteria by *Acromyrmex* leaf-cutting ants. *The ISME Journal* 1 (4): 313–30. doi: 10.1038/ismej.2007.41.
70. Currie, C. R., et al. (1999). Fungus-growing ants use antibiotic-producing bacteria to control garden parasites. *Nature* 398: 701–4. doi: 10.1038/19519.
71. Vasconcelos, H. L. & Fowler, H. G. (1990). Foraging and fungal substrate selection by leaf-cutting ants. In: Vander Meer, R. K., et al. (eds.) *Applied Myrmecology—a World Perspective*, pp. 411–9. Westview Press: Boulder, CO.
72. Wirth, R., et al. (2003). Herbivory of leaf-cutter ants: a case study of *Atta columbica* in the tropical rainforest of Panama. *Ecological Studies* 164: 1–233.
73. Urbas, P., et al. (2007). Cutting more from cut forests: edge effects on foraging and herbivory of leaf-cutting ants in Brazil. *Biotropica* 39: 489–95.

74. Costa, A. N., et al. (Dec., 2008). Do herbivores exert top-down effects in Neotropical savannas? Estimates of biomass consumption by leaf-cutter ants. *Journal of Vegetation Science* 19 (6): 849–54. doi: 10.3170/2008-8-18461.

75. Silva, P. S. D., et al. (2009). Decreasing abundance of leaf-cutting ants across a chronosequence of advancing Atlantic forest regeneration. *Journal of Tropical Ecology* 25: 223–27.

76. Correa, M. M., et al. (Jan., 2010). How leaf-cutting ants impact forests: drastic nest effects on light environment and plant assemblages. *Oecologia* 162 (1): 103–15.

77. Meyer, S. T., et al. (Feb., 2011). Ecosystem engineering by leaf-cutting ants: nests of *Atta cephalotes* drastically alter forest structure and microclimate. *Ecological Entomology* 36 (1): 14–24. doi: 10.1111/j.1365-2311.2010.01241.x.

78. Leal, I. R., et al. (Sept., 2014). The multiple impacts of leaf-cutting ants and their novel ecological role in human-modified neotropical forests. *Biotropica* 46 (5): 516–28. doi: 10.1111/btp.12126.

79. Moreira, A. A., et al. (April, 2004). External and internal structure of *Atta bisphaerica* Forel (Hymenoptera: Formicidae) nests. *Journal of Applied Entomology* 128 (3): 204–11. doi: 10.1111/j.1439-0418.2004.00839.x.

80. Cherrett, J. M. (1989). Leaf-cutting ants. In: Lieth, H. & Werger, M. J. A. (eds.), *Ecosystems of the World*, pp. 473–86. Elsevier: New York.

81. Farji-Brener, A. G. & Illes, A. E. (2000). Do leaf-cutting ant nests make "bottom-up" gaps in neotropical rain forest? A critical review of the evidence. *Ecology Letters* 3: 219–27.

82. Hart, A. G. & Ratnieks, F. L. W. (2002). Waste management in the leaf-cutting ant *Atta colombica*. *Behavioural Ecology* 13 (2): 224–31. doi: 10.1093/beheco/13.2.224.

83. Bot, A. N. M., et al. (2001). Waste management in leaf-cutting ants. *Ethology Ecology & Evolution* 13 (3): 225–37. doi: 10.1080/08927014.2001.9522772.

84. Dalling, J. W. & Wirth, R. (1998). Dispersal of *Miconia argentea* seeds by the leaf-cutting ant *Atta colombica* (L.). *Journal of Tropical Ecology* 14: 705–10.

85. Farji-Brener, A. G. & Ghermandi, L. (2000). Influence of nests of leafcutting ants on plant species diversity in road verges of northern Patagonia. *Journal of Vegetation Science* 11: 453–60.

86. Moutinho, P., et al. (2003). Influence of leaf-cutting ant nests on secondary forest growth and soil properties in Amazonia. *Ecology* 84: 1265–76.

87. Pinto-Tomás, A. A., et al. (2009). Symbiotic nitrogen fixation in the fungus gardens of leaf-cutter ants. *Science* 326 (5956): 1120–3.

88. Meyer, S. T., et al. (2013). Leaf-cutting ants as ecosystem engineers: topsoil and litter perturbations around *Atta cephalotes* nests reduce nutrient availability. *Ecological Entomology* 38: 497–504. doi: 10.1111/een.12043.

89. Farji-Brener, A. G., & Silva, J. F. (1995). Leaf-cutting ants and forest grooves in a tropical parkland savanna of Venezuela: facilitated succession? *Journal Tropical Ecology* 11: 651–69.

90. Garrettson, M., et al. (1998). Diversity and abundance of understory plants on active and abandoned nests of leaf-cutting ants (*Atta cephalotes*) in Costa Rica rain forest. *Journal Tropical Ecology* 14: 17–26.

91. Farji-Brener, A. G. (Dec., 2005). The effect of abandoned leaf-cutting ant nests on plant assemblage composition in a tropical rainforest of Costa Rica. *Ecoscience* 12 (14): 554–60.

92. Sousa-Souto, L., et al. (2007). Leaf-cutting ants, seasonal burning and nutrient distribution in Cerrado vegetation. *Austral Ecology* 32: 758–65.

93. Kitajima, K. (1996). Ecophysiology of tropical tree seedlings. In Mulkey, S. S., et al. (eds.), *Tropical Forest Plant Ecophysiology*, pp. 559–96. Chapman and Hall: New York.

94. Chomicki, G. & Renner, S. S. (27 Nov., 2016). Obligate plant farming by a specialized ant. *Nature Plants* 2, Article number 16181. doi: 10.1038/nplants.2016.181.

95. Li, H., et al. (2006). Arbuscular mycorrhizal fungi contribute to phosphorus uptake by wheat grown in a phosphorus-fixing soil even in the absence of positive growth responses. *New Phytologist* 172 (3): 536–43. doi: 10.1111/j.1469-8137.2006.01846.x. PMID 17083683.

96. Hogan, C. M. (2011). Phosphate. In Jorgensen, A. (ed.), *Encyclopedia of Earth*. Topic. Ed.-in-Chief C. J. Cleveland. National Council for Science and the Environment: Washington, DC.

97. Rodriguez, R. J., et al. (2009). In White, J. F., Jr. & Torres, M. S. (eds.), *Mycology* V. 27: *Defensive Mutualism in Microbial Symbiosis*. Chapter 20. Habitat-adapted symbiosis as a defense against abiotic and biotic stress. CRC Press, Boca Raton, FL; London; New York.

98. Stamets, P. (12 Jan., 2012). 6 ways mushrooms can save the world. TED Talk. TED.com. https://www.ted.com/talks/paul_stamets_on_6_ways fungus can save the world.

99. Lehto, T. (1992). Mycorrhizas and drought resistance of *Picea sitchensis* (Bong.) Carr. I. In conditions of nutrient deficiency. *New Phytologist* 122 (4): 661–8. JSTOR 2557434.

100. Nikolaou, N., et al. (2003). Effects of drought stress on mycorrhizal and non-mycorrhizal Cabernet Sauvignon grapevine, grafted onto various rootstocks. *Experimental Agriculture* 39 (3): 241–52. doi: 10.1017/S001447970300125X.

101. Frey-Klett, P., et al. (2007). The mycorrhiza helper bacteria revisited. *New Phytologist* 176: 22–36. doi: 10.1111/j.1469-8137.2007.02191.x.

102. Klironomos, J. N., & Hart. M. M. (5 April, 2001). Food-web dynamics: animal nitrogen swap for plant carbon. *Nature* 410: 651–65. doi: 10.1038/35070643.

103. Harrison, M. J. (2005). Signaling in the arbuscular mycorrhizal symbiosis. *Annual Review Microbiology* 59: 19–42. doi: 10.1146/annurev.micro.58.030603.123749. PMID 16153162.

104. Smith, S. E. & Read, D. (2008). *Mycorrhizal Symbiosis*. Third Edn. Elsevier Ltd.: Philadelphia, PA. ISBN 978-0-12-370526-6.

105. Wang, B., & Qiu, Y. L. (2006). Phylogenetic distribution and evolution of mycorrhizas in land plants. *Mycorrhiza* 16 (5): 299–363. doi: 10.1007/s00572-005-0033-6. PMID 16845554.

106. Taylor A. F. S. & Alexander I. (2005). The ectomycorrhizal symbiosis: life in the real world. *Mycologist* 19: 102–12.

107. Simard, S. W., et al. (2012). Mycorrhizal networks: Mechanisms, ecology and modeling. *Fungal Biology Review* 26: 39–60.

108. Simard, S. W., et al. (1997). Net transfer of carbon between ectomycorrhizal tree species in the field. *Nature* 388 (6642): 579–82. doi: 10.1038/41557.

109. Simard, S. W. (2011). Do Trees Communicate? A documentary.

110. Babikova, Z., et al. (July, 2013). Underground signals carried through common mycelial networks warn neighbouring plants of aphid attack. *Ecology Letters* 16 (7): 835–43. doi: 10.1111/ele.12115.

111. Barto (Morris), E. K., et al. (Nov., 2011). The fungal fast lane: common mycorrhizal networks extend bioactive zones of allelochemicals in soils. *PLoS ONE* 6 (11): 1–7. e27195.

112. Achatz, M., et al. (2014). Soil hypha-mediated movement of allelochemicals: arbuscular mycorrhizae extend the bioactive zone of juglone. *Functional Ecology* 28: 1020–9.

113. Andrade, G., et al. (1998). Bacterial associations with the mycorrhizosphere and hyphosphere of the arbuscular mycorrhizal fungus *Glomus mosseae*. *Plant Soil* 202: 79–87.

114. Andrade, G., et al. (1998). Soil aggregation status and rhizobacteria in the mycorrhizosphere. *Plant Soil* 202: 89–96.

115. Founoune, H., et al. (2002). Interactions between ectomycorrhizal symbiosis and fluorescent pseudomonads on *Acacia holosericea*: isolation of mycorrhiza helper bacteria (MHB) from a Soudano-Sahelian soil. *FEMS Microbiology Ecology* 41: 37–46.

116. Garbaye, J. (1991). Biological interactions in the mycorrhizosphere. *Experientia* 47: 370–5.

117. Garbaye, J. (1994). Helper bacteria: a new dimension to the mycorrhizal symbiosis. *New Phytologist* 128: 197–210.

118. Garbaye, J. & Bowen, G. D. (1989). Stimulation of mycorrhizal infection of *Pinus radiata* by some microorganisms associated with the mantle of ectomycorrhizas. *New Phytologist* 112: 383–8.

119. Duponnois, R. & Plenchette, C. (2003). A mycorrhiza helper bacterium (MHB) enhances ectomycorrhizal and endomycorrhizal symbiosis of Australian *Acacia* species. *Mycorrhiza* 13: 85–91. Medline, ISI.

120. Mamatha, G., et al. (2002). Innoculation of field-established mulberry and papaya with arbuscular mycorrhizal fungi and a mycorrhiza helper bacterium. *Mycorrhiza* 12: 313–6.

121. Budi, S. W., et al. (Nov., 1999). Isolation from the *Sorghum bicolor* Mycorrhizosphere of a bacterium compatible with arbuscular mycorrhiza development and antagonistic towards soilborne fungal pathogens. *Applied and Environmental Microbiology* 65 (11): 5148–50.

122. Frey-Klett, P., et al. (2005). Ectomycorrhizal symbiosis affects functional diversity of rhizosphere fluorescent pseudomonads. *New Phytologist* 165: 317–28.

123. Xie, Z. P., et al. (1995). Rhizobial nodulation factors stimulate mycorrhizal colonization of nodulating and nonnodulating soybeans. *Plant Physiology* 108: 1519–25.

124. Xavier, L. J. C. & Germida, J. J. (2003). Bacteria associated with *Glomus clarum* spores influence mycorrhizal activity. *Soil Biology and Biochemistry* 35: 471–8.

125. Bending, G. D., et al. (2002). Characterisation of bacteria from *Pinus sylvestris–Suillus luteus* mycorrhizas and their effects on root–fungus interactions and plant growth. *FEMS Microbiology Ecology* 39: 219–27.

126. Duponnois, R. (1992). Les bacteries auxiliaires de la mycorhization du Douglas (*Pseudotsuga menziessii* (Mirb.) Franco) par *Laccaria laccata* souche S238. *PhD thesis*, University of Nancy 1, France.

127. Poole, E. J., et al. (2001). Bacteria associated with *Pinus sylvestris–Lactarius rufus* ectomycorrhizas and their effects on mycorrhiza formation in vitro. *New Phytologist* 151: 743 –51.

128. Schrey, S. D., et al. (2005). Mycorrhiza helper bacterium *Streptomyces* AcH 505 induces differential gene expression in the ectomycorrhizal fungus *Amanita muscaria*. *New Phytologist* 168: 205–16.

129. Frey-Klett, P., et al. (1997). Location and survival of mycorrhiza helper *Pseudomonas fluorescens* during establishment of ectomycorrhizal symbiosis between *Laccaria bicolor* and Douglas fir. *Applied and Environmental Microbiology* 63: 139–44.

130. Gamalero, E., et al. (2004). Impact of two fluorescent pseudomonads and an arbuscular mycorrhizal fungus on tomato plant growth, root architecture and P acquisition. *Mycorrhiza* 14: 185–92.

131. Lopez, M. F., et al. (2007). Increased trehalose biosynthesis in the Hartig net hyphae of ectomycorrhizas. *New Phytologist* 174: 389–98.

132. Wiemken, V. (2007). Trehalose synthesis in ectomycorrhizas a driving force of carbon gain for fungi? *New Phytologist* 174: 228–30.

133. Izumi, H., et al. (2006). Endobacteria in some ectomycorrhiza of Scots pine (*Pinus sylvestris*). *FEMS Microbiology and Ecology* 56: 34–43.

134. Uroz, S., et al. (2007). Effect of the mycorrhizosphere on the genotypic and metabolic diversity of the soil bacterial communities involved in mineral weathering in a forest soil. *Applied and Environmental Microbiology* 73: 3019–27.

135. Duponnois, R. & Kisa, M. (2006). The possible role of trehalose in the mycorrhiza helper effect. *Canadian Journal of Botany* 84: 1005–8.

136. Lumini, E., et al. (2007). Presymbiotic growth and sporal morphology are affected in the arbuscular mycorrhizal fungus *Gigaspora margarita* cured of its endobacteria. *Cellular Microbiology* 9: 40–53.

137. Brown, J. (2003). Ancient horizontal gene transfer. *Nature Review Genetics* 4: 121–32.

138. Buades, C. & Moya, A. (1996). Phylogenetic analysis of the isopenicillin-N-synthetase horizontal gene transfer. *Journal of Molecular Evolution*. 42: 537–42.

139. Yang, C.-H. & Crowley, D. E. (2000). Rhizosphere microbial community structure in relation to root location and plant iron nutritional status. *Applied and Environmental Microbiology* 66: 345–51.

140. Herschkovitz, Y., et al. (2005). Inoculation with the plant growth promoting rhizobacterium *Azospirillum brasilense* causes little disturbance in the rhizosphere and rhizoplane of maize (*Zea mays*). *Microbial Ecology* 50: 277–88.

141. Lerner, A., et al. (2006). Effect of *Azospirillum brasilense* on rhizobacterial communities analyzed by denaturing gradient gel electrophoresis and automated intergenic spacer analysis. *Soil Biology and Biochemistry* 38: 1212–8.

142. Grayston, S. J., et al. (1998). Selective influence of plant species on microbial diversity in the rhizosphere. *Soil Biology and Biochemistry* 30: 369–78.

143. Smalla, K., et al. (2001). Bulk and rhizosphere soil bacterial communities studied by denaturing gradient gel electrophoresis: plant-dependent enrichment and seasonal shifts revealed. *Applied and Environmental Microbiology* 67: 4742–51.

144. Van Overbeek, L., & Van Elsas, J. D. (2008). Effects of plant genotype and growth stage on the structure of bacterial communities associated with potato (*Solanum tuberosum* L.). *FEMS Microbiology Ecology* 64: 283–96.

145. Andreote, F. D., et al. (2009). Endophytic colonization of potato (*Solanum tuberosum* L.) by a novel competent bacterial endophyte, *Pseudomonas putida* strain P9, and the effect on associated bacterial communities. *Applied and Environmental Microbiology*. doi: 10.1128/AEM.00491-09.

146. Dunfield, K. E. & Germida, J. J. (2003). Seasonal changes in the rhizosphere microbial communities associated with field-grown genetically modified Canola (*Brassica napus*). *Applied and Environmental Microbiology* 69: 7310–8.

147. Márquez. L. M., et al. (26 Jan., 2007). A virus in a fungus in a plant: three-way symbiosis required for thermal tolerance. *Science* 315 (5811): 513–5. doi: 10.1126/science.1136237.

148. Redman, R.S., et al. (22 Nov., 2002). Thermotolerance generated by plant/fungal symbiosis. *Science* 298 (5598): 1581. doi: 10.1126/science.1078055.

149. Bordenstein S. R. & Wernegreen, J. J. (2004). Bacteriophage flux in endosymbionts (*Wolbachia*): infection frequency, lateral transfer, and recombination rates. *Molecular Biology and Evolution* 21 (10): 1981-91. doi: 10.1093/molbev/msh211.

150. Moran, N. A., et al. (2005). The players in a mutualistic symbiosis: Insects, bacteria, viruses, and virulence genes. *PNAS USA* 102 (47): 16919–26. doi: 10.1073/pnas.0507029102.

151. Rasmann, S. (Aug., 2016). Root signals that mediate mutualistic interactions in the rhizosphere. *Current Opinion in Plant Biology* 32: 62–8. doi: 10.1016/j.pbi.2016.06.017.

152. Frank, J., et al. (2006). Oaks below ground: Mycorrhizas, truffles, and small mammals. *General Technical Report PSW-GTR* 217: 131–38.

153. Smith, S. S. & Smith, F. A. (Jan./Feb., 2012). Fresh perspectives on the roles of arbuscular mycorrhizal fungi in plant nutrition and growth. *Mycologia* 104 (1): 1–13. Published online before print: September, 2011. doi: 10.3852/11-229.

154. Pirozynski, K. A. & Malloch, S. W. (March, 1975). The origin of land plants: a matter of mycotrophism. *Biosystems* 6 (3): 153–164. doi: 10.1016/0303-2647(75)90023.

155. Read, D. J., et al. (2000). Symbiotic fungal associations in "lower" land plants. *Philosophical Transactions of the Royal Society of London B* 355: 815–31.
156. Remy, W., et al. (1994). Four hundred million-year-old vesicular arbuscular mycorrhizae. *PNAS USA* 91: 11841–3.
157. Simon, L., et al. (1993). Origin and diversification of endomycorrhizal fungi and coincidence with vascular land plants. *Nature* 363: 67–9.
158. Smith, M. R. (April, 2016). Cord-forming Palaeozoic fungi in terrestrial assemblages. *Botanical Journal of the Linnean Society* 180 (4): 452–60. doi: 10.1111/boj.12389.
159. Walters, D. (2017). *Fortress Plant. How to Survive When Everything Wants to Eat You.* Oxford University Press: Oxford, UK.
160. Gagliano, M., et al. (2012). Towards understanding plant bioacoustics. *Trends in Plant Science* 954: 1–3. doi: 10.1016/j.tplants.2012.03.002.
161. Crepet, W. L. (1984). Advanced (constant) insect pollination mechanisms: pattern of evolution and implications vis-a-vis angiosperm diversity. *Annals of the Missouri Botanical Garden. Historical Perspectives of Angiosperm Evolution* 71 (2): 607–30.
162. Regal, P. J. (1977). Ecology and evolution of flowering plant dominance. *Science* 196: 622–9.
163. Stebbins, G. L. (1981). Why are there so many species of flowering plants? *BioScience* 31: 573–7.
164. McPeek, M. A. (1996). Linking local species interactions to rates of speciation in communities. *Ecology* 77: 1355–66.
165. Koh, L. P., et al. (10 Sept., 2004). Species coextinctions and the biodiversity crisis. *Science* 305 (5690): 1632–4. doi: 10.1126/science.1101101.
166. Thorne, R. F. (2002). How many species of seed plants are there? *Taxonomy* 51 (3): 511–22. doi: 10.2307/1554864. JSTOR 1554864.
167. Scotland, R. W. & Wortley, A. H. (2003). How many species of seed plants are there? *Taxonomy* 52 (1): 101–4. doi: 10.2307/3647306. JSTOR 3647306.
168. Govaerts, R. (2003). How many species of seed plants are there?—A response. *Taxonomy* 52 (3): 583–4.
169. Crane, P. R., et al. (1995). The origin and early diversification of angiosperms. *Nature* 374: 27–33.
170. Christenhusz, M. J. M. & Byng, J. W. (2016). The number of known plants species in the world and its annual increase. *Phytotaxa* 261 (3): 201–17. doi: 10.11646/phytotaxa.261.3.1.
171. Lidgard, S. & Crane, P. R. (1988). Quantitative analyses of the early angiosperm radiation. *Nature* 331: 344–6. doi: 10.1038/331344a0.
172. Proctor, M., et al. (1996). *The Natural History of Pollination.* Timber Press: Portland, OR.
173. Dafni, A., et al. (eds.) (2005). *Practical Pollination Biology.* Enviroquest, Ltd.: Cambridge, Ontario, Canada.
174. Berendse, F. & Scheffer, M. (Sept., 2009). The angiosperm radiation revisited, an ecological explanation for Darwin's 'abominable mystery'. *Ecology Letters* 12 (9): 865–72. doi: 10.1111/j.1461-0248.2009.01342.x.
175. Brodribb, T. J. & Field, T. S. (2010). Leaf hydraulic evolution led a surge in leaf photosynthetic capacity during early angiosperm diversification. *Ecology Letters* 13: 175–83. doi: 10.1111/j.1461-0248.2009.01410.
176. Baker, H. G. & Hurd, P. D. (1968). Intrafloral ecology. *Annual Review Entomology* 13: 385–414.
177. Ollerton, J., et al. (March, 2011). How many flowering plants are pollinated by animals? *Oikos* 120 (3): 321–6. doi: 10.1111/j.1600-0706.2010.18644.x.
178. Landry, C. L. (2010). Mighty mutualisms: the nature of plant-pollinator interactions. *Nature Education Knowledge* 3 (10): 37.

179. Erwin, T. (1982). Tropical forests: their richness in Coleoptera and other arthropod species. *The Coleopterists Bulletin* 36 (1): 74–5.
180. Fenton, M. B. & Simmons, N. B. (2015). *Bats: A World of Science and Mystery*, p. 115. University of Chicago Press: Chicago, IL. ISBN 978-0226065120.
181. Hodgkison, R., et al. (2003). Fruit bats (Chiroptera: Pteropodidae) as seed dispersers and pollinators in a lowland Malaysian rain forest. *Biotropica* 35 (4): 491–502. doi: 10.1111/j.1744-7429.2003.tb00606.x.
182. National Research Council. (2007). *Status of Pollinators in North America*. National Academies Press: Washington, DC.
183. Landry, C. L., et al. (2005). Flower visitors to white mangrove: a comparison between three Bahamian islands and Florida. In *The Proceedings of the 10th Symposium on the Natural History of the Bahamas*. Buckner, S. & McGrath, T. (eds.). pp. 84–94. San Salvador Island: Gerace Research Center.
184. Michener, C. D. (2007). *The Bees of the World*, Second edn. The John Hopkins University Press: Baltimore, MD.
185. Zimmermann, Y., et al. (2006). Species-specific attraction to pheromonal analogues in orchid bees. *Behavioral Ecology and Sociobiology* 60: 833–43.
186. Pellmyr, O. (2003). Yuccas, yucca moths, and coevolution: a review. *Annals of the Missouri Botanical Garden* 90: 35–55.
187. Clarke, D., et al. (5 April, 2013). Detection and learning of floral electric fields by bumblebees. *Science* 340 (6128): 66–9. doi: 10.1126/science.1230883.
188. Atamian, H. S., et al. (2016). Circadian regulation of sunflower heliotropism, floral orientation, and pollinator visits. *Science* 353: 587–90.
189. Cook, J. M. & Rasplus, J. Y. (2003). Mutualists with attitude: coevolving fig wasps and figs. *Trends in Ecology & Evolution* 18: 241–8.
190. Labandeira, C. C. & Sepkoski, J. J., Jr. (16 July, 1993). Insect diversity in the fossil record. *Science* 261 (5119): 310–5. doi: 10.1126/science.11536548.
191. Hammond, P. M. (1992). Species inventory. In Groombridge, B. (ed.). *Global Biodiversity, Status of the Earth's Living Resources*, pp. 17–39. Chapman and Hall: London, UK. ISBN 0412472406.
192. Capinera, J. L. (2008). *Butterflies and Moths. Encyclopedia of Entomol. 4*. Second edn. Springer: New York, pp. 626–72. ISBN 9781402062421.
193. Mallet, J. (12 June, 2007). *Taxonomy of Lepidoptera: The Scale of the Problem. The Lepidoptera Taxome Project*. University College: London, UK.
194. Van der Pijl, L. (1982). *Principles of Dispersal in Higher Plants, Third Revised and Expanded Edition*. Springer-Verlag: Berlin, Heidelberg, Germany; New York, N. Y., U. S. A.
195. Janzen, D. H. (Nov.–Dec., 1970). Herbivores and the number of tree species in tropical forests. *The American Naturalist* 104 (940). doi: 10.1086/282687.
196. Connell, J.H. (1970). On the role of natural enemies in preventing competitive exclusion in some marine animals and in rain forest trees. In Den Boer, P.J. & Gradwell, G.R. (eds.), *Dynamics of Population*. Wageningen: Pudoc.
197. Stoner, K. E. & Henry, M. (2009). Seed dispersal and frugivory in tropical ecosystems. *Encyclopedia of Life Support Systems (EOLSS). International Commission on Tropical Biology and Natural Resources* V: 176–190.
198. Foster, R. B., et al. (1986). Dispersal and the sequential plant communities in Amazonian Peru floodplain. In Estrada, A. & Fleming, T. H. (eds.), *Frugivores and Seed Dispersal*, pp. 357–70. Dr. W. Junk Publsher: Dordrecht, the Netherlands.
199. Ridley, H. N. (1930). *The Dispersal of Plants Throughout the World*. L. Reeve & Co.: Ashford, KY.
200. Darwin, C. (1859). *The Origin of Species*. The New American Library, Inc.: New York, N. Y., U. S. A.

201. Dawson, E. (1962). The giants of Galapagos. *Natural History* 71: 52–7.

202. Rick, C. M. & Bowman, R. J. (1961). Galapagos tomatoes and tortoises. *Evolution* 15: 407–17.

203. Klimstra, W. D. & Newsome, F. (1960). Some observations on the food coactions of the common Box Turtle (*Terrapene c. caroline*). *Ecology* 41: 637–47.

204. Barquin, D. E. & de la Torre, W. W. (1975). Disseminacion de plantas canarias. *Datos iniciales. Vieraca (Tenerife)* 5: 38–60.

205. Buckley, R.C. (1982). Ant-plant interactions: a world review. In Buckley, R.C. (ed.), *Ant-Plant Interactions in Australia*, pp. 111–41, Dr. W. Junk Publishers: The Hague, the Netherlands.

206. Lengyel S., et al. (2009). Ants sow the seeds of global diversification in flowering plants. *PLoS ONE* 4 (5): e5480. doi: 10.1371/journal.pone.0005480. PMC 2674952. PMID 19436714.

207. Lengyel S., et al. (2010). Convergent evolution of seed dispersal by ants, and phylogeny and biogeography in flowering plants: a global survey. *Perspectives in Plant Ecology, Evolution and Systematics* 12 (1): 43–55. doi: 10.1016/j.ppees.2009.08.001.

208. Beattie, A. J. (1985). *The Evolutionary Ecology of Ant-Plant Mutualisms*. Cambridge University Press: Cambridge, U.K.

209. Hanzawa, F. M., et al. (1988). Directed dispersal: demographic analysis of an ant-plant mutualism. *The American Naturalist*, 131 (1): 1–13.

210. Culver, D. C., & A. J. Beattie. (1978). Myrmecochory in *Viola*: Dynamics of seed-ant interactions in some West Virginia species. *Journal of Ecology* 66 (1): 53–72.

211. Heithaus, E. R., et al. (1980). Models of some ant-plant mutualisms. *The American Naturalist* 116 (3): 347–61.

212. Giladi, I. (2006). Choosing benefits or partners: a review of the evidence for the evolution of myrmecochory. *Oikos* 112 (3): 481–92. doi:10.1111/j.0030-1299.2006.14258.x.

213. Restrepo C., et al. (2002). The role of vertebrates in the diversification of New World mistletoes. In Levey, D., et al. (eds.), *Seed Dispersal and Frugivory: Ecology, Evolution and Conservation*. CABI Publishing: New York.

214. Regan, B. C., et al. (March, 2001). Fruits, foliage and the evolution of primate colour vision. *Philosophical Transactions B* 356 (1407). doi: 10.1098/rstb.2000.0773.

215. Isbell, L. A. (July, 2006). Snakes as agents of evolutionary change in primate brains. *Journal of Human Evolution* 51 (1): 1–35. doi: 10.1016/j.jhevol.2005.12.012.

216. Bello, C., et al. (18 Dec., 2015). Defaunation affects carbon storage in tropical forests. *Science Advances* 1: e1501105: pp. 1–10.

217. UNEP (2001). UNEP-WCMC World *Atlas of Coral Reefs Coral Reef Unit.*

218. Sherman, C. D. H. (2009).The Importance of fine-scale environmental heterogeneity in determining levels of genotypic diversity and local adaption. University of Wollongong, Ph.D. Thesis.

219. Marshall, P. & Schuttenberg, H. (2006). *A Reef Manager's Guide to Coral Bleaching*. Great Barrier Reef Marine Park Authority: Townsville, Australia. ISBN 1-876945-40-0.

220. Spalding, M. D. & Grenfell, A. M. (1997). New estimates of global and regional coral reef areas. *Coral Reefs* 16 (4): 225–30. doi: 10.1007/s003380050078.

221. Sorokin, Y. I. (1993). *Coral Reef Ecology*. Springer-Verlag: Berlin, Heidelberg; Germany; New York. ISBN 978-0-387-56427-2.

222. Hatcher, B. G. (1 May, 1988). Coral reef primary productivity: a beggar's banquet. *Trends in Ecology & Evolution* 3 (5): 106–11. doi: 10.1016/0169-5347(88)90117-6.

223. McClellan, K. & Bruno, J. (2008). Coral degradation through destructive fishing practices. *Encyclopedia of Earth* 1–5.

224. Vermeij, M. J. A., et al. (14 May, 2010). Coral larvae move toward reef sounds. *PLoS One* 5 (5): e10660. doi: 10.1371/journal.pone.0010660.

225. Kaplan, M. (2009). How the sponge stays slim. *Nature*. doi: 10.1038/news.2009.1088.

226. McKeon, C. S. & Moore, J. M. (30 Sept., 2014). Species and size diversity in protective services offered by coral guard-crabs. *PeerJ* 2: e574. doi: 10.7717/peerj.574. PubMed 25289176.

227. Hay, M. E. & Stachowicz, J. J. (1 Sept., 1999). Mutualism and coral persistence: the role of herbivore resistance to algal chemical defense. *Ecology* 80: 2085–101. doi: 10.2307/176680.

228. Castro, P. & Huber, M. (2000). *Marine Biology*. Third edn. McGraw-Hill: Boston, MA.

229. Achituv, Y. & Dubinsky, Z. (1990). *Evolution and Zoogeography of Coral Reef Ecosystems of the World*. Vol. 25: pp. 1–8. Elsevier: Amsterdam, the Netherlands.

230. Mallet, J. & Joron, M. (1999). Evolution of diversity in warning color and mimicry: polymorphisms, shifting balance, and speciation. *Annual Review of Ecology, Evolution, and Systematics* 30: 201–33.

231. Futuyma, D. J. (2005). *Evolution*. Sinauer Assoc., Inc.: Sunderland, MA. p. 286, Fig. 12.19 and p. 445, Fig.18.24.i

232. Rudman, W.B. (24 July, 2004). Symbiosis, commensalism, mutualism and parasitism. [In] *Sea Slug Forum*. Australian Museum, Sydney. Available at http://www.seaslugforum.net/factsheet.cfm?base=symbio.

233. Hutchison, V. H. & and Hammen, C. S. (1958). Oxygen utilization in the symbiosis of embryos of the salamander, *Ambystoma maculatum* and the alga, *Oophila amblystomatis*. *Biology Bulletin* 115: 483–489. doi: 10.2307/1539111.

234. Kerney, R., et al. (2011). Intracellular invasion of green algae in a salamander host. *PNAS USA* 108: 6497–6502.

235. Lago, J., et al. (2015). Tetrodotoxin, an extremely potent marine neurotoxin: distribution, toxicity, origin and therapeutical uses. *Marine Drugs* 13 (10): 6384–406. doi: 10.3390/md13106384. PMC 4626696. PMID 26492253.

236. Homann, M., et al. (2016). Evidence for cavity-dwelling microbial life in 3.22 Ga tidal deposits. *Geology*. 44 (1): 51–4. doi: 10.1130/G37272.1.

237. de Fouw, J., et al. (25 April, 2016). Drought, mutualism breakdown, and landscape-scale degradation of seagrass beds. *Current Biology* 26 (8): 1051–6. doi: 10.1016/j.cub.2016.02.023.

238. Taylor, P. D., et al. (Sept., 1989). Symbiotic associations between hermit crabs and bryozoans from the Otago region, southeastern New Zealand. *Journal of Natural History* 23 (5): 1059–85. doi: 10.1080/00222938900770971.

239. Heinrich, B., et al. (2002). Common ravens preferentially associate with grey wolves as a foraging strategy in winter. *Animal Behaviour* 64: 283–90. doi: 10.1006/anbe.2002.3047.

240. Vucetich, J. A., et al. (2003). Raven scavenging favours group foraging in wolves. *Animal Behaviour* 67: 1117–26. doi: 10.1016/j.anbe hav.2003.06.018.

241. Nell, L. A., et al. (2 Mar., 2016). Presence of breeding birds improves body condition for a Crocodilian nest protector. *PLOS One*. doi: 10.1371/journal.pone.0149572.

242. Nardi, J.B., et al. (2002). Could microbial symbionts of arthropod guts contribute significantly to nitrogen fixation in terrestrial ecosystems? *Journal of Insect Physiology* 48: 751–63.

243. Armitage, D. W. (1 Nov., 2016). Bacteria facilitate prey retention by the pitcher plant *Darlingtonia californica*. doi: 10.1098/rsbl.2016.0577.

244. Matthews, D. M. & Jenks, S. M. (June, 2003). Ingestion of *Mycobacterium vaccae* decreases anxiety-related behavior and improves learning in mice. *Behavioural Processes* 96: 27–35. doi: 10.1016/j.beproc.2013.02.007.

245. Raine, N. E., et al. (Nov., 2002). Spatial structuring and floral avoidance behavior prevent ant-pollinator conflict in a Mexican ant-acacia. *Ecology* 83 (11): 3086–96. doi: 10.1890/0012-9658(2002)083[3086:SSAFAB]2.0.CO;2.

246. González-Teuber, M., et al. (April, 2014). Mutualistic ants as an indirect defense against leaf pathogens. *New Phytologist* 202 (2): 640–50. doi: 10.1111/nph.12664.

247. Brock, D. A., et al. (20 Jan., 2011). Primitive agriculture in a social amoeba. *Nature* 469: 393–6. doi: 10.1038/nature09668.

248. Hata, H., et al. (2010). Geographic variation in the damselfish-red alga cultivation mutualism in the Indo-West Pacific. *BMC Evolutionary Biology* 10, Article number 185.

249. Silliman, B.R. & Newell, S.Y. (2003). Fungal farming by a snail. 100 (26): 15643–8. doi: 10.1073/pnas.2535227100.

250. Gittman, R. K. & Keller, D. A. (Dec., 2013). Fiddler crabs facilitate *Spartina alterniflora* growth, mitigating periwinkle overgrazing of marsh habitat. *Ecology* 94 (12): 2709–18.

251. Canestrari, D., et al. (21 Mar., 2014). From parasitism to mutualism: unexpected interactions between a cuckoo and its host. *Science* 343 (6177): 1350–2. doi: 10.1126/science.1249008.

252. Ingram, W. M., et al. (18 Sept., 2013). Mice infected with low-virulence strains of *Toxoplasma gondii* lose their innate aversion to cat urine, even after extensive parasite clearance. *PLoS ONE* 8, e75246. doi: 10.1371/journal.pone.0075246. Also see *Nature* doi: 10.1038/nature.2013.13777.

253. Greig, E. I. & Pruett-Jones, S. (2010). Danger may enhance communication: predator calls alert females to male displays. *Behavioral Ecology* 21 (6): 1360. doi: 10.1093/beheco/arq155.

254. Lovas-Kiss, Á., et al. (7 July, 2020). Experimental evidence of dispersal of invasive cyprinid eggs inside migratory waterfowl. *PNAS USA* 117 (27): 15397–15399. doi: 10.1073/pnas.2004805117.

255. Grman, E. (Aug., 2009). Plant sanctions cannot prevent parasitism by cheating mycorrhizal fungi. *94th ESA Annual Convention*.

256. Michalakis, Y., et al. (Feb., 1992). Pleiotropic action of parasites: how to be good for the host. *Trends in Ecology & Evolution* 7 (2): 59–62. doi: 10.1016/0169-5347(92)90108-N.

257. Matabos, M., et al. (Feb., 2014). High-frequency study of epibenthic megafaunal community dynamics in Barkley Canyon: a multi-disciplinary approach using the NEPTUNE Canada network. *Journal of Marine Systems* 130: 56–68. doi: 10.1016/j.jmarsys.2013.05.002.

258. Dorenbosch, M., et al. (2005). Indo-Pacific seagrass beds and mangroves contribute to fish density and diversity on adjacent coral reefs. *Marine Ecology Progress Series* 302: 63–76.

259. Lamb, J. B., et al. (17 Feb., 2017). Seagrass ecosystems reduce exposure to bacterial pathogens of humans, fishes, and invertebrates. *Science* 355 (6326): 731–3. doi: 10.1126/science.aal1956.

260. Maheswari, R. U., et al. (June, 2011). Interrelation among coral reef and sea-grass habitats in the Gulf of Mannar. *International Journal of Biodiversity and Conservation* 3 (6): 193–205.

261. Schlichting, P. E., et al. (2019). Efficiency and composition of vertebrate scavengers at the land-water interface in the Chernobyl Exclusion Zone. *Food Webs* 18: e00107. doi: 10.1016/j.fooweb.2018.e00107.

262. Drymon, J. M., et al. (Sept., 2019). Tiger sharks eat songbirds: scavenging a windfall of nutrients from the sky. *Ecology* 100 (9). doi: 10.1002/ecy.2728.

263. Bristow, C.S., et al. (2010). Fertilizing the Amazon and equatorial Atlantic with West African dust. *Geophysics Research Letters* 37 (14): L14807. doi: 10.1029/2010GL043486.

264. Washington R., et al. (10 Feb., 2006). Dust and the low level circulation over the Bodélé Depression, Chad: observations from BoDEx 2005. *Journal of Geophysics Research Atmospheres* 111 (D3): D03201. doi: 10.1029/2005JD006502.
265. Pett-Ridge, J. C. (2009). Contributions of dust to phosphorus cycling in tropical forests of the Luquillo Mountains, Puerto Rico. *Biogeochemisty* 94: 63–80.
266. Aciego, S. M., et al. (2017). Dust outpaces bedrock in nutrient supply to montane forest ecosystems. *Nature Communications* 8, Article number 14800. doi: 10.1038/ncomms14800.

5 Commensalism Is Ubiquitous, and Maintains and Increases Diversity

INTRODUCTION

Commensalism is symbiosis in which one species benefits and the other is unaffected. I will refer to the species that helps as the benefactor, and the species that is helped as the beneficiary. Commensalism is a major mechanism by which the Autocatalytic Biodiversity Hypothesis (ABH) is carried out, as it greatly increases species diversity. One species either benefits from or cannot exist without the unaffected species. Countless species are able to persist only because of benefactor species. Indirect commensalism occurs when a species benefits from the beneficiary, and thus indirectly from the benefactor. As in indirect mutualism, commensalism is much more common when indirect commensalism is taken into account. The more common commensalism is, the more important it is to diversity and the ABH. It is not known precisely how common it is, although the majority view that it is very common. It is almost certainly the most common interspecific interaction when one takes the microbiome into account. The microbiome is a diverse ecosystem of microbes in all eukaryotes, most of them probably commensal with their host. Commensalism maintains biodiversity because the beneficiary is aided. Keystone species help many species commensally. Examples of commensalism are listed in this chapter to demonstrate its prevalence and importance. As with all ecological interactions, but more so on average than in other such interactions, commensalisms vary in strength and duration from intimate, long-lived relationships to brief, weak interactions. In some cases, the beneficiary cannot survive without its host; in others, it can.

TYPES OF COMMENSALISM

INQUILINISM

Inquilinism is commensalism in which an animal, the inquiline, lives in the nest, burrow, or other dwelling of another species of animal. It is a special case of the provision of habitat to other species, examples of which will be discussed in the section on keystone species.

The most widely distributed types of inquiline are those found in association with the nests of social insects, especially ants and termites, where a single colony may support dozens of different species. However, some species that inhabit social insect nests are parasites, not inquilines. There are several species that exploit ant colonies for protection from predators, to sleep, and obtain food. It is thought there are many species that exploit ants commensally; this area has been little explored.

The Lace Monitor, a large Australian lizard, lays its eggs in termite nests. Since the temperature in the nests are well regulated by the termites to be neither too hot nor too cold, the eggs do well. When the babies hatch, they are too small to burrow out of the nest. The female returns at this time and digs them out. Cuckoo flies lay their eggs in the nests of wasps and solitary bees commensally, although some are parasites. Guest bees are common. They do not collect pollen or have organs to collect or carry it, but enter the nests of solitary and social bees and lay their eggs. Their young eat the pollen of their host bees. Some guest bees are considered inquiline because there is plenty of pollen for hosts and guests, but they show that there can be a fine line between inquiline and parasite. In fact, the distinctions between parasites and inquilines are subtle. However, parasites are not inquilines, because they have a deleterious effect on the host species.

The abandoned nest holes of many woodpecker species are used by several other bird species. Famously, in the southwest United States, the Elf Owl (*Micrathene whitneyi*) uses the hole in a Saguaro Cactus (*Carnegiea gigantea*) of the Gila Woodpecker (*Melanerpes uropygialis*) for nesting after the woodpecker is done nesting in it.

The term inquiline has also been applied to aquatic invertebrates that spend all or part of their life cycles in water-filled structures in plants. These include the liquid in pitcher plants that normally digests small animals, but which some invertebrates have adapted to live in, and water in bromeliads, tree hollows, bamboo internodes, and water accumulated at the base of leaves, petals, or bracts of plants.

The Mariana Trench in the western Pacific Ocean is the deepest part of the seas. In it are the largest individual cells known in the deep sea at over four inches in length. They are called xenophyophores, and are 35,000 feet beneath the sea surface. They usually build their dwelling structures with sediments. These structures are also used by sea stars, crustaceans, clams, and worms.

The Sociable Weaver (*Philetairus socius*) of southern Africa builds huge, spectacular compound community nests. These nests are used by several other bird species, such as the Red-headed Finch and Rosy-faced Lovebird, which use the nests for breeding. Other bird species roost on them, including the Acacia Pied Barbet, Familiar Chat, and Ashy Tit. Larger birds use the nests as a platform to build their nests. This includes vultures and owls. The Pygmy Falcon is the most common user of these nests. Its presence discourages predators of the weaver, such as mongoose and snakes, from approaching the colony. The falcon eats some weaver young, although it mostly eats other prey. The falcon has to limit its predation on the weavers, lest they abandon the colony and start it again elsewhere. The falcon may thus act as a prudent predator (Chapter 9). It is not a commensal. It is both mutualist and predator, showing the complexity of interspecific relationships.

The Pygmy Blue-tongued Skink (*Tiliqua adelaidensis*) of Australia lives in trap door spider burrows and lays its eggs in them. The young stay there a while after hatching, and then leave.

In Argentina, lizards lay eggs inside the nests of the caimans. This is safe because the mother caiman does not feed during incubation of her eggs. The lizard's eggs get protection and a nest.

USE OF ANOTHER ORGANISM AS HABITAT

Many beneficiary species use other species, or parts of them, as habitat.

Some unicellular organisms use silicon or calcium shells of phytoplankton (such as those of coccolithophors) to construct their own armor. Hermit crabs use the shells of dead sea snails as homes. They are also mutualists with sea anemones that grow on the snail's shell. The anemone is a beneficiary commensal of the sea snail whose shell the hermit crabs lives in.

Epiphytes benefit from living on trees mainly by gaining height and access to sunlight for photosynthesis. They also gain access to a greater number of canopy animal pollinators. They can more easily disperse their seeds via wind. Host plants also benefit epiphytic plants by providing them with moisture. Epiphytes do obtain moisture from rain and the air, but they also get it from the surface of the host plant. Finally, they are out of reach of ground-dwelling herbivores. They achieve all this without spending the enormous amount of energy that a tree expends in growing to a great height. For the most part, the host trees are not significantly affected by this, even when supporting a great number of epiphytes. Occasionally, epiphytes are so numerous that their weight damages their host, but this is not the rule. Epiphytes are found in all types of forests in the world, especially in rainforests, cloud forests, and tropical dry forests.

Some plant taxa are specialized as epiphytes or have many species that are so specialized; these include the bromeliads (family Bromeliaceae) and orchids (family Orchidaceae). There are over 2,000 species of epiphytes that are in the bromeliad family. Seventy percent of orchids are epiphytes, and orchids are the most diverse family of flowering plants, with over 18,000 species, representing about 8% of all flowering plants on Earth. Furthermore, it is estimated that there are 10,000–12,000 orchid species that have yet to be described. This yields about 12,600 described species of epiphytic orchid, and an estimate of about 19,600–21,000 epiphytic orchid species, when undescribed species are added! There are a number of cacti that are epiphytes, although they represent a minority of this family. Although the term epiphyte most often refers to higher plants, many species of ferns (phylum or division Pterophyta) are also epiphytic species. And many bacteria, algae, lichens, mosses, and liverworts also grow on trees. Additionally, there are several species of algae that are epiphytic on aquatic plants, both in fresh and saltwater, the most common hosts being aquatic angiosperms and seaweeds.

Sometimes grasses, small bushes, and small trees grow on other trees. Overall, there are well over 15,000 epiphytic species in the Neotropical realm alone, and over 30,000 worldwide, among the described species. There is a large, unknown number of uncataloged epiphytic species. There are epiphytes in 83 plant families. Striking evidence for the prevalence of commensalism comes from the estimate that 24%–25% of all land plants are epiphytes (Almeda, personal comm., 2012). Thus, assuming there are about as many host plant species as there are epiphytes, about half of all land plants are in at least one commensal relationship, either as host or beneficiary, because of epiphytes alone.

Some epiphytes act commensal benefactor species. They create niches that are exploited by a wide range of species. One of the best examples of a tiny ecosystem in an epiphyte is the tank bromeliad of South America, whose stiff, upturned leaves can hold more than 2 gallons (8 L) of water. These reservoirs of water not only provide a drinking supply for many canopy animals, but also create an entire habitat which species use for living and breeding. A number of species of aquatic insects, insect larvae, salamanders, and frogs exist in these pools and are fed upon by other animals. The water catchments of the tank bromeliad serve as a nursery for poison-arrow frog tadpoles. The female frog lays its eggs on a leaf or in burrows on the forest floor. When the tadpoles hatch, she allows them to climb upon her back and she climbs to a great height up a tree to a bromeliad where she deposits the tadpoles into one of the plant's pools that appear free of potential predators. The tadpole feeds on the developing insect larvae of the catchment. The females of some frog species actually return to the bromeliad every few days to deposit an infertile egg into the water. The tadpole then feeds on the yolk of this egg. Bromeliads, especially those with interconnecting chambers, are often colonized by mutualistic stinging ants, which provide the plant with nutrients produced by ant waste and their collection of decaying debris. The ants get a habitat and water. And water-filled insectivorous pitcher plant leaves are inhabited by animals that have adapted to the hostile environment of the leaves' digestive fluids. Thus epiphytes, commensal beneficiaries of trees, help many other species survive by both mutualism and commensalism. In some cases, this involves indirect commensalism. For example, the tree benefits the bromeliad, which benefits the frog, so the tree is an indirect commensal benefactor of the frog.

Freshwater mussel larvae must live on fish gills. They usually do not harm the fish, although sometimes they do, in which case it is parasitism. In some cases the mussels use lures that mimic fish prey such as small fish or crayfish to entice the fish into ingesting the larvae.

On the rocky shores of western North America, *Mytilus californianus* forms extensive mussel beds. Hundreds of invertebrate species, including barnacles, polychaetes, and snails live on and in the interstices of the mussels, which function analogously to trees in a rainforest. Mussels provide a three-dimensional habitat and substrate upon which to live, but they also provide a wet microclimate with food for commensals.

Pearlfish, fish family Carapidae, are small fish that in most species live as adults inside invertebrates, mainly clams, starfish, or sea squirts, as commensals, not harming their hosts. Some species are parasites, living in sea cucumbers, bivalves, and starfish, eating their insides. The ones that live in sea cucumbers live in the anus and eat the gonads.

Many species of fish, often juveniles, swim around jellyfish, gaining a safe haven from potential predators. It is thought that the jellyfish is not affected by the relationship because it is not eaten by the fish nor does it eat the fish. Small invertebrates and fish utilize the areas between sea urchin spines as protective habitat.

BATESIAN MIMICRY

Batesian mimicry is a relationship where one species, the mimic, resembles an unpalatable species, the model. The model may be poisonous to eat, or have a venomous bite or sting. Model species have aposematic color patterns, sounds, or smells that they may reinforce with behaviors. The mimic is harmless, and benefits from the model by imitating its warning colors, sounds, smell, and/or behaviors. The model is mostly unaffected, although it acquires a small cost because the predator has a harder time learning that the model is unpalatable, since it encounters the palatable mimic once in a while. Sometimes the predator does not need to learn to avoid the unpalatable species, but instinctively avoids it. The model species must outnumber the mimic, so that the predator does not encounter the mimic so often that it does not learn to, or evolve to instinctively, avoid the model. The percentage of all animal species that are Batesian mimics is not known. It is thought to be more common in the tropics. An informal, unpublished estimate is 5%–10% of Neotropical butterflies are either Batesian or Müllerian mimics, but this needs more research to confirm.

Mimics may have different models during different stages of their life cycles. Models may have more than one mimic, although the need for models to outnumber mimics limits this.

Mimicry is common in the Lepidoptera. There are mimetic polymorphs, in which there are several morphs in the mimic species, each morph resembling a different model species. Since the model must always be more abundant than the mimic for the system to function and persist, polymorphic mimicry allows the mimic to have a much larger population size than it otherwise would. This type of mimicry occurs especially in beetles and butterflies. Sometimes the female of a butterfly species is polymorphic, mimicking more than one species, while the male is monomorphic, lacking this variation. An example is the Mocker Swallowtail (*Papilio dardanus*), in which all males look identical, but the female has several morphs. These correspond to a variety of different unpalatable models, including *Amauris niavius*, *Amauris echeria*, *Acraea poggei*, and *Danaus chrysippus*. The reason only females are polymorphic is poorly understood. Some biologists postulate that if males were also polymorphic, then certain morphs would be at a disadvantage in male-to-male territorial conflicts, so would in time disappear.

Eresia pelonia produces several differently-colored morphs, each mimicking a different toxic model species.It even has different subspecies that mimic different model species. This may be causing speciation in the mimic. The nominate subspecies *E. pelonia pelonia* is a mimic of the Ithomiine *Callithomia alexirrhoe thornax*; while subspecies *E. pelonia callonia* is a very good mimic of *Hypothyris mansuetus meterus*.

The tiger complex is a group of about 200 Neotropical Lepidopteran species which all share a similar pattern of orange and yellow stripes on a black ground color. The complex includes many unpalatable members. These include species in the tribe Ithomiini (subfamily Danainae, family Nymphalidae, the brush-footed butterflies), such as *Tithorea harmonia*, *Tithorea tarricina*, *Melinaea marsaeus*, and *Forbestra equicola*; passion-vine butterflies (genus *Heliconius*, family Nymphalidae), which show a huge diversity of wing patterns and number of species; *Lycorea pasinuntia* (subfamily Danainae); and several highly toxic day-flying moths from the Arctiid subfamily Pericopinae. Palatable mimics include many species, many not close relatives of the models, such as *Heliconius ismenius, Heliconius hecale*, and *Eueides isabella* of subfamily Heliconiinae of family Nymphalidae; *Eresia eunice* of subfamily Nymphalinae; *Stalachtis calliope* of family Riodinidae (metalmarks); *Consul fabius* of Nymphalid subfamily Charaxinae (leafwings); and *Pterourus zagreus* of subfamily Papilioninae of family Papilionidae (swallowtails). Many of the unpalatable species are Müllerian mimics as well.

Members of the tiger complex habitually come together in great numbers in moist depressions in forests at the end of the dry season, when they are easily caught by birds. At this time, mimicry is most important for the defense of species in this complex. Any bird that has the unpleasant experience of attempting to ingest an unpalatable member of the tiger complex quickly learns in a single trial to avoid attacking any species that resembles it. Some believe birds can communicate the unpalatable nature of sampled species to other birds.

Coincidentally similarly-named lepidopterans in this and the next paragraph are not in the tiger complex. Several palatable butterfly species mimic different species from the highly noxious genus *Battus*, which belongs to the subfamily Papilioninae of the family Papilionidae. Many palatable moth species produce ultrasonic click calls that resemble the unpalatable tiger moths, a case of Batesian mimicry by sound.

The Plain Tiger (*Danaus chrysippus*), widespread in Asia, Australia, and Africa, is a butterfly that is an unpalatable model that is polymorphic, and is mimicked by the females of two butterfly species, the Indian Fritillary (*Argyreus hyperbius*) and Danaid Eggfly (*Hypolimnas misippus*). And the following three harmless butterflies have a general resemblance to both the Plain Tiger and Common Tiger (*Danaus genutia*), which is another unpalatable butterfly: Leopard Lacewing (*Cethosia cyane*), males and females; Indian Tamil Lacewing (*Cethosia nietneri mahratta*), males and females; and Common Palmfly (*Elymnias hypermnestra*), females. The Common Crow (*Euploea core*) is a butterfly found in India, Southeast Asia, and parts of Australia that is an unpalatable model that the inexperienced predator will attack, but soon learn to avoid, because the alkaloids in its body are so strong that ingesting one butterfly causes vomiting. It has a strong scent that also helps predators to identify it as inedible. It has the following Batesian mimics: Common Mime (*Chilasa clytia*) form *dissimilis*, both male and female; Malabar Raven (*Papilio dravidarum*), both male and female; Common Raven (*Papilio castor*), female; Great Eggfly (*Hypolimnas bolina*), female; Ceylon Palmfly (*Elymnias singala*), male and female (not in India). However, due to evolution of the predators, not all birds are sensitive to the toxins, and common invertebrate predators and parasites such as spiders, dragonflies, flies, and wasps suffer no ill effects from exploiting these butterflies. Thus, the system is very complex.

Some caterpillars and some chrysalises look like feces. For example, caterpillars of the moth *Apochima juglansiaria*, which dwells in Japan, takes on a posture and has a coloration that causes it to resemble bird droppings, and predators avoid it. Many moths have eyespots, thought to resemble larger animals' eyes.

The system of both Batesian and Müllerian mimicry is complicated by the fact that there is three-way antagonistic coevolution between the host plant the caterpillar feeds on, the butterfly, and its predator. The plant evolves toxins in response to larval herbivory. But the caterpillar evolves a defense that results in its being largely immune to the toxin, and can even store the toxin in its tissues, including when it is an adult, protecting it from predators, and turning a defense evolved against it into its ally. The insect then evolves aposematic coloration. Their predators respond by avoiding them, either through learning or evolving instinctive avoidance of the warning color patterns. Then palatable insects evolve Batesian mimicry of the toxic model insects. And the unpalatable models coevolve Müllerian mimicry with other unpalatable species. The system maintains biodiversity by aiding both Batesian and Müllerian mimics. It increases diversity when it leads to speciation. This demonstrates the importance of coevolution.

There are many types of insect mimics, and mimics may resemble models that are wholly unrelated. There are flies and moths that mimic bees. Bee flies (family Bombyliidae) are true flies that tend to eat nectar from flowers. Harmless, many of them closely resemble stinging bee models. There are several species of hoverfly which mimic stinging species of wasps. The Ash Borer (*Podosesia syringae*), a moth of the Clearwing family (Sesiidae), mimics the Common Wasp (*Vespula vulgaris*). I observed a moth of a species I could not identify in Monte Verde National Park in Costa Rica, which, when disturbed, folded and pulled its wings over its head, curling and exposing its abdomen, which was banded yellow, red, and black. It seemed to be mimicking a venomous coral snake.

Most known mimics are insects, but mimicry occurs in many animal groups, including all classes of vertebrates, and numerous invertebrate taxa that are not insects. Plants and fungi may also be mimics.

In the Amazon, the helmeted woodpecker (*Dryocopus galeatus*), a rare species which lives in the Atlantic Forests of Brazil, Paraguay, and Argentina, has a similar red crest, black back, and barred underside to two larger woodpeckers: *Dryocopus lineatus* and *Campephilus robustus*. This mimicry reduces attacks on *Dryocopus galeatus* from predators, which think it is a larger species than it is.

Londoño et al. [1] found that baby Cinereous Mourners (*Laniocera hypopyrra*), Amazon rainforest birds, have long, modified downy feathers with long, orange barbs and white tips all over their bodies that cause them to resemble a poisonous, aposematic, hairy caterpillar, and thus be avoided by snake and monkey predators. Six days after hatching, they move their heads very slowly from side to side, mimicking the caterpillar's behavior. They even do this in the presence of their parent unless the parent makes a specific vocalization that causes them to open the mouth and beg for food. The resemblance ceases at about two weeks, when the birds have the strength to jump from the nest if a predator appears.

In snakes, the False Cobra (*Malpolon moilensis*) is a mildly venomous but harmless snake of the common snake family Colubridae which mimics the characteristic hood (large, flattened neck) of the highly venomous Indian Cobra's threat display. The Eastern Hognose Snake (*Heterodon platirhinos*) of the United States displays a hood, hisses, and strikes, imitating the threat display of venomous snakes, though it never bites. If the bluff fails, it lies on its back with its mouth open and plays dead. The Eastern Coral Snake (*Micrurus fulvius*) of the southeastern United States is a highly venomous member of the cobra family (Elapidae), with beautiful bands of red, yellow, and black that warn predators not to attack, thus protecting the snake. It is mimicked by the Scarlet Snake (*Cemophora coccinea*), and three subspecies of the same species of kingsnake: the Louisiana Milk Snake (*Lampropeltis triangulum amaura*), the Mexican Milk Snake (*L. t. annulata*), and the Scarlet Kingsnake (*L. t. elapsoides*). All mimics are nonvenomous, and have the gorgeous tricolor banding pattern. In the tropics, there are many species of coral snakes and their harmless mimics, with the same tricolor banding pattern or patterns similar to it. Many of the venomous ones are also Müllerian mimics. Potential predators have evolved to instinctually avoid the tricolor pattern with no need for learning.

The Mimic Octopus (*Thaumoctopus mimicus*), a Batesian mimic of the Indo-Pacific is likely the best mimic of all, capable of changing color and texture to blend cryptically with the environment and to mimic up to 15 species, many poisonous, more than any other animal can. It can change shape and take on the appropriate behavior to reinforce the mimicry. In many cases, it does not mimic a venomous or poisonous animal, so it is not Batesian mimicry, but cryptic coloration. Some of the common animals it mimics are lion fish, where its legs resemble spines; sea snake, by waving two arms in opposite directions, with the yellow and black markings of the snake; flatfish, by flattening out its body while moving forward along the seafloor; tunicates; sponges; jellyfish, imitating their motions; and even algae-covered rock coral. It mimics the animal appropriate to the predator it encounters.

OTHER KINDS OF MIMICRY

Some predators mimic the food of their prey to increase their chances of capturing prey. For example, anglerfishes of the teleost order Lophiiformes wiggle a fleshy growth from the head that resembles a worm and lures predators of worms near to it. Several turtle species, and the Frogmouth Catfish (*Chaca* sp.) of Southeast Asia, have worm-like tongue extensions that are used to attract prey. The benefactor commensal in these cases is the worm being mimicked.

Cryptic coloration overlaps mimicry. It often involves commensalism, because the animal blends in by resembling an organism it is benefiting from, often a part of a tree or plant. Flower mantises and some spiders look like flowers or parts of them. Planthoppers and geometer moth caterpillars look like sticks, twigs, bark, leaves, or flowers. There are chrysalids that resemble twigs, and some that look like leaves. Some katydids are excellent leaf mimics. They are nocturnal, and normally remain perfectly still during the day, often in a position that helps them resemble leaves even more. Some have a body coloring and shape that perfectly matches leaves, including half-eaten leaves, dying leaves, and leaves with bird droppings. There are moths that resemble hummingbirds, and ants that look like tufts of thistle fluff.

The number of cryptic species that actually mimic all or parts of plants, trees, or other organisms is astronomically high. There are 360,000 species of beetle (order Coleoptera) described, with new species frequently found. It is estimated their total species number is five to eight million. A good estimate is that beetles comprise about 40% of all insect species. Many beetle species live on and blend in with plants and trees. The largest family of beetles is Curculionidae, the "true" weevils, or snout weevils, with over 60,000 described species, or one-sixth the described species of beetles. As a whole, snout weevils live on plants and are cryptically colored, with the color of their host plant. The nine families of the order Mantodea, the Praying Mantises, comprising about 2,200 described species, all tend to be green or brown, blending in with the plants they live on. The order Neuroptera, family Chrysophidae, the Green Lacewings, with about 85 described genera and about 1,300–2,000 described species, tend to be cryptically green like the leaves and plants they eat and inhabit. True Bugs, order Hemiptera, have suborder Homoptera, the cicadas, leaf-hoppers, and aphids, with about 45,000 described species, which as a general rule resemble the trees and plants they occupy and feed on as adults. The final instar of the cicada nymph tends to be the color of the trunk of the tree upon which it crawls and molts into an adult.

The order Orthoptera, with more than 20,000 described species, has the suborder Caelifera, the locusts and grasshoppers, with 2,400 genera and about 11,000 known species, almost all of which are camouflaged on the grasses and other plants such as forbs that they live on and eat. Katydids (family Tettigoniidae) have about 6,400 species, essentially all of which resemble the vegetation they live on, many species being green.

In the Lepidoptera, a large plurality of the total number of species are microlepidoptera, a subgroup of about 20 families that does not make up a true taxonomic group. These are small moths. They comprise about 77,000 described species, out of about 180,000 species of Lepidoptera. In addition, it is estimated that about 20% of the microlepidoptera have not been described. The vast majority of the microlepidoptera are brown, gray, or white, and roost on trees or other plants, blending in with the bark or other tree or plant parts. A great many of the larvae and pupae of both butterflies and moths are cryptically colored, matching the trees, forbs, or other plants they are found on.

Walking Sticks, insects of the order Phasmatodea, are elongated, with some resembling sticks or twigs, others leaves. The vast majority mimic the plants they live on and consume. Their body is often further modified to resemble vegetation, with ridges resembling leaf veins, bark-like tubercles, and other plant parts. A few species can even change their pigmentation to match their surroundings. Their camouflage is extraordinarily effective. The eggs are also typically camouflaged, resembling plant seeds. There are in excess of 3,000 described species, with many more yet to be described. They have a worldwide distribution. Most species are tropical, and the tropical species vary from stick-like species to those resembling bark, leaves, and even lichens. They sit still like twigs, or move back and forth, resembling branches blowing in the wind.

PHORESY

Phoresy is commensalism in which the larger host species transports the smaller species, called the guest, without affecting the host. Unless otherwise stated, the host is not significantly affected in any of the following examples.

Phoresy mainly occurs in arthropods. The beetle *Nymphister kronaueri* uses its mandibles to attach between the thorax and abdomen of the army ant *Eciton mexicanum* to obtain transport in the rainforests of Costa Rica [2]. The beetle mimics the ant's abdomen. Other hitchhikers ride on army ants' backs, follow in their wake on foot, or stow themselves on top of prey the ants carry from nest to nest.

Pink rays hitch rides on stingrays. Remoras have dorsal fins modified as suckers to attach to and get transport on sharks and other large fish. They eat food scraps from the prey of the sharks and likely get protection from them.

Various mite species get from flower to flower in the beaks of pollinating hummingbirds, or get transport from dragonflies, beetles, flies, and bees. Pseudoscorpions use their large pincers to hitchhike on bats. Some pseudoscorpion species disperse by riding under the wing covers of large beetles, such as those of the cerambycid family, others on mammals. The pseudoscorpions gain the advantage of being dispersed over large areas while being protected from predators. Baby lobsters attach to jellyfish for transport. Millipedes get rides from birds. Flatworms, snails, and other invertebrate eggs, larvae, and adults are often transported many miles between ponds on the legs of birds. This means of dispersal can possibly lead to range increases and speciation.

Though not strictly phoresy, many plants produce fruits that adhere to fur and feathers, and are thereby dispersed by mammals and birds. Some North American examples of such plants are the Greater Burdock (*Arctium lappa*), Common Beggarticks (*Bidens frondosa*), and Showy Tick-trefoil (*Desmodium canadense*). The fruits of these plants have special anatomical adaptations for adhering to fur. Burdock fruit is the botanical model from which Velcro was developed as a fastening material. Rarely, individual animals become so loaded with these sticky fruits that their fur peels off excessively, but this is very uncommon, so the relationship is usually commensal.

OTHER COMMENSALISMS

The competitively dominant mussel *Perumytilus purpuratus* can settle and establish itself more effectively if there are barnacles, filamentous algae, or clumps of mussels present in the mid rocky intertidal zone in central Chile. *P. purpuratus* recruits only on the walls of adult barnacles that form the patch borders, and never on bare rock. The barnacles thus provide habitat for the mussels, which in turn provide habitat for many species that live between or on the mussels. So the barnacles are benefactors in indirect commensalism with the species that benefit directly from the mussels. The algae and mussels themselves also help mussel recruitment [3].

The vast majority of birds use material from other species, such as twigs and leaves from plants, to build their nests. Tree and Violet Green Swallows build nests of dry grasses lined with feathers from water birds, chickens, and turkeys. The vast majority of birds are commensal beneficiaries of trees and plants because they build their nests in them. There are about 18,000 species of birds [4], so there are conservatively at least 15,000 commensal beneficiary bird species and a somewhat smaller number of tree and plant benefactor species that they nest in.

About 18 of the 200 species of antbirds, family Thamnophilidae, specialize in following columns of army ants to eat the small invertebrates flushed by the ants.

Pirt and Lee [5] found that methane-generating Methanobacteria of the genera *Methanobacterium*, *Methanococcus*, and *Methanosarcina*, found in certain ponds and swamps, are commensal with and dependent on motile aerobic bacteria that consume O_2 and produce CO_2. The Methanobacteria are obligate anaerobes. So the aerobic bacteria help the Methanobacteria by removing oxygen from their environment and providing them with the energy source, CO_2, which they use for the anaerobic production of CH_4. The aerobic bacteria do not benefit from the Methanobacteria. This happens in wetlands of various depths and sediment compositions. Methane is created at the same rate O_2 is consumed, showing conclusively that the two processes are linked.

Cats and dogs, both domestic and some wild species, commensally eat grass, not as herbivores. They do get folic acid, a B vitamin, from grass. But thick-leaved grass is a laxative, and may add fiber and bulk to the diet, helping animals eliminate worms and fur from the gut. Grass with thin leaves can induce vomiting, since dogs and cats cannot digest its cellulose. This clears the digestive tract of fur balls from grooming, feathers and bones from birds the dogs and cats ate, and other such undigestible debris. This does not cause any significant effect on the grass population.

The Western Fence Lizard has evolved a protein in its blood that kills the *Borrelia* bacteria that cause Lyme disease, making the lizard immune to this disease. Deer ticks (genus *Ixodes*) carry the disease, spreading the bacteria that causes it with their bite. The lizard provides the tick with a place to live and blood meals, usually at little to no cost to the lizard. The lizard greatly reduces the number of ticks with bacteria in them that carry Lyme disease, and thus aids other hosts of the tick by greatly reducing the chance that they will get the disease. These include wood rats, White-footed Mice, deer, and humans, all of which do not affect the lizard, so are commensal with it. This system works in the western United States, but not the eastern part of the country, because this lizard does not occur there.

Western Honeybees collect chemicals from mushroom mycelia that protect them against viral infections [6].

Some insects acquire compounds from their host plants and use them as sex pheromones or precursors to them ([7], and references therein). Other insects produce or release sex pheromones in response to specific host plant cues. Host chemicals often enhance the response of an insect to sex pheromones of its species. The courtship pheromones of the Bella Moth, *Utetheisa ornatrix* (Arctiidae), are manufactured from pyrrolizidine alkaloids (PAs) ingested by the larvae from its host plant, the herbaceous *Crotalaria spectabilis* (Fabaceae). The alkaloids are retained into the adult stage, and provide protection to the next generation of eggs. Also, PA-sequestering species are found in many butterflies and moths, certain aphids, some leaf beetles, and some grasshoppers. They make members of these taxa unpalatable to both invertebrate and vertebrate predators. These relationships are commensalism in addition to the herbivore-plant relationship. The insects evolved to use compounds in their host plants to their advantage.

The cicada *Cryptotympana facialis* lives in high densities in urban parks in central Japan in summer. They pierce holes in Keyaki Trees (*Zelkova serrate*) with their stylets (mouthparts), then feed on the exuding xylem sap. Three ant species, two wasp species, and two species of flower chafer beetles feed on the sap after the cicada finishes feeding [8]. The flower chafers also displace the cicadas from the sap to feed on it. Thus, the chafers have a negative impact on the cicadas, so are not commensals. They are kleptoparasites.

Butterflies in the western Amazon land on and drink the liquid in the eyes of Yellow-spotted River Turtles to obtain sodium, perhaps other minerals, and possibly amino acids. The carnivorous turtles get plenty of salt and minerals in their diet, and sometimes exude salt from their eyes, but the herbivorous butterflies often find their diet short of all these nutrients. This area of the Amazon is very low in salt and minerals. In addition, the butterflies get salt and minerals commensally from animal urine and the sweat of people. Bees also drink the tears for sodium and other minerals, with no likely effect on the turtles.

In the shallow waters of Lake Tanganyika, the fish *Xenotilapia boulengeri*, in the highly diverse cichlid family, lives in the benthic zone, and rests in a cluster around the nesting sites of two species of cichlids that build nests on the lake bottom, *Lepidiolamprologus attenuatus* and *L. elongatus*. The two nesting species guard their nests from other fish that might eat their young. They do not guard against *X. boulengeri* because it does not eat their eggs or young. Two species of cichlid fish that eat the scales of other fish, *Perissodus microlepis* and *Plecodus straeleni*, are chased away by the guarding cichlid species, providing protection for *X. boulengeri*, accounting for why it stays in proximity to these species when they are guarding their nests [9].

The salt marsh plant species of rushes (genus *Juncus*) and marsh elders (genus *Iva*) are unable to tolerate the high soil salinities when evaporation rates are high. Thus, they depend on neighboring plants to shade the sediment, slow evaporation, and help maintain tolerable soil salinity levels [10].

In syntrophy, or cross feeding, one species lives off the products of another species. House dust mites live off human skin flakes, of which a healthy human being produces about 1 g/day. These mites can also produce chemicals that stimulate the production of skin flakes, and people can become allergic to these compounds. So sometimes the host is harmed, but usually this relationship is commensal.

Gordon [11] showed that sounds of a healthy coral reef attract fish to the reef, helping them find a habitat, in a "group commensalism". Acoustic enhancement in a degraded reef enhanced fish community development across all major trophic guilds, resulting in a doubling of overall abundance and a 50% greater species richness. This is likely not only commensalism, since some of the recruited fish probably benefit some of the species already there.

Even nature as a whole is commensal with humans, where there are esthetic preferences for views containing trees and other vegetation, and a strong preference for nature over urban and unspectacular natural views. Esthetic preference is central to a landscape observer's thoughts, conscious experience, and behavior. Views of nature have been shown to have positive influences on emotional and physiological states. These benefits appear greatest for individuals experiencing stress or anxiety. Responses to vegetation are directly linked to health [12].

INDIRECT COMMENSALISM

Indirect commensalism occurs when the beneficiary species benefits indirectly from the benefactor through an intermediary species that benefits directly from the benefactor. For example, Gila Woodpeckers benefit Elf Owls that nest in the holes they create in cacti. The owls benefit various plant species by regulating the population of rodents that eat these plants and their seeds. Thus, the woodpeckers are indirect commensal benefactors of the plants. When one considers the great number of indirect commensal relationships, the number of commensal relationships in nature is phenomenal.

All higher organisms host tens of thousands of species of commensal microbes that benefit from the habitat the host provides without significantly affecting the host. If a host species is the commensal beneficiary of another species, all the microbes in the beneficiary host indirectly benefit as commensals of the species benefiting their host. Any keystone species is an indirect commensal benefactor to tens to hundreds of thousands of microbial species residing in the multicellular organisms it benefits. So when microbiomes are taken into account, there are at least hundreds of thousands of indirect commensalisms in every ecosystem.

KEYSTONE SPECIES

A keystone species is one that has a disproportionately large positive effect on other species relative to its abundance and compared to the effect an average species has on other species. If it is removed, the ecosystem loses many species or even collapses. It is the commensal benefactor of a great many species, as well as being in mutualistic relationships with several species. Keystone species play a key role in determining their community's structure, and greatly increase and maintain their community's diversity. Many are keystone species because they provide a habitat for immense numbers of species. The concept has been rightly criticized for oversimplifying complex ecological systems [13]. Still, the concept has some validity.

Decomposers are keystone commensal species that benefit many animal and plant species by removing and recycling feces and dead organisms. They are also commensal beneficiaries of countless animal and plant species, which provide them with nutrition when they die, and, in the case of animals, defecate. Since the species they benefit are often the same ones that benefit them, they are also mutualists with countless species. Their species numbers are spectacular. There are tremendous numbers of species of bacteria and protozoa that decompose. The number of decomposing species in the Fungus kingdom is likely in the tens of thousands. Most species in the subfamily Scarabaeinae, the true dung beetles, of the family Scarabaeidae (scarab beetles) feed exclusively on feces. This subfamily has over 5,000 species. There are dung-feeding beetles in other families, such as the Geotrupidae. Lorquin's Admiral and the California Sister are butterfly species that eat animal feces to obtain needed minerals. Scavengers are a type of decomposer. There are many scavengers that are keystone species, such as some species of vultures.

Nitrogen-fixing bacteria provide useable nitrogen for multitudes of organisms. Though mutualistic with their plant allies, they commensally benefit tens of thousands of species indirectly. Cyanobacteria fix nitrogen in the sea, helping countless species.

Mistletoe is a particularly important symbiont that merits discussion because it can greatly increase the diversity of its symbionts, and is now recognized as a keystone species [14]. Many mistletoe species are benefactor species of several species they are commensal with. Mistletoe is an order of plants (Santalales), whose largest family, Loranthaceae, has 73 genera and over 900 species. Mistletoe species grow on a wide range of host tree species. European Mistletoe (*Viscum album*) can grow on more than 200 tree and shrub species. All mistletoe species are hemiparasites, since they photosynthesize at least a small amount for at least short periods of their life cycle, although some species produce almost no carbohydrate. They draw nutrients from the tree's xylem. They can cause reduced growth, stunting, and loss of distal branches from the mistletoe's weight [15]. Mistletoe can grow in very large, heavy clumps, causing tree branches to break and fall in heavy storms, because of the increased weight. It can reduce the growth rate of a tree, stunt it, and can kill it if there is a very heavy infestation of it. Mistletoe may be one of the many factors that collectively take 300 years to kill a big oak tree. Heavy infestations can kill the host more quickly. Numerous dwarf mistletoe species (genus *Arceuthobium*) cause disease in trees. Severe dwarf mistletoe infection can reduce tree growth, reduce seed and cone development, increase the susceptibility of the tree to pathogen and/or insect attack, and even kill the tree. However, mistletoe often causes no significant harm to the host. Trees generally grow at about the same rate with and without mistletoe. So it is likely more often commensal than parasitic, though it can act as either. It ranges across the full spectrum from almost fully commensal to fully parasitic with its host. It can even be a direct or indirect mutualist with its host.

Almost all mistletoe species have their seeds dispersed by animals, and many are animal-pollinated, thus being in mutualistic relationships with both seed dispersers and pollinators. And many mistletoe species have other mutualisms. It evolved sticky berries that force birds to rub their beaks on branches to free themselves of the berries that stick to the outer part of their beaks. This generally plants it on a new tree. The sticky coating also allows the seeds to stick to branches and sprout. It also remains after birds defecate the seeds. And birds carry seeds stuck to their bodies, feathers, and feet, including in mud stuck to their legs. In its early life stages, mistletoe sometimes gets its nutrients from the bird feces. On oaks, such birds as Western Bluebirds, American Robins, and Cedar Waxwings, eat and disperse mistletoe fruits. The Phainopepla (*Phainopepla nitens*) famously disperses the fruit of the Desert Mistletoe, *Phoradendron californicum*, in the western US and Mexican deserts.

In general, areas with greater mistletoe densities support higher animal species diversity. Many species depend on mistletoe for food, eating the leaves, young shoots, berries, and sometimes pollen. However, Pacific Mistletoe (*Phoradendron villosum*) is poisonous, indicating intense past herbivory selected for this defense. It may have herbivores, since animals are good at evolving defenses to plant toxins. Dwarf mistletoe, which has no leaves and has 42 recognized species mainly in North America, may protect trees from Douglas Fir Beetle attacks [16]. It can cause witches' brooms to form in the trees they grow on. These are deformities or diseases resulting in a dense mass of shoots growing from a single point. Witches' brooms are excellent roosting and nesting sites for two rare, protected species, the Northern Spotted Owl and Marbled Murrlet [17]. There is increased foraging by wildlife, including birds and mammals, and more nesting by mammals in Douglas Fir trees with a specific species of Dwarf Mistletoe (*Arceuthobium douglasii*) brooms than without (ibid). Parks and Bull [18] showed that American Martens (*Martes americana*) in the Pacific northwest seek out in dwarf mistletoe as well as brooms made by rust fungi (*Chrysomyxa* sp. and *Melampsora* sp.) to keep warm during cold nights. Since it protects its host from beetles and the witches brooms are a manifestation of disease, dwarf mistletoe is a both parasite and mutualist to its host. A study of mistletoe in juniper trees showed that more than twice as many juniper seedlings sprout in stands with mistletoe than without, since mistletoe attracts birds that not only eat and disperse its berries, but also juniper berries [19]. Since birds disperse the seeds of both juniper and mistletoe, this is a system with a three-way mutualism, with the juniper, mistletoe, and bird each benefiting the other two species. There are several bird species in this system, so many more than three species are involved. The mistletoe provides a stable resource of berries, while the juniper berry crop varies considerably and is not stable, varying 10- to 15-fold over the three-year study. Birds were attracted to juniper in most years and mistletoe in all years, and mistletoe was the best predictor of avian seed dispersers. Here, mistletoe benefits directly from its host, and benefits its host indirectly through a bird intermediary. The bird can act at times as a parasite of the tree when the mistletoe is so abundant it is a parasite to its host, and the bird's dispersal of mistletoe seeds thus harms the tree. This is another example of the complexity on interspecific relationships.

Wildlife often use mistletoe when it grows in large clumps. It has been shown that there is a greater diversity of bird species in trees with more mistletoe than those with less ([20], and references therein). Some birds of prey, such as the Mexican Spotted Owl and Goshawk, nest preferentially in trees with certain types of mistletoe. Other birds use it, though not necessarily preferentially, as nest sites. In Australia, more than 240 bird species have been recorded nesting in mistletoe, representing over three quarters of the resident avifauna. Examples include the Diamond Firetail and Painted Honeyeater. Some species of possum require mistletoe and specialize on it in this country. In other countries as well as Australia, it provides habitat for a broad range of animals, including squirrels.

Several animal species pollinate mistletoe. Some species of mistletoe have small flowers pollinated by insects, while others have spectacular flowers pollinated by birds. It is likely each species of mistletoe benefits at least one species of animal, probably several. The defecation of these animals in the soil next to the host tree is an indirect benefit of the mistletoe to the tree, fertilizing the soil. Thus, mistletoe can be a parasite, commensal, or mutualist, or any combination of these, to its host tree.

I will now coin a term for groups of related species in the same ecosystem that provide similar benefits and behave collectively as keystone species. They are keystone species groups. For example, thousands of coral reef species benefit from the three-dimensional habitat provided by several keystone coral species. This includes hundreds of fish and thousands of invertebrate species.

Many tree species, including figs, oaks, and rainforest species, are keystone species, providing habitat and/or food for a multitude of bacteria, fungi, plants, invertebrates, birds, and mammals, and some amphibians and reptiles. The beneficiary species use and/or inhabit the trees' shade, leaves, branches, trunk, and/or roots, depending on the beneficiary. Trees provide a three-dimensional habitat by adding height. They provide shade and shelter for ground dwellers, and food for herbivores. Predators that eat these herbivores benefit indirectly from trees. Dead trees provide food and habitat for many small vertebrate and invertebrate species, some of which live under rotting logs. The trees of forests are keystone species groups.

Insects likely have the most commensal beneficiary species of any animal class. Their benefactor species are most often trees. The canopy of tropical rainforests is the most diverse terrestrial ecosystem on Earth, mainly because of its insects. The number of insect species in an acre of tropical rainforest canopy could approach tens of thousands or more, and sampling of a tropical rainforest canopy found such high insect diversity that it was proposed the total estimated number of animal species on Earth be increased from 5 to 10 million to 30 to 60 million [21]. There are an estimated 16,000 tree species in the Amazon rainforest. There can be a hundred species of tree per acre in rainforests. Each tree species provides habitat to overlapping but primarily different commensal and mutualistic species living on it or otherwise benefiting from it. So insect species symbiotic with rainforest trees number at least in the hundreds of thousands, and every tree species on Earth has insect symbionts, some commensal beneficiaries. Every rainforest tree species has many insect mutualist and beneficiary commensal species. Thus, the numbers of both tree and insect species in commensal relationships is enormous.

Every full-grown tree of every species is a small ecosystem, with amazingly far-reaching impacts on its community and ecosystem, greatly increasing species richness in its community. Many of the beneficiaries aid trees and are thus mutualists, but many are commensal. The number of species that benefit from any given tree species is astronomical.

Perhaps the vast majority of tree species are keystone species and/or are in keystone species groups, since trees typically support small ecosystems. I will discuss only oak trees, a genus (*Quercus*) of about 400 species, and only gall wasps to illustrate the importance of trees as keystone species. Bear in mind that many plant species are keystone species, and essentially all natural forests consist of trees that are keystone species groups, and each of these plants and forests aids a multitude of species. The vast majority of tree species and forests consist of several of benefactor species. Thousands of plant species have a complex, diverse network of commensalism and mutualism.

Tiny wasps of the family Cynipidae, known as gall wasps, lay their eggs on buds, leaves, or new stems, depending on the gall wasp species, of oak trees. Growths of tree tissue called galls induced by the larvae provide a home for the larvae. Any given gall wasp species lays its eggs on only one of the three subgenera of oak tree, and never the other two oak subgenera. The 22 species of oak in California support over 200 cynipid wasp species of the total described 800 cynipid wasp species in several genera (new species of this wasp are discovered each year). A single, large Blue Oak of about 30 feet in height could support approximately 120,000 gall wasp larvae of the species *Andricus kingi*. Inquiline wasp species may lay their eggs in the galls, so there can be more than one species per gall. Trees other than oaks host some of these species, but oaks host more species of this wasp than any other group of trees in North America. When the female wasp deposits an egg, she injects a chemical into the tree that causes it to form a nutritive shell around the egg. Upon hatching, the larva feeds on the nutritive shell, and in the process, releases a chemical that induces the host tree to produce a larval nursery of specific size, shape, and color for each species of wasp. This is the gall. Although scientists suspect that the wasp larvae manipulate genes in specific plant tissues, inducing the formation of galls characteristic of their species, the precise mechanism of gall formation is unknown. The wasp larvae generally do not significantly harm the oak, which provides them food and a home. The gall protects the larva, but it is not fool proof. Some species of predators and parasites can attack the larva during its early development. For example, wasps slightly larger than gall wasps seek and find the galls and lay their eggs next to the wasp larvae in the gall. The predatory wasp larva then eats the gall wasp larva. Some wasp larvae have evolved further protection besides the gall; they convert normal plant starches to sugar and make a honeydew that the gall exudes in copious quantities. This sweet liquid attracts large numbers of ants and bees of various species, and yellow jackets. While feeding, these insects prevent insect predators of the larvae from accessing them, greatly reducing predation. Thus, the gall wasp benefits from the oak and honeydew eaters; the honeydew eaters benefit from oak and gall wasp; and the oak is not normally significantly negatively affected by any of these species. The species that eat the honeydew are mutualists with the gall wasps, and indirect commensal beneficiaries of the oak tree. Gall wasp predators are likewise indirect commensal beneficiaries of the oak tree. The Urchin Gall Wasp, which inhabits the Blue Oak, stays in a prepupal form in a gall that hardens and drops to the ground. Some galls fall with the leaves, and some drop off with the help of the wind or birds. They stay in the leaf litter through the winter. Though their galls are especially hard, some of them end up providing food to mice and wood rats, which are thus indirect commensals of the oaks.

Gall wasp larvae are attacked by parasites, which are attacked by parasitoids. These in turn are attacked by hyperparasites. That represents four trophic levels of hosts and parasites! There are also insects that eat gall tissues, and they often kill the larva inside of the galls. These parasites and predators are indirect commensal beneficiaries of the oaks, and illustrate the complexity of the system and its tendency toward diversity. One 30-foot tall Blue Oak could directly and indirectly support 200,000 insects from the groups listed in this paragraph alone!

Some species are "indirect keystone species"—keystone species because they directly benefit keystone species, indirectly benefiting many other species. For example, some species of sea star prey on sea urchins, controlling their populations and keeping them from decimating coral reefs. Many species of kelp (order Laminariales) are keystone species because they provide three-dimensional habitat to many fish and invertebrate species. Sea Otters eat sea urchins, which eat the holdfasts of kelp, the anchors that hold kelp to the seafloor. When sea urchins eat the holdfast, the entire kelp frond floats away and dies. Sea urchins thus can remove huge kelp fronds from the ecosystem by eating only a tiny portion of the kelp's biomass. These kelp fronds can be as long as 61 m. So sea urchins can destroy entire kelp forests if not controlled. Sea Otters and some sea star species preserve kelp forests by controlling sea urchin populations. They are thus indirect keystone species because they indirectly benefit hundreds of species that benefit from kelp. Their relationship to kelp is mutualism, because Sea Otters wrap up in kelp fronds to keep from drifting away when they sleep. More than 800 animal species have been documented to live in kelp forests, essentially all benefiting directly or indirectly from the kelp, and hence indirectly from the Sea Otter. These include California Sea Lions, Harbor Seals, the Garibaldi (a damselfish), Jack Mackerel, Giant Sea Bass, Opal Eye, Blacksmith Fish, gobies, blennies, sculpin, Cabezon, Lingcod, moray eels, more than 60 species of rockfishes, cleaner shrimps, sea slugs, octopuses, California Spiny Lobsters, occasionally Barracuda, and many species of sea anemones, sea stars, brittle stars, and snails. And thousands of invertebrates live in the holdfast, including brittle stars, sea stars, sea anemones, sponges, and tunicates. Colonies of hydroids and bryozoans grow on the kelp blades. Certain fish species eat these, benefiting from and aiding the kelp by cleaning its fronds, indirectly helping many other species. All the fish and invertebrates defecate and eventually die, and their feces and decaying bodies provide nutrient for the kelp. They are thus to some extent directly mutualistic with the kelp, and indirectly with Sea Otters. The system is a complex web of direct and indirect commensalisms and mutualisms.

Species that are keystone species as a result of being predators need not be apex predators. Sea stars are prey for sharks, rays, and sea anemones. Yet they can be keystone species by controlling sea urchins. They can also maintain high diversity among their prey by selectively eating the superior competitor.

The Jaguar is a keystone species in Central and South America because it has an extremely varied diet, helping to balance the mammalian jungle ecosystem with 87 different prey species. This not only benefits the prey populations that are regulated, but many mammals, birds, invertebrates, and plants that benefit because their predators and herbivores are regulated.

The cassowary, a large, flightless bird of the tropical forests of New Guinea, nearby islands, and northeastern Australia, of which there are three living species, eats fruit. It spreads the seeds of several hundred rainforest plant and tree species from at least 26 families; some will not grow unless they have been through a cassowary's digestive system. It can spread the seeds over 1 km, depositing them in large, dense scats. This keystone species indirectly aids all the species that benefit from the many tree species whose seeds it disperses.

The Grizzly Bear is a keystone species because it transfers nutrients from the sea and rivers to forests by depositing partly-eaten salmon and defecating on forest floors, benefiting many tree species and, indirectly, their beneficiaries, as described in Chapter 4. There are keystone predators, such as wolves and some sharks, whose removal results in catastrophic consequences for their ecosystems, as discussed in Chapter 9.

Some Ecosystem Engineers Are Keystone Species

A central part of the ABH states that all species are ecosystem engineers, but some ecosystem engineers have an extraordinary impact, and these ones are keystone species.

Many species of earthworms are ecosystem engineers, since they help the soil. In many soils, worms pull fallen leaves and manure lying on the ground down below the surface, for food or to plug their burrows. In their burrows, underground, the worms nibble the leaf and partially digest it, saturate it with intestinal secretions, and mix it into the soil. Thus, the worms convert organic matter such as leaves and manure on the surface into rich humus in the soil, greatly improving its fertility.

Earthworms also eat soil particles. These are ground in the gizzard by tiny pieces of grit into a fine paste which the intestine then digests. The material is excreted in the form of casts, on the surface or deeper in the soil. The casts contain minerals and nutrients that are accessible to plants, but were not before the worms put them through this process. Earthworm casts are five times richer in available nitrogen, seven times richer in available phosphates, and eleven times richer in available potash than the surrounding upper six inches (150 mm) of soil. They can contain 40%

more humus than the top 9 inches (23 cm) of soil near where they live. In conditions with abundant humus, one worm can produce 10 lbs (4.5 kg) of casts per year. The worms mix the soil, and their activity helps plants take up minerals and nutrients.

The worms create a great deal of channels by burrowing. This allows aeration and drainage of the soil, essential to its health and useful to life. They even secrete mucous that lubricates the soil, making movement easier. The worms act as pumps, pumping air in and out of the soil, more rapidly at night. So the worms not only create passages for air and water to move through, but are also active pistons to move the air.

Their impact is great, for their numbers are huge. Even poor soil may support 250,000 earthworms/acre (62/m²), while rich, fertile farmland may have up to 1,750,000/acre (432/m²). So the weight of earthworms beneath the soil of a farm is likely greater than that of the livestock on its surface.

Thus, earthworms help all species that benefit from soil. This includes myriad species of plants, especially trees, the mycorrhizal fungi, and nitrogen-fixing bacteria associated with the roots of trees and some plants, thousands of species of soil invertebrates, and burrowers like moles and gophers. It also includes huge numbers of species that benefit indirectly from the soil, such as various species of birds and squirrels that benefit in one way or another from trees or plants, and the predators of these birds and squirrels, such as birds of prey, coyotes, foxes, Pine Martins, Fishers, raccoons, snakes, and so on. Other earthworm beneficiaries include insects and other invertebrates above ground that feed on the plants, and their predators. I estimate that this amounts to hundreds to thousands of species in temperate regions, and thousands to tens of thousands of species in the tropics.

In addition to greatly aiding soil and life dependent on it as commensal benefactors, earthworms are extremely important as prey in many food chains. Many animal species eat them, including several birds, such as starlings, gulls, crows, thrushes, and European Robins; some snake species; several mammals, including bears, foxes, hedgehogs, and moles; and several species of invertebrate, such as ground beetles and other beetles, and some snails and slugs. They also help their many internal parasites, including Protozoa, flat worms, and round worms. Many species benefit earthworms directly and indirectly, most notably the many plant species that supply them with leaves, so many of their relationships are also mutualistic.

There are several other invertebrate species that benefit the soil and thus plants, and so are in direct and indirect commensalism and mutualism with several animal, plant, and microbial species.

Juergens [22] found the likely cause of fairy circles. In South Africa's extremely arid Namib desert, sand termites (*Psammotermes allocerus*) remove short-lived vegetation that appears after rain, leaving circular barren patches. This increases local water retention by reducing evapotranspiration and causing rapid percolation. This produces the fairy circles, local ecosystems supported by circles of perennial vegetation that grow within otherwise mostly barren desert. Once generated, the circles collect water, which sustains the growth of perennial vegetation at the edges of the circles, allowing for long-term persistence of the termites and a local increase in biodiversity. This termite-generated ecosystem persists through prolonged droughts lasting many decades. The termites are ecosystem engineers that maintain a favorable habitat for themselves and the many species in the fairy circle ecosystem.

Seagrasses are ecosystem engineers and many or all of the approximately 60 species are keystone species. Lamb et al. [23] showed that seagrass meadows, the most widespread coastal ecosystem on Earth, reduce bacterial pathogens. When they were present, there was a 50% reduction in the relative abundance of bacterial pathogens of humans and marine organisms. There were twofold reductions in disease levels in coral reefs adjacent to seagrass meadows compared to those not adjacent to them. This help they provide to a different ecosystem, the coral reef, greatly enhances diversity because of the high diversity of reefs and their ability to protect coastal habitats. Seagrasses have natural biocides. Seagrasses are also great at removing excess nitrogen and phosphorous from coastal waters. They are one of the largest stores of carbon in the sea. Seagrasses provide habitat for large numbers of animals.

Oysters, mussels, and some other bivalve mollusks are ecosystem engineers. They filter out pollutants, including industrial waste, and algae, cleaning the water. A single oyster can filter and clean 50 gallons of water per day. Freshwater mussels clean rivers; a single one can filter over 50 liters of water per day. Bivalves anchor sediment, preventing erosion. Freshwater mollusk shells stabilize river beds. Bivalves protect coastal habitats from daily tides and storm surges. They provide a three-dimensional habitat for several fish and invertebrate species. They are food for sea stars, some fish, some snails, and other species.

Of course angiosperms are ecosystem engineers, holding soil in place, weathering rocks, fertilizing soil, providing animal habitat, and more. It is interesting that Schneider et al. [24] used molecular data to show that polypod ferns (over 80% of living fern species) diversified in the Cretaceous, after angiosperms, suggesting these ferns had an ecologically opportunistic response to the diversification of angiosperms, as angiosperms came to dominate terrestrial ecosystems. They suggest angiosperms helped polypod ferns diversify by providing structurally complex forests (many polypod ferns are epiphytes), and causing low-light subcanopy conditions. (In fact, ferns needed high light levels when angiosperms first appeared, but adapted to low light levels created by angiosperms.) Angiosperms have high transpiration rates that create locally higher relative humidities and cloud cover—especially noticeable over the Amazon—that could enhance reproduction of ferns, especially epiphytic ones. I will add that fertilization of the soil via leaf litter, and higher and faster nutrient cycling, no doubt helped polypod ferns, although this ecosystem engineering by angiosperms would have helped all plants in the ecosystem.

Mangroves are ecosystem engineers. They protect coastal habitat by reducing erosion from tides and dampening the energy of storm waves. They prevent erosion and hold soil in place. They even accumulate and build soil in their tangled roots, creating habitat for themselves. They filter out toxins. They provide habitat for many fish and invertebrate species. They provide nurseries for larvae of coral and some coral reef fish. They reduce atmospheric CO_2 by drawing it out of the air and burying the carbon.

Echidnas, Australian monotreme mammals, are ecosystem engineers because they till and aerate the soil, helping a multitude of species. They spend about 12% of their day digging, mainly for invertebrates, but at times to escape predation. One echidna churns up $204\,m^3$ of soil in one year [25].

Prairie dogs of several species in the western United States dig burrows for themselves that are employed as nesting sites for Burrowing Owls and Mountain Plovers, and provide habitat for various species of snakes, lizards, salamanders, frogs, toads, mice, other rodents, and invertebrates, allowing them to escape the heat and cold. This is inquilinism. Burrowing Owls that use the burrows flee predators in response to the prairie dogs warning whistles. Their tunnel systems help channel rainwater into the water table, preventing runoff and water loss, as well as erosion. Their tunnels also act as channels for the air to go into the soil, increasing aeration. This is good for the soil and its ability to support life. Prairie dogs till and loosen the soil, reversing soil compaction, a deleterious effect on the soil that can be caused by grazing ungulates. The soil compaction is not a counterexample to the Autocatalytic Biodiversity Hypothesis (ABH) because this effect is always ameliorated by burrowing animals that coexist with the ungulates, and the ungulates improve the soil for life by tilling it. Prairie dogs trim the vegetation around their colonies, perhaps to better see their predators. They also eat back the vegetation, trimming it short as they graze. The grasses and other vegetation grow back with much higher nutritional content after prairie dog trimming. This greatly helps grazing hoofed mammals, such as American Bison, and Mule Deer. These two species have shown a proclivity for grazing areas trimmed by prairie dogs. This is interesting, because these species are competitors with the prairie dog, eating many of the same plants. Yet, they benefit more from the prairie dog than they are harmed by it. Thus, Mule Deer are both a commensals and competitors with prairie dogs. Bison are competitors and mutualists with prairie dogs because when they graze, they leave behind ideal habitat for prairie dogs. Black-tailed Prairie Dogs prefer using heavily grazed area for habitat. Experiments show that foraging by prairie dogs and bison both improve the nutritional value of plants for the other species [26].

Gophers, Desert and Gopher Tortoises, American Badgers, ground squirrels, mongooses, and other animals that dig burrows have many animals that use their burrows temporarily or permanently, including insects, scorpions, other arthropods, salamanders, frogs, toads, lizards, snakes, and rodents. The burrows also house fungi, which some of the insect inquilines often eat. The California Tiger Salamander (*Ambystoma californiense*) utilizes and is largely dependent on the burrows of the California Ground Squirrel (*Otospermophilus beecheyi*). These are all cases of inquilinism. As with prairie dogs, the benefactor species in this paragraph till the soil, and their burrows aerate the soil and help channel rainwater. They are all keystone species, although possibly usually with less effect than prairie dogs.

All of the benefactor species discussed in the two previous paragraphs help cool the planet to life's benefit because their burrows aerate the soil. This inhibits the growth of methanogens, which cannot grow in the presence O_2 and produce the powerful greenhouse gas, methane.

Bioturbation is the reworking of soils and sediments, including aquatic ones, by animals or plants. These include burrowing, and ingestion and defecation of sediment. Many organisms that do this are ecosystem engineers and keystone species, since it has profound positive effects on the environment and is a primary driver of biodiversity [27]. Bioturbers improve habitat for themselves and myriad other species. Bioturbation is a major component in the cycling of elements, including magnesium, nitrogen, calcium, strontium, and molybdenum.

This chapter has discussed bioturbation by terrestrial fauna, such as earthworms and prairie dogs. It is also done by tree roots. The evolution of trees during the Devonian Period enhanced soil weathering and increased the spread of soil due to bioturbation by their roots [28]. Root penetration and uprooting also enhanced soil carbon storage by causing mineral weathering and the burial of organic matter [28]. Bioturbation on land helps soil formation [27].

Bioturbation in the sea is done by walruses, polychaetes, ghost shrimp, mud shrimp, and several other fauna, and in freshwater systems by salmon, midge larvae, and many other fauna. Bioturbation in aquatic systems increases nutrients in aquatic sediment and overlying water; mixes water and solutes with sediments (called bioirrigation), enhancing habitat; provides burrows; and improves sediment texture. The transport of solutes like dissolved O_2, and enhancement of organic matter decomposition and sediment structure, on land and in water, by bioturbators aids the survival and colonization by other macrofaunal and microbial communities [29]. Increased microbial activity increases organic matter decomposition and sediment oxygen uptake [30] in a positive feedback loop, helping much of the community. Nutrients released from enhanced microbial decomposition of organic matter cause increased growth of marine phytoplankton and bacteria, aiding all species in the food web which they are the basis of. And Canfield and Farquhar [31] argued convincingly that bioturbation has resulted in a several-fold increase in seawater sulfate concentration. Sulfur is, of course, essential to life.

Bioturbation by walrus feeding is a significant source of nutrient flux, community structure, and species richness in the Bering Sea [32]. Walruses feed by digging their muzzles into the sediment and extracting clams through powerful suction [32].

A lack of bioturbating organisms in coastal ecosystems results in clogging of the sediment with fine particles, a drastic reduction in sediment permeability, inability of O_2 to penetrate deeply into the sediment, and accumulation of reduced mineralized products in pore water. As a result, species richness decreases greatly.

Bioturbation is crucial in the deep sea because this habitat has little to no abiological forces for sediment movement for nutrient cycling. In areas with relatively still water, it is the only force distributing minerals in the sediment [33]. There is positive feedback. Bioturbation causes higher diversity, which causes more bioturbation, until a stable state is reached. Animals on the seafloor facilitate carbon incorporation into the sediment, where it is consumed by sediment-dwelling animals and bacteria, sequestering some of it [34]. In some deep-sea sediments, manganese and nitrogen cycling are aided by intense bioturbation [33]. Sea cucumbers are among the many bioturbators in the deep sea.

Bioturbation is thought to have been an important factor of the Cambrian Explosion [35,36]. It increased at that time, for Seilacher, et al. [37] pointed out trace fossils show as great a diversification in the Ediacaran-Cambrian transition as metazoan body fossils. In Ediacaran-age shallow-marine deposits, trace fossils are horizontal, simple, and rare, and indicate feeding on microbial mats. Arthropod tracks and sinusoidal nematode trails are lacking. In the Early Cambrian, there was a dramatic increase in the diversity of distinct ichnotaxa followed by the onset of vertical bioturbation and the disappearance of a matground-based ecology. On deep-sea bottoms,

Ediacaran fauna had low diversity and was dominated by the horizontal burrows of animals that fed underneath mats. There was a great diversification of behaviors during the Early Cambrian. The precise time of the increase in bioturbation and diversity in the deep sea is uncertain, but bioturbation appeared at approximately the beginning of the Cambrian explosion; the timing was about right for it to have aided the explosion.

Microbial mats—in particular, stromatolites—were the main biological structures in the sea before the explosion, providing habitat for other forms and driving many of the ecological functions, consistent with the ABH, but the habitat services were limited. Prokaryotes on the seafloor caused organic matter to accumulate in the sediments. The bacteria and this organic matter were food sources for bottom-feeding metazoans. Predators were a selective force favoring hard skeletons and parts, including bristles, spines, and shells, in their prey. This allowed burrowing, which the prey did both for protection and to feed on the organic matter in the sediments. Predators then evolved to burrow into the sediments to seek their prey. Thus, bioturbation came about, and burrowing animals disturbed the microbial mat system and created a mixed sediment layer with greater biological and chemical diversity, which is thought to have led to the evolution and diversification of seafloor-dwelling species and the Cambrian Explosion [35]. In the Ediacaran period, microbial mats provided food for heterotrophs and benefited many species by keeping sediments out of the water, but hardened the seafloor, making it difficult for many animal species to utilize it. There was an expansion of burrowing 550 to 540 mya [38], possibly by primitive worms, just prior to the Cambrian explosion. The burrowing seems to have opened up new living spaces for other animal species. Churning the mud would soften it and free it of the stiff microbial mat that covered much of the seafloor, making it easier to colonize. This would have had a planet-wide impact, creating habitat for many species, which could then diversify. Fossils from the Ediacaran Dengying Formation (551 to 541 Ma) in South China likely represent bilateran animal activities related to under-mat feeding and epibenthic locomotion on the microbial mats to exploit nutrients and O_2 resources [39]. These animals heralded a new age in ecosystem engineering, animal-sediment interaction, and biogeochemical cycling. This indicates the Cambrian explosion was made possible by bioturbers in the form of burrowers creating habitat for more complex seafloor dwellers.

It is reasonable to assume that bioturbation has had a profound effect on evolution throughout the Phanerozoic, for since its onset about 541 mya, which is the start of the Panerozoic, it has been important in nutrient cycling [40]. Pillay [36] pointed out that bioturbators potentially play a big role in evolution, especially at the microevolutionary level, in which they influence the selective pressure on coexisting species, and, as a result, the evolution of novel morphologies and behavioral interactions. Bioturbating animals likely affected the cycling of sulfur in the early oceans in a way helpful to life.

Logan et al. [41] found evidence suggesting a radiation of zooplankton with a gut capable of forming large fecal pellets resulted in great quantities of these pellets sinking to the seafloor, at the beginning of the Cambrian. This delivered large quantities of nutrients, such as nitrate, phosphorus, and carbon to the seafloor, and precipitated a positive feedback loop that aided the Cambrian explosion. The increased nutrients on the seafloor allowed metazoans to colonize it. They obtained larger sizes and diversified there, with arthropods being the major phylum. The impact of these metazoans on carbon and nutrient cycles was tremendous and pushed the biosphere into a new biogeochemical regime (ibid). Some seafloor metazoans joined others at the surface. Then, metazoans in shallow water sent much greater quantities of nutrients to the seafloor, helping further diversification there. The increased nutrient flow from surface waters in the Neoproterozoic-Cambrian allowed increased activity in the benthic environment for sediment-processing metazoans and heterotrophic bacteria. The nutrient increase allowed greater bacterial populations and diversity. Some of the zooplankton's fecal pellets sank to the sea floor, carrying reduced carbon with them. This increased O_2 levels in the ocean, and this also played a key role in the Cambrian explosion.

The metazoans altered the sediments by processing them through their guts and burrowing in them [42]. This introduced water into the sediments, changing their rheological properties and increasing their suitability as a habitat [43]. Deeper burrowing became easier. Bioturbation enhances habitat in several ways. Alteration of the redox interface within sediments by bioturbation in modern sediments has been shown to change pH, calcium fluorapatite, pyritization, and authigenic calcite precipitation, relative to unbioturbed sediments [44]. And the early Cambrian bioturbation increased the supply of organic matter to bacteria due to particle movement [45]. It aerated the seafloor sediments, enabling metazoans and aerobic bacteria to extend their habitat to deeper levels, and forcing anaerobic bacteria deeper. This is significant because O_2 is rapidly depleted by aerobic bacteria in the upper sediments. Grazing of microbes by metazoans increases diversity and biomass of bacteria by reducing competition for nutrients [46,47]. Churning of the upper layer produced a more three-dimensional habitat and thus distribution of bacterial communities [48–50]. These effects of bioturbation are essentially absent in Proterozoic compared to Phanerozoic strata, showing a major change in the biosphere across the Proterozoic-Cambrian boundary. This was a mutualism between metazoa and bacteria that increased the diversity of both. The metazoans were bioengineers, and their activity increased their own diversity. It is interesting that the sinking of zooplankton fecal pellets likely precipitated a chain reaction and positive feedback that led to perhaps life's greatest diversification.

There are many species of bioturbators, they occur in all ecosystems, and they are ecosystem engineers that profoundly improve habitat and increase species richness.

Elephants are ecosystem engineers. They dig water holes to get access to underground water when surface water is in short supply. A host of other animal species then use this water that was formerly unavailable to them. This includes all the major predators, herbivores, and scavengers that share the ecosystem with elephants. In Africa, for example, this is a tremendous number of species, including lions, cheetahs, leopards, various species of gazelle, giraffe, wildebeest, jackal, small mammals like mongoose, and diverse species of birds, reptiles, amphibians, aquatic insect larvae, flatworms, and other invertebrates. Elephants dig up Earth to obtain salt and other minerals. Their tusks are used to churn the ground. The elephant eats dislodged pieces of soil, ingesting nutrients. The digging creates holes that are several feet deep, making salt and other minerals available to many other species. Animals such as rodents and reptiles use these holes as habitat. The footprints of African Elephants (*Loxodonta Africana*) in Kibale's tropical rainforest in Uganda fill with rainwater and are used in the dry season by water beetles, mosquito larvae, dragonfly larvae, and other invertebrates, representing 61 distinct taxa, including 27 families or orders [51]. Colonization of water-filled footprints is fast. They constitute important habitats with high diversity, and they act as stepping stones for invertebrate dispersal. Platt et al. [52] found water-filled tracks of the Asian Elephant (*Elephas maximus*) in Myanmar had anuran eggs and larvae. Their findings suggest that the tracks persist for over a year and function as small lentic waterbodies that provide temporary, predator-free breeding habitat for anurans during the dry season when alternate sites are unavailable, and trackways could also function as stepping stones that connect anuran populations. Asian Elephant footprints are also used by mosquitoes to breed in India. All this suggests that sauropod footprints made pools that were habitats for bacteria, archaea, protists, algae, fungi, insects and their larvae, amphibians and their larva, and perhaps other organisms in the time of dinosaurs. Elephants also eat and destroy trees, maintaining grasslands. This increases biodiversity, because it maintains a mosaic of two habitats, forest and savanna, even though forest is more diverse than savanna. If elephants did not remove trees, there would be only one type of habitat, forest. Elephant defecation adds tremendous nutrient to the savanna, fertilizing it and keeping it healthy. This allows better growth of the grasses, and hence healthier herbivores and their predators. All three trophic levels are thus made more species-rich.

On the deep-sea floor, many mounds and depressions are formed by benthic animals, such as worms, mollusks, crustaceans, starfish, brittle stars, shrimps, fishes, sea cucumbers, and sea urchins. Some of these mounds are large and called seamounts. Seamounts are excellent habitat for high concentrations of plankton. This food source attracts many small fish and invertebrate species, which in turn attract predatory fish and invertebrates. Marine mammals, sharks, tuna, and cephalopods all congregate over the seamounts to feed.

Various species of copepods; krill; shrimp, such as brine shrimp; and other small, oceanic taxa are ecosystem engineers that cause great turbulence in the sea when they swim in great numbers, often by migrating from shallow to deeper waters and back. Brine shrimp can cause fluid velocities of about 1–2 cm/s. These various taxa mix shallow waters with deeper, saltier waters. Without this mixing, waters of different densities and temperatures would remain separated in distinct layers, becoming stratified and stagnated. Parts of the sea would become anoxic. The swarms mix ocean layers, and deliver nutrients from deep waters to sunlit shallow waters with phytoplankton, just as upwelling does. This increases phytoplankton and hence all members of the food web above phytoplankton. Phytoplankton growth helps sequester carbon, and hence regulate Earth's temperature. The swimmers also aid the exchange of gases between sea and atmosphere. The animals are tiny, but extremely numerous. As one swims upward, it kicks the water downward. That parcel of water is then kicked down by another animal, and then another. This results in a downward flow of water that strengthens as the migration continues, and eventually extends as far as the entire migrating group, which can be as much as 100 m.

Various termite and leafcutter ant species are exceptionally influential ecosystem engineers (Chapter 4).

Beavers Are Good Examples of Keystone Species and Ecosystem Engineers That Increase Species Richness Tremendously

The beaver is a major ecosystem engineer that transforms parts of rivers to ponds or small lakes, creating habitat for great numbers of species. As a beaver family moves into new territory, they drop a large tree across a stream to begin a new dam, which creates a pond for their lodge. They cover the tree with sticks, mud, and stones. As the water level rises behind the dam, it submerges the entrance to their lodge and protects the beavers from predators.

This pooling of water leads to a cascade of ecological changes. The pond nourishes young trees of the Salicaceae family (includes willows, aspens, poplars, cottonwood), the beaver's preferred trees for food and lodging. So the beaver's engineering makes the habitat better for it. The pond aids growth of grass and shrubs alongside of it, which improves habitat for songbirds, deer, and elk. Since dams raise underground water levels, they increase water supplies.

The dam increases habitat heterogeneity and diversity by creating a pond and a wetland, resulting in three habitats—river, pond, and wetland—whereas before there was only the river. Thus, beavers increase biodiversity by increasing the number of habitat types, creating a mosaic of habitats. Species that can exist as a result of beaver ponds include several species of each of these: algae, plants that thrive in still water, insects with aquatic larvae, other invertebrates, fish that prefer still or slow-moving water, frogs, salamanders, some snakes, and water-loving birds. Indirect benefactors are all those species that are helped by the species the beaver benefits, such as snakes that eat frogs. This amounts to a huge number of beneficiary species that are commensal with the beaver, both directly and indirectly.

Beaver dams can both create wetlands and restore human-destroyed ones. Benefits of wetlands include flood control, provision of habitat for many common as well as rare species, and water purification. The latter is achieved by the ability of small dams to break down toxins, and because the roots of many wetland plants remove toxins. Beaver dams retain silt. The accumulation of silt over several years produces good topsoil. Beaver dams reduce erosion by holding the silt in place. They reduce water turbidity, which is a limiting factor for much aquatic life. Almost 50% of endangered and threatened North American species rely on wetlands. Beaver dams are small, so do not have the destructive effects of human-built megadams.

Beaver dams raise the water table alongside a stream, aiding the growth of trees and plants that stabilize the banks and prevent erosion. The trees provide shade that keeps the water temperature from getting very hot, allowing the survival of fish such as salmon.

Rivers with beaver dams in their head waters are less variable, with lower high water and higher low water levels. Many bacterial species break down the cellulose in the branches in beaver dams. These bacteria use phosphorous and nitrogen compounds, removing them from the water, preventing eutrophication, a problem more prone to occur in agricultural than natural areas. Excess nitrogen and phosphorous compounds due to human activity are a serious environmental problem today. Some bacteria in beaver dams denitrify, converting excess nitrogen compounds to nitrogen gas [54].

Wright et al. [55] showed that by increasing habitat heterogeneity, beavers increase the number of species of herbaceous plants in the riparian zone by over 33% at a scale encompassing "both beaver-modified patches and patches with no history of beaver occupation, in the Adirondacks, in New York". They suggested that ecosystem engineers increase species richness whenever there are species present in a landscape that is restricted to engineered habitats during at least some stages of their life cycle.

Beaver dams create nurseries for salmon, trout, and other fish. When beavers were made extinct in the Columbia River watershed in the early 1800s, salmon populations fell steeply in the following years. None of the other factors associated with the decline of salmon runs were extant at that time. Beaver dams increase salmon runs. The ponds they create are sufficiently deep for juvenile salmon to not be easily visible to predatory wading birds. They trap nutrients the juvenile salmon consume after losing their yolk sacs, including the bonanza nutrient pulse from the mass die-off from the migration of the adult salmon upstream. They create calm water with the result that the young salmon need not fight currents and can thus use their energy for growth. They keep the water clean. All salmonoids are sensitive to water quality.

Kemp et al. [53] found the number of benefits of beaver dams to fish outweighed the number of costs, 184 to 119. (But I find it hard to believe there are as many as 119 deleterious effects of beaver ponds.) The dams increase rearing and overwintering habitat, refuges, and invertebrate production. The latter increases species richness in its own right, and also provides food for fish and other taxa.

Beaver dams help frogs and toads, because they provide areas with deep, warm, well-oxygenated water, and places for concealment, for their larvae. Warm, oxygen-rich water enhances the development and growth of frog and toad larvae. Amphibian traps in beaver ponds in Alberta, Canada, captured 5.7 times more newly metamorphosed wood frogs, 29 times more western toads, and 24 times more boreal chorus frogs than on nearby free-flowing streams without beavers [56].

Beaver dams help migrating songbirds, including ones in decline, by stimulating the growth of plants and trees that provide food and habitat for them. Beaver dams are associated with increased songbird diversity [57]. Shrubs increase, in the long run increasing moose and hare populations, and thus wolf populations. When wolves eat beaver, they excrete less intestinal parasites called cestodes (tapeworms are in this taxon), which are harmless to wolves, but can be picked up by caribou (in tundra habitats) and moose, which they do harm.

If sediment accumulates, making the water too shallow, or if trees become too few, beavers leave their pond. In time, the dam breaks and water drains freely. The rich, thick layer of branches, dead leaves, and silt behind the dam forms a wetland that supports a highly diverse community. Ecological succession follows, and the wetland slowly dries out, and is eventually replaced by a meadow, then a forest of trees adapted to riverine environments, including the beaver's preferred Salicaceae. Beavers recolonize nearby rivers, and the cycle repeats. Each successive cycle lays down another layer of rich organic soil. This builds a valley that gets wider with each cycle, building rich habitat that can become grassland or, more likely as the final ecosystem, diverse forest.

CONCLUSION

Careful observation and analysis indicates the number and percentage of keystone species and ecosystem engineers appears larger than commonly thought, indicating commensalism and mutualism account for a great amount of the tremendous biodiversity we observe on the planet.

Commensalism is ubiquitous. Every metazoan has a microbiome with thousands of microbial species using it as a habitat, many not affecting its fitness, and many of these microbial species are commensal with each other. Even if we do not consider the microbiome, it is likely every species is a commensal benefactor to at least one species and a commensal beneficiary of at least one species. From barnacles on sea turtles to epiphytes on trees, every large organism almost for certain has one or more other species, probably several, living on it, without significantly affecting it.

REFERENCES

1. Almeda, F. (2012). Personal communication.
2. Londono, G. A., et al. (2014). Morphological and behavioral evidence of Batesian mimicry in nestlings of a lowland Amazonian bird. *The American Naturalist* 185 (1): 135–41.
3. von Beeren, C. & Tishechkin, A. K. (2017). *Nymphister kronaueri* sp. nov., an army ant-associated beetle species (Coleoptera: Histeridae: Haeteriinae) with an exceptional mechanism of phoresy. *BMC Zoology* 2 (1): 3. doi: 10.1186/s40850-016-0010-x.
4. Navarrete, S. A. & Castilla, J. C. (1990). Barnacle walls as mediators of intertidal mussel recruitment: effects of patch size on the utilization of space. *Marine Ecology Progress Series* 68: 113–9.
5. Barrowclough, G. F., et al. (23 Nov., 2016). How many kinds of birds are there and why does it matter? *PLOS ONE*. http://journals.plos.org/plosone/article?id=10.1371/journal.pone.0166307.
6. Pirt, S. J. & Y. K. Lee. (1983). Enhancement of methanogenesis by traces of oxygen in bacterial digestion of biomass. *FEMS Microbiology Letters* 18 (1–2): 61–3.
7. Stamets, P. E., et al. (2018). Extracts of polypore mushroom mycelia reduce viruses in honey bees. *Scientific Reports* 8, Article number 13936. doi: 10.1038/s41598-018-32194-8.
8. Reddy, G. V. P. & Guerrero, A. (May, 2004). Interactions of insect pheromones and plant semiochemicals. *Trends in Plant Science* 9 (5): 253–261. doi: 10.1016/j.tplants.2004.03.009.
9. Yamazaki, K. (2007). Cicadas "dig wells" that are used by ants, wasps and beetles. *European Journal of Entomology* 104 (2): 347–9.
10. Ochi, H. & Yanagisawa, Y. (May, 1998). Commensalism between cichlid fishes through differential tolerance of guarding parents toward intruders. *Journal of Fish Biology* 52 (5): 985–96. doi: 10.1111/j.1095-8649.1998.tb00598.x.
11. Bertness, M. D., & Hacker, S. D. (1994). Physical stress and positive associations among marsh plants. *The American Naturalist* 144: 363–72.
12. Gordon, T. A. C. (29 Nov., 2019). Acoustic enrichment can enhance fish community development on degraded coral reef habitat. *Nature Communications* 10, Article number 5414. https://doi.org/10.1038/s41467-019-13186-2.
13. Ulrich, R. S. (1986). Human responses to vegetation and landscapes. *Landscape and Urban Planning* 13: 29–44. doi: 10.1016/0169-2046(86)90005-8.
14. Mills, L. S., et al. (1993). The keystone-species concept in ecology and conservation. *BioScience* 43 (4): 219–24.
15. Watson, D. M. (2001). Mistletoe: a keystone resource in forests and woodlands worldwide. *Annual Review of Ecology, Evolution, and Systematics* 32: 219–49.

16. Hadfield, J. S. (Jan., 1999). Douglas-fir dwarf mistletoe infection contributes to branch breakage. *Western Journal of Applied Forestry* 14 (1): 5–6.

17. Hadfield, J. S. & Flanagan, P. T. (1 Jan., 2000). Dwarf mistletoe pruning may induce Douglas-Fir beetle attacks. *Western Journal of Applied Forestry* 15 (1): 34–6. doi: 10.1093/wjaf/15.1.34.

18. Parks, C. G., et al. (1 April, 1999). Wildlife use of Dwarf Mistletoe brooms in Douglas-Fir in northeast Oregon. *Western Journal of Applied Forestry* 14 (2): 100–5. doi: 10.1093/wjaf/14.2.100.

19. Parks, C. G. & Bull, E. L. (1997). American marten use of rust and dwarf mistletoe brooms in northeastern Oregon. *Western Journal of Applied Forestry* 12 (4): 131–3.

20. van Ommeren, R. J. & Whitham, T. G. (2002). Changes in interactions between juniper and mistletoe mediated by shared avian frugivores: parasitism to potential mutualism. *Oecologia* 130: 281–8.

21. Watson, D. M. (26 Aug., 2002). Effects of mistletoe on diversity: a case-study from southern New South Wales. *EMU* 102 (3) 275–81. doi: 10.1071/MU01042.

22. Erwin, T. L. (1983). Tropical forest canopies: the last biotic frontier. *Bulletin of the Entomological Society of America* 29: 14–9.

23. Juergens, N. (29 March, 2013). The biological underpinnings of Namib Desert fairy circles. *Science* 339 (6127): 1618–21. doi: 10.1126/science.1222999.

24. Lamb, J. B., et al. (17 Feb., 2017). Seagrass ecosystems reduce exposure to bacterial pathogens of humans, fishes, and invertebrates. *Science* 355 (6326): 731–3. doi: 10.1126/science.aal1956.

25. Schneider, H., et al. (1 April, 2004). Ferns diversified in the shadow of angiosperms. *Nature* 428: 553–7. doi: 10.1038/nature02361.

26. Clemente, C. J., et al. (2016). The private life of echidnas: using accelerometry and GPS to examine field biomechanics and assess the ecological impact of a widespread, semi-fossorial monotreme. *Journal of Experimental Biology* 219: 3271–83. doi: 10.1242/jeb.143867.

27. Krueger, K. (1 June, 1986). Feeding relationships among bison, pronghorn, and prairie dogs: an experimental analysis. Ecology 67 (3): Pages 760–70. https://doi.org/10.2307/1937699.

28. Wilkinson, M. T., et al. (1 Dec., 2009). Breaking ground: pedological, geological, and ecological implications of soil bioturbation. *Earth-Science Reviews* 97 (1): 257–72. Bibcode: 2009ESRv...97..257W. doi: 10.1016/j.earscirev.2009.09.005.

29. Algeo, T. J. & Scheckler, S. E. (29 Jan., 1998). Terrestrial-marine teleconnections in the Devonian: links between the evolution of land plants, weathering processes, and marine anoxic events. *Philosophical Transactions of the Royal Society of London B* 353 (1365): 113–30. doi: 10.1098/rstb.1998.0195. ISSN 0962-8436. PMC 1692181.

30. Braeckman, U., et al. (28 Jan., 2010). Role of macrofauna functional traits and density in biogeochemical fluxes and bioturbation. *Marine Ecology Progress Series* 399: 173–86. Bibcode: 2010MEPS..399..173B. doi: 10.3354/meps08336. ISSN 0171-8630.

31. Andersen, F. O. & Kristensen, E. (1991). Effects of burrowing macrofauna on organic matter decomposition in coastal marine sediments. *Symposia of the Zoological Society of London* 63: 69–88.

32. Canfield, D. E. & Farquhar, J. (19 May, 2009). Animal evolution, bioturbation, and the sulfate concentration of the oceans. *PNAS USA* 106 (20) 8123–7. doi: 10.1073/pnas.0902037106.

33. Ray, G. C., et al. (2006). Pacific walrus: benthic bioturbator of Beringia. *Journal of Experimental Marine Biology and Ecology* 330 (1): 403–19. doi: 10.1016/j.jembe.2005.12.043.

34. Aller, R. C., et al. (1998). Biogeochemical heterogeneity and suboxic diagenesis in hemi-pelagic sediments of the Panama Basin. *Deep Sea Research Part I: Oceanographic Research Papers* 45 (1): 133–65. Bibcode: 1998DSRI...45..133A. doi: 10.1016/s0967-0637(97)00049-6.

35. Vardaro, M. F., et al. (1 Nov., 2009). Climate variation, carbon flux, and bioturbation in the abyssal North Pacific. *Limnology and Oceanography* 54 (6): 2081–8. Bibcode: 2009LimOc..54.2081V. doi: 10.4319/lo.2009.54.6.2081. ISSN 1939-5590.

36. Meysman, F., et al. (2006). Bioturbation: a fresh look at Darwin's last idea. *Trends in Ecology and Evolution* 21 (12): 688–95. doi: 10.1016/j.tree.2006.08.002.

37. Pillay, D. (23 June, 2010). Expanding the envelope: linking invertebrate bioturbators with micro-evolutionary change. *Marine Ecology Progress Series.* 409: 301–3. Bibcode: 2010MEPS..409..301P. doi: 10.3354/meps08628. ISSN 0171-8630.

38. Seilacher, A., et al. (10 Nov., 2005). Trace fossils in the Ediacaran–Cambrian transition: behavioral diversification, ecological turnover and environmental shift. *Palaeogeography, Palaeoclimatology, Palaeoecology* 227 (4): 323–56. doi: 10.1016/j. palaeo.2005.06.003.

39. Grazhdankin, D. (March, 2014). Patterns of evolution of the Ediacaran soft-bodied biota. *Journal of Paleontology* 88 (2): 269–83. doi: 10.1666/13-072. Published online by Cambridge University Press: Cambridge, MA, 15 Oct., 2015.

40. Chen, Z., et al. (Jan., 2013). Trace fossil evidence for Ediacaran bilaterian animals with complex behaviors. *Precambrian Research* 224: 690–701. doi: 10.1016/j. precamres.2012.11.004.

41. Rhoads, D. C. & Young, D. K. (1970). The influence of deposit-feeding organisms on sediment stability and community trophic structure. *Journal of Marine Research* 28 (2): 150–78.

42. Logan, G. A., et al. (4 July, 1995). Terminal Proterozoic reorganization of biogeochemical cycles. *Nature* 376 (6535): 53–6. doi: 10.1038/376053a0.

43. McIlroy, D. & Logan, G. A. (1999). The Impact of bioturbation on infaunal ecology and evolution during the Proterozoic–Cambrian transition. *PALAIOS* 14: 58–72.

44. Rhoads, D.C. (1970). Mass properties, stability and ecology of muds related to burrowing activity. In Crimes, T. P. & Harper, J. C., (eds.), *Trace Fossils: Geological Journal Special Issue 3*, pp. 391–406. Seel House Press: Liverpool, UK.

45. Reimers, C. E. (1996). Porewater pH and authigenic phases formed in the uppermost sediments of the Santa Barbara basin. *Geochmica et Cosmochimica Acta* 60: 4037–57.

46. Moriaty, D. J. W. (1991). Heterotrophic bacterial activity and their growth in sediments of the continental margin of eastern Australia. *Deepsea Research* 38: 693–712.

47. White, D. C. (1983). Analysis of microorganisms in terms of quantity and activity in natural environments. *Microbes in Their Natural Environments* 34: 37–66.

48. White, D. C. (1988). Validation of quantitative analysis for microbial biomass, community structure and metabolic activity. *Archiv fur Hydrobiologie Beiheft* 31: 1–18.

49. Aller, R. C. (1978). Experimental studies of changes produced by deposit feeding on pore water, sediment and overlying water chemistry. *American Journal Science* 278: 1185–234.

50. Aller, R. C. (1982). The effects of macrobenthos on chemical properties of marine sediments and overlying water. In McCall, P. L. & Tevesz, M. J. S., (eds.), *Animal-Sediment Relations*, pp. 53–102. Plenum: New York, N. Y.

51. Aller, R. C. (1994). Bioturbation and remineralization of sedimentary organic matter: effects of redox oscillation. *Chemical Geology* 114: 331–45.

52. Remmers, W., et al. (23 Sept., 2016). Elephant (*Loxodonta africana*) footprints as habitat for aquatic macroinvertebrate communities in Kibale National Park, south-west Uganda. *African Journal of Ecology* 55 (3): 342–35. doi: 10.1111/aje.12358.

53. Platt, S. G., et al. (9 Dec., 2019). Water-filled Asian elephant tracks serve as breeding sites for anurans in Myanmar. Published online. doi: 10.1515/mammalia-2017-0174.

54. Kemp, P. S., et al. (2012). Qualitative and quantitative effects of reintroduced beavers on stream fish. *Fish and Fisheries* 13 (2): 158–81. doi: 10.1111/j.1467-2979.2011.00421.

55. Lazar, J., et al. (16 Sept., 2015). Beaver ponds: resurgent nitrogen sinks for rural watersheds in the northeastern United States. *Journal of Environmental Quality* 44 (5): 1684–93. doi: 10.2134/jeq2014.12.0540.
56. Wright, J. P., et al. (2002). An ecosystem engineer, the beaver, increases species richness at the landscape scale. *Oecologia* 132 (1): 96–101. doi: 10.1007/s00442-002-0929-1. http://www.springerlink.com/index/0637GF0979LRU90J.pdf.
57. Stevens, E., et al. (Jan., 2007). Beaver (*Castor canadensis*) as a surrogate species for conserving anuran amphibians on boreal streams in Alberta, Canada. *Biology Conservation* 134 (1): 1–13.
58. Cooke, H. A. & Zack, S. (2008). Influence of beaver dam density on riparian areas and riparian birds in shrubsteppe of Wyoming. *Western North Amer. Naturalist* 68 (3): 365–73. doi: 10.3398/1527-0904(2008)68[365:IOBDDO]2.0.CO;2.

6 Interspecific Competition Increases Species Richness

The competitive exclusion principle or Gause principle states that no two species can coexist and occupy the same niche. Said another way, no two species can coexist indefinitely on a single limiting resource. The idea is that the better competitor will drive the other species locally extinct. It is named for Georgy Gause, who did not formulate it, but demonstrated it using two species of ciliated protozoa, *Paramecium aurelia* and *P. caudatum*, in the laboratory [1]. *Caudatum* initially dominated, but *aurelia* quickly recovered and drove *caudatum* extinct via superior exploitative resource competition. A number of theoretical and mathematical models also predict competitive exclusion. Since the Gause principle means that one of two competing species will tend to go extinct, it implies competition leads to a decrease in biodiversity and a contradiction of the Autocatalytic Biodiversity Hypothesis (ABH). Yet Gause was able to cause *caudatum* to survive by varying environmental conditions, such as water chemistry and food availability, showing the rule only applies if the environment is constant. And in fact, a huge literature accumulated over the years showing that the Gause principle rarely applies in nature, mainly because of the ecological and evolutionary responses of the competitors, and sometimes due to the influence of other species, especially predators. In short, the ABH is supported by the responses of species to competitive interactions, since biodiversity is maintained by them. The literature now has a good body of evidence supporting the idea that evolution in response to interspecific competition is one of the major causes of adaptive radiation [2].

MacArthur [3] showed that five species of the wood warbler (genus *Setophaga*) in the coniferous forests of Maine and Vermont coexisted with overlapping niches by niche partitioning. Although they overlapped in what they ate, each species ate somewhat different diets and sometimes from different areas, such as different parts of the same trees, than the other warblers. We now know niche partitioning is a general principle, as shown by the examples below. The examples are of closely related species, the most common form of niche partitioning, but it sometimes occurs between unrelated species.

Lack [4] showed the White-breasted Cormorant (*Phalacrocorax carbo*) and European Shag (*P. aristotelis*) differ markedly in both nesting sites and food. In one case he found isolation by breeding season. He stated the main types of ecological isolation found in birds are also found in mammals.

De León et al. [5] studied the diets of four sympatric Darwin's ground finch species on Santa Cruz Island of the Galápagos Islands. They were generalists with overlapping diets, but each species had aspects of its diet specific to it for which its morphology was best suited. The use of these private resources increased considerably, and diet overlap decreased, when the availability of preferred shared foods, such as arthropods, was reduced during drought conditions. The birds have overlapping resources under benign conditions in space or time, but then at least partially shifted to resources for which they are best adapted during periods of food limitation. These behaviors allow coexistence.

Three species of nectar-feeding honeyeaters (genus *Melidectes*) in the mountains of New Guinea occur such that only two of the species occur together in the same mountain range, and they partition the habitat by altitude, one species always occurring higher than the other. Interestingly, which two species occur on any given mountain range appears to be a matter of chance [6].

There are 130 hummingbird species in Ecuador's cloud forests, many sympatric with overlapping niches. They require twice their body weight in nectar daily. Some species coexist stably by obtaining their nectar in different ways. The Green Violetear chases away other species. The Booted Racket-tail uses agility and speed to outmaneuver its opposition. The tiny Purple-throated Woodstar mimics the sound of bees to scare off rivals. And the Wedgebill pierces a hole in the flower's base and steals nectar, obtaining it too quickly for the other species to thwart it. Many competing species of hummingbird coexist in the Andes and Panama [7].

Four Neotropical bat species of similar morphology honed in on different acoustic features of the male mate-attraction song of 12 sympatric katydid species to locate their prey [8]. The authors stated that this niche partitioning not only allows the coexistence of the bat species, but contributes to the substantial diversity in the mate attraction signals of male katydids.

Niche partitioning can be temporal. In Isreal's southern deserts, the Common Spiny Mouse (*Acomys cahirinus*) is nocturnally active while the Golden Spiny Mouse (*A. russatus*) is diurnal [9]. Both feed on insects and they are sympatric. Three cryptic bumblebee species in one location in northern Scotland were found to partition niches by time of flight throughout the season, interspecific temporal variation according to thermal conditions, and use of forage plants [10]. This also shows that morphological similarity does not necessarily equate to ecological equivalence. It is also known that different bee species can gather nectar from the same flowers at different times of day if the flowers continuously replenish their nectar.

It is generally true that specialist species are better competitors than generalists, and this allows competing species to coexist, yielding higher diversity than if the reverse were true. An example of this was provided by Bovbjerg [11], who showed the two crayfish, *Orconectes virilis* and *O. immunis*, have similar ranges but are ecologically isolated within these ranges, the former species inhabiting lake margins and streams, the latter ponds and sloughs. *Immunis* is a generalist and better burrower, able to live in all these habitats, while *virilis* cannot inhabit ponds and sloughs, being unable to withstand summer drying and periodic low oxygen periods in these habitats. But *virilis* is more aggressive than and physically excludes *immunis*, evicting it from crevices in the lake margins and streams. Similarly, Miller [12] showed four species of pocket gophers of family Geomydae in Colorado all had the same habitat preferences of deep, light soils. The generalist species with the greatest habitat tolerances were displaced by the specialists, resulting in the ranges of the better competitors being subsets of the potential ranges of the generalists.

The barnacle, *Semibalanus balanoides* (called *Balanus balanoides* at the time of the research), outcompetes and displaces the barnacle *Chthamalus stellatus* on rocky intertidal substrates at lower tide heights [13]. It is more of a specialist than its generalist competitor, so cannot live at higher locations, where it is subject to drying out when exposed on dry rock for too long. *Chthalamus*, the generalist, can live at higher tide limits and rock levels. The lower limit of survival of *S. balanoides* is set by predation on it by a snail.

Coexistence also occurs because inferior competitors tend to be better colonizers. This competition-colonization trade-off is seen in plants in ecological succession, where pioneer species arrive first and are displaced later by better competitors. For example, herbaceous plants colonize unoccupied areas first, but are displaced by trees. Body lice are better competitors than wing lice on their hosts, Rock Pigeons (*Columba livia* Gmelin). Wing lice are better dispersers, both vertically to nestlings and by phoretic hitchhiking on parasitic flies (*Diptera*: Hippoboscidae). Body lice cannot do the latter [14]. Wing lice arrive earlier, but are eventually displaced by body lice, so the species coexist.

Competing species can coexist as a result of their consumer selectively eating the better competitor or keeping their populations too low for competition to complete exclusion of the weaker competitors, in both herbivores eating plants and carnivores eating prey (Chapters 8 and 9).

Interspecific competition can lead to two competing species evolving increasing differences between them. Such character displacement may be behavioral, morphological, ecological, physiological, or any combination of these. For a general hypothetical example, suppose beak sizes in two bird species are associated with a resource characteristic, such as size of seeds that the two species are competing for. The two seed-eating bird species have overlapping normal curves of bill sizes and sizes of seeds they eat. The mean seed and beak size is larger in one species than the other. Of the birds of the species that eats larger seeds, those that eat the largest seeds will have a selective advantage over their conspecifics because they will experience little to no interspecific competition. Likewise, of the birds of the species that eats smaller seeds, those that eat the smallest seeds will have a similar advantage. So the bill sizes of the two species will diverge by selection. The most extreme bill sizes and preferences for the most extreme seed sizes will be the most fit. Interspecific competition will decrease or disappear as a result of this selection. It results in the two species evolving less overlap in their resource use and more difference in their relevant phenotypes [15,16]. There is good evidence for this model [17]. It is true that during the 1970s and 1980s, character displacement was questioned and its importance downgraded [18], but this resulted in theoretical and methodological advances and more rigorous criteria for testing it [18]. Character displacement has provided strong evidence for evolutionary divergence due to competition in nature [16,17]. And Dayan and Simberloff [19] pointed out that research has provided sound statistical support for character displacement over a wide variety of taxa, although with a phylogenetically skewed representation. Many studies demonstrate morphologically-based resource partitioning. Theoretical studies and experimental work have added further support for the hypothesis (see discussion below).

Two species of Galapagous Island Darwin's ground finches, *Geospiza fortis* and *G. fuliginosa* differ more in bill size where they occur together than where they occur alone [20]. Competition between the two species where they are sympatric caused them to diverge in bill size, resulting in the two species eating seeds of different sizes, thus minimizing competition.

Closely related species of bumblebees (genus *Bombus*) in Colorado partition by taking nectar and pollen from flowers of different corolla lengths, and these flowers are generally different species. The length of the proboscis of the bee species is correlated with the length of the flower's corolla it feeds on, as a result of character displacement and coevolution of each flower species with its pollinator. In some cases, more than one species use flowers with similar length. In that case, they have spatial niche partitioning, living at different altitudes [21].

Where the two Appalacian salamanders, *Plethodon hoffmani* and *P. cinereus*, are sympatric, *hoffmani* has a faster-closing jaw needed for its larger prey, while *cinereus* has a slower, stronger jaw for the smaller prey it eats. But when they are allopatric, they have no significant differences in prey size or morphology [22].

In two other salamander species, *P. jordani* and *P. teyahalee*, head shape differed because of aggressive interactions, demonstrating a direct link between interference competition and morphology [23]. This is evidence that morphological variation can be generated by mechanisms other than resource exploitation, which has profound implications for interpreting patterns of biological diversity.

Introduced species have provided unintended experiments in nature that further support character displacement and evolutionary rates associated with it [24]. The American Mink (*Mustela vison*) was introduced to Belarus, with the result that it decreased in size while the native European Mink (*Mustela lutreola*) increased in size over a period of only ten years [25]. This shows competition can drive very rapid evolutionary change and divergence. Stuart et al. [26] showed the native Green Anole (*Anolis carolinensis*) perches higher and evolves toes better adapted for climbing when the non-native Brown Anole (*A. sagrei*) is present than when it is not on islands off Florida.

Character displacement has also been shown in many other groups, such as mammalian carnivores; snails; and many pairs of bird species, including rock nuthatches in Asia, shearwaters in the Cape Verde Islands, Australian parrots, Australian honey eaters of the genus *Myzantha*, and flycatchers of the Bismarck Archipelago [27]. Brown and Wilson [27] viewed character displacement as important in speciation. In fact, it is now largely accepted that interspecific competition and character displacement are often fundamental to the speciation process. This shows interspecific competition can increase biodiversity.

Niche partitioning with character displacement can lead to ecotypes. For example, lizards of the genus *Anolis*, known by the common name anoles, display a fascinating pattern of spatial niche partitioning on Caribbean islands [28–33]. About 110 species of them occur on the Greater Antilles, which are Cuba, Jamaica, Puerto Rico, and Hispaniola. There are different ecotypes that occupy specific niches they are adapted to. All four islands have a tree-canopy giant that lives in tree tops and is very large, an arboreal specialist, and a specialist that dwells on the tree's base. Three islands have grass specialists, and two have tree trunk specialists that have flatter bodies and shorter tails than specialists on the tree's base. Each ecotype is specifically adapted to its environment. The pattern of ecotypes is basically the same on each island, a repeating pattern that implies that there is a deterministic process that drives the evolution of the anoles to repeat the same pattern over and over again, shaping a consistent community structure on all islands. Thorough lab experiments showed that *Anolis* species with long legs could run faster. The long-legged Puerto Rican species, *Anolis gundlachi*, which specializes on the base of the trunk, can run at double the speed of the short-legged Jamaican species, *A. valencienni*, which is a twig specialist. But the advantage of greater speed diminishes as the diameter of the trunk or twig the lizard is on decreases. And short-limbed species are sure-footed on branches and trunks of all diameters, while long-limbed ones trip or fall more often on small-diameter surfaces like twigs. So long-legged specialists on the broad base of tree trunks that can run fast on those wide surfaces coexist on all the islands with short-legged twig specialists that are adept on branches of very small diameter. Did each ecotype evolve and spread from island to island, or did one ancestor on each island evolve into all ecotypes on that island? Over 20 years of study, looking at fossils records, field observations, DNA sequences, and experiments, have shown conclusively that each island had its own colonizing species that evolved into all the ecotypes on that island, and that the same ecotypes on different islands are not as closely related as different ecotypes on the same island. It is impossible to show competition among the ancestors of these species molded the ecotypes, but experiments show that competition occurs among the lizards today, selecting for specialization to their various niches. Where the Brown Anole, *A. sangrei*, occurs in the absence of other *Anolis* species in the Bahama Islands, it employs a wider array of perching sites, including higher in the tree, than where it occurs sympatrically with other anole species. Competition with more arboreal species seems to force it to use a smaller part of the tree, shrinking its habitat, causing it to specialize. Researchers removed all *A. gundlachi*, a long-legged anole, from areas of the Puerto Rican rainforest, with the result that the Evermann's Tree-canopy Anole, *A. evermanni*, had significantly higher densities eight weeks after this removal than in control areas where *A. gundlachi* was not removed. Similar experiments that introduced the Green Anole, *A. carolinensis*, to small islands in the Bahamas, found they did much better alone than when they coexisted and had to compete with *A. sangrei*. Some may still insist other factors explain these results. For example, one species may be eating the other. But competition for limited resources is clearly the most plausible explanation. And other experiments showed that the more ecologically and morphologically similar the species are, the more they affect each other. Smaller islands in the Caribbean than the four large islands of the Greater Antilles have similar patterns of specialization and niche partitioning, but with less ecotypes. Thus, competition led to niche partitioning in Caribbean islands, and with a consistent, repeating pattern of specialized forms for specific niches.

In fish, evolution of ecotypes within a species and their divergence by *intraspecific* competition can lead to sympatric speciation, resulting in a benthic and limnetic form. For example, sympatric speciation was shown in cichlid fish in Nicaraguan crater lakes, where the ancestral benthic species (*Amphilophus citrinellus*) gave rise to the reproductively isolated limnetic species (*A. zaliosus*) in less than ~10,000 years [35]. The two species are eco-morphologically distinct. The process of creating two distinct ecotypes was likely completed by interspecific competition.

Beardmore et al. [36] showed in experiments with bacteria that even in homogeneous environments, species with overlapping niches can coexist by metabolic and physiological trade-offs and negative frequency-dependent selection. The better competitor that utilizes food more efficiently coexists with the species that is more resilient to mutations. The more trade-offs there are, the wider is the maintenance zone of coexistence. The principle applies both to species within a community and lineages within a species.

There is also conditional niche differentiation, in which one competing species does better in one set of environmental conditions, and the other in another. In a fluctuating environment, the species alternate as to which is the better competitor, and they thus coexist. If an environment that favors one species persists for too long in this situation, one species will be extirpated.

A study of a community with a radish, aphid, and parasitoid on the aphid demonstrated that it was not simply an increase in number of species, but the niche partitioning itself that led to increased resource use in the community [37].

Evidence is now accumulating that species can coexist with overlapping niches with no niche partitioning. There is a group of species of hispine beetles which all seem to eat the same food and live in the same habitat, coexisting with no clear evidence of niche partitioning, or even aggression within or between species [38]. It is possible that food and habitat are not limiting, or high rates of predation or parasitism, or any combination of these factors, allows the coexistence, but none of these have been demonstrated.

Shmida and Ellner [39] presented studies and a mathematical model that derived several mechanisms whereby plant species can coexist in a community without differing in their relationships with habitats, resources, and exploiters. The model is based on the dynamics of species turnover in microsites, and incorporates localized competition, nonuniform seed dispersal, and aspects of spatiotemporal environmental heterogeneity. They proposed that these mechanisms contribute to the dissimilarity of within-community replicate samples and the maintenance of many rare species in plant communities. They showed plants can coexist with great niche overlap and little to minimal niche partitioning.

Rastetter and Ågren [40] showed in a model that many species can coexist at an asymptotically stable state, even if there is but one limiting resource.

There are many cases of great numbers of species coexisting with niches that apparently overlap tremendously. The cichlids of the African lakes, rainforest trees of the Amazon and Congo Basins, insects in rainforests canopies in the Amazon, and hummingbirds are examples. Up to 25 hummingbird species are sympatric in the Andes, some with overlapping niches [41]. The African Great Lakes around the East African Rift, which include Lake Victoria, Lake Tanganyika, and Lake Malawi, have tremendous numbers of fish species. More than 1,500 cichlid fish species live in the lakes [42] as well as other fish families. They contain 10% of the world's fish species. Many are endemic. Lake Malawi has more fish species than any other lake, including at least 700 species of cichlids [42]. In these lakes, many of the species are tightly packed [43], and many have overlapping niches.

I must concede that sometimes interspecific competition results in the extinction of the inferior competitor. Placental mammals displaced marsupials in most parts of the world, except Australia and for a time South America, where marsupials were isolated from placentals. However, when North and South America were joined by the Isthmus of Panama, not all marsupials were driven extinct by invading placentals. And the North American Opossum (*Didelphis virginiana*) survived in the face of placental competition. Coexistence, niche partitioning, and character displacement are the rules in competing species. Introduced species driving local species extinct is an unnatural situation. The introduced species has unnatural advantages, such as being free of competitors.

There is a phenomenon known as the paradox of the phytoplankton, which is that phytoplankton show high biodiversity at several taxonomic levels, including species, even though the resources available to them are very limited. Several resources for which they compete, including light, nitrate, phosphate, silicic acid, and iron, are in short supply. The competitive exclusion principle would suggest much lower diversity than we see. G. Evelyn Hutchison first noted it, and proposed explaining it by vertical gradients of light or turbulence, which would involve niche partitioning; mutualism; commensalism; differential predation; or constantly changing environmental conditions [44]. It is clear that niche partitioning by living at different depths and thus receiving different levels of sunlight is a factor in their coexistence. With more recent research, others have proposed or provided evidence for chaotic fluid motion [45]; size-selective grazing [46]; spatiotemporal heterogeneity [47]; and environmental fluctuations [48]. All of these are either purely or partly biological, and the third implies possible niche partitioning. One may think chaotic fluid motion is physical, but various species of copepods, krill, shrimp, and other small, oceanic taxa cause great turbulence in the sea when they swim in great numbers. Such swimmers may increase phytoplankton diversity by creating fluid motion. Scheffer et al. [49] posited that the continual interaction of environmental factors has the result that equilibrium is never reached, and so no species can dominate. Organisms can act as one or more such environmental factor(s). We now know that many phytoplankton both ingest other organisms and photosynthesize, and this could play a role in their coexistence. Niche partitioning is part of the solution to this paradox, and the proposed explanations are all plausible, although we do not know the full answer. It appears from this paradox that organisms cause the coexistence of greater species numbers by more means than niche partitioning, and not all of these mechanisms are understood.

Interspecific competition can cause the evolution of major breakthroughs and innovations. These would be followed by adaptive radiation, resulting in many new species.

Competition with dinosaurs probably contributed to the evolution of mammals by forcing their ancestors to live as small, mainly nocturnal insectivores. A nocturnal lifestyle likely caused selection favoring the development of fur and higher metabolic rates [50]. It also could have selectively favored endothermy.

Competition for light has selected for taller trees in the rainforest, where trees can reach 150 feet in height. This is generally between species, since two rainforest trees in close proximity are rarely the same species.

Competition from birds for the diurnal niche led to bats evolving into night-flying animals, with many adaptations to a nocturnal niche, such as sonar. As a result, a new adaptive zone was invaded by bats, which then radiated into two suborders and over 1,200 species.

REFERENCES

1. Gause, G. F. (1934). *The Struggle for Existence*. Williams and Wilkins: Baltimore, MD.
2. Schluter, D. (2000). *The Ecology of Adaptive Radiation*. Oxford University Press: Oxford, UK.
3. MacArthur, R. H. (1 Oct., 1958). Population ecology of some warblers of northeastern coniferous forests. doi: 10.2307/1931600.
4. Lack, D. (May, 1945). The ecology of closely related species with special reference to Cormorant (*Phalacrocorax carbo*) and Shag (*P. aristotelis*). *Journal of Animal Ecology* 14 (1): 12–6.
5. De León, L. F., et al. (18 April, 2014). Darwin's finches and their diet niches: the sympatric coexistence of imperfect generalists. *Journal of Evolutionary Biology* 27 (6): 1093–104. doi: 10.1111/jeb.12383.
6. Diamond, J. M. (1975). Assembly of species communities. In Cody, M. L. & Diamond, J. M. (eds.), *Ecology and Evolution of Communities*, pp. 342–444. Harvard University Press: Cambridge, MA.
7. McGuire, J. A., et al. (5 May, 2014). Molecular phylogenetics and the diversification of hummingbirds. *Current Biology* 24 (9): 1038. doi: 10.1016/j.cub.2014.03.016.
8. Falk, J. J., et al. (7 June, 2015). Sensory-based niche partitioning in a multiple predator–multiple prey community. *Proc. of the Royal Soc. B. Biological Sciences*. doi: 10.1098/rspb.2015.0520.
9. Kronfeld-Schor, N. & Dayan, T. (1999). The dietary basis for temporal partitioning: food habits of coexisting *Acomys* species. *Oecologia* 121: 123–8. doi: 10.1007/s004420050913.
10. Scriven, J. J., et al. (28 Jan., 2016). Niche partitioning in a sympatric cryptic species complex. *Ecology and Evolution*. doi: 10.1002/ece3.1965.
11. Bovbjerg, R. V. (1 March, 1970). Ecological isolation and competitive exclusion in two crayfish (*Orconectes virilis* and *Orconectes immunis*). *Ecology* 51 (2): 225–36. doi: 10.2307/1933658.
12. Miller, R. S. (1 April, 1964). Ecology and distribution of pocket gophers (Geomyidae) in Colorado. *Ecology* 45 (2): 256–72. doi: 10.2307/1933839.
13. Connell, J. H. (1961). The influence of interspecific competition and other factors on the distribution of the barnacle *Chthamalus stellatus*. *Ecology* 42: 710–23.
14. Harbison, C. W., et al. (Nov., 2008). Comparative transmission dynamics of competing parasite species. *Ecology* 89 (11): 3186–94. doi: 10.1890/07-1745.1.
15. Slatkin, M. (1980). Ecological character displacement. *Ecology* 61: 163–77.
16. Taper, M. L. & Case, T. J. (1992). Coevolution among competitors. *Oxford Surveys in Evolutionary Biology* 8: 63–109.
17. Schluter, D. (2000). *The Ecology of Adaptive Radiation*. Oxford University Press: Oxford, UK.
18. Losos, J. B. (2000). Ecological character displacement and the study of adaptation. *PNAS USA* 97: 5693–5.
19. Dayan, T., & Simberloff, D. (2005). Ecological and community-wide character displacement: the next generation. *Ecology Letters* 8: 875–94.
20. Grant, P. R. (1986). *Ecology and Evolution of Darwin's Finches*. Princeton University Press: Princeton, NJ.
21. Pyke, G. (1982). Local geographic distribution of bumblebees near Crested Butte, Colorado; competition and community structure. *Ecology* 63: 555–73.
22. Adams, D. C., & Rohlf, F. J. (2000). Ecological character displacement in Plethodon: biomechanical differences found from a geometric morphometric study. *PNAS USA* 97: 4106–11.

23. Adams, D. C. (2004). Character displacement via aggressive interference in Appalachian salamanders. *Ecology* 85: 2664–70.
24. Schluter, D. & J. D. McPhail (1992). Ecological character displacement and speciation in sticklebacks. *The American Naturalist* 140: 85–108.
25. Sidorovich, V., et al. (1999). Body size and interactions between European and American mink (*Mustela lutreola* and *M. vison*) in Eastern Europe. *Journal of Zoology* 248: 521–7.
26. Stuart, Y. E., et al. (2014). Rapid evolution of a native species following invasion by a congener. *Science* 346: 463–6.
27. Brown, W. L., Jr. & Wilson, E. O. (1956). Character displacement. *Systematic Zoology* 5: 49–64.
28. Losos, B., et al. (1 May, 1997). Adaptive differentiation following experimental island colonization in *Anolis* lizards. *Nature* 387: 70–3.
29. Irschick, D. J. & Losos, J. B., (Feb., 1998). A comparative analysis of the ecological significance of maximal locomotor performance in Caribbean *Anolis* lizards. *Evolution* 52 (1): 219–26.
30. Losos, J. R., et al. (Mar. 27, 1998). Contingency and determinism in replicated adaptive radiation of island lizards. *Science* 279: 2115–8.
31. Leal, M., et al. (Nov. 17, 1998). An experimental study of interspecific interactions between two Puerto Rican *Anolis* lizards. *Oecologica* 117 (1/2): 273–8.
32. Mahler, D. L., et al. (Sept., 2016). Discovery of a giant chameleon-like lizard (*Anolis*) on Hispaniola and its significance to understanding replicated adaptive radiations. *The American Naturalist* 188 (3): 357–64. doi: 10.1086/687566.
33. Losos, J. B. (March, 2001). Evolution: a lizard's tale. *Scientific American* 284 (3): 64–69. doi: 10.1038/scientificamerican0301-64. PMID: 11234508. Secondary literature.
34. Beardmore, R. E., et al. (21 April, 2011). Metabolic trade-offs and the maintenance of the fittest and the flattest. *Nature* 472: 342–6. doi: 10.1038/nature09905.
35. Finke, D. L. & Snyder, W. E. (12 Sept., 2008). Niche partitioning increases resource exploitation by diverse communities. *Science* 321 (5895): 1488–90.
36. Strong, D. R. J. (1982). Harmonious coexistence of hispine beetles on *Heliconia* in experimental and natural communities. *Ecology* 63 (4): 1039–49. doi: 10.2307/1937243. JSTOR 1937243.
37. Shmida, A. & Ellner, S. (1984). Coexistence of plant species with similar niches. *Vegetatio* 58: 29–55.
38. Rastetter, E.B. & Ågren, G.I. (Dec., 2002). Changes in individual allometry can lead to coexistence without niche separation. *Ecosystems* 5 (8): 789–801.
39. McGuire, J. A., et al. (5 May, 2014). Molecular phylogenetics and the diversification of hummingbirds. *Current Biology* 24 (9): 1038. doi: 10.1016/j.cub.2014.03.016.
40. Turner, G. F., et al. (2001). How many species of cichlid fishes are there in African lakes? *Molecular Ecology* 10: 793–806.
41. Seehausen, O. (2000). Explosive speciation rates and unusual species richness in haplo-chromine cichlid fishes: effects of sexual selection. *Advances in Ecology Research* 31: 237–74. doi: 10.1016/S0065-2504(00)31015-7.
42. Hutchinson, G. E. (1961). The paradox of the plankton. *The American Naturalist* 95: 137–45. doi: 10.1086/282171.
43. Károlyi, G., et al. (2000). Chaotic flow: the physics of species coexistence. *PNAS USA* 97: 13661–5.
44. Wiggert, J. D., et al. (2005) The role of feeding behavior in sustaining copepod populations in the tropical ocean. *Journal of Plankton Research* 27: 1013–31.
45. Miyazaki, T., et al. (2006). Spatial coexistence of phytoplankton species in ecological timescale. *Population Ecology* 48 (2): 107–12.

46. Descamps-Julien, B. & Gonzalez, A. (2005). Stable coexistence in a fluctuating environment: an experimental demonstration. *Ecology* 86: 2815–24. doi: 10.1890/04-1700.
47. Scheffer, M., et al. (2003). Why plankton communities have no equilibrium: solutions to the paradox. *Hydrobiologia* 49: 9–18.
48. Hallam, A. & Wignall, P. B. (1997). *Mass Extinctions and their Aftermath*. Oxford University Press: Oxford, UK. ISBN 978-0-19-854916-1.

7 Plants Are Ecosystem Engineers That Aid Other Life and Are Linked to It

Major contributions of trees, plants, and other photosynthetic organisms to biodiversity such as O_2 production, carbon sequestration, providing habitat for nitrogen-fixing bacteria, providing carbohydrate to mutualistic fungi and helper bacteria, and coevolution and mutualism with pollinators and seed dispersers are discussed in other chapters.

Plants provide the carbohydrate through photosynthesis that is the basis of all terrestrial ecosystems on Earth's surface. Phytoplankton are the basis of the sea's food webs, including most of the deep sea. So photosynthetic organisms are the engines that run essentially all of Earth's food webs and ecosystems, except such things as sea vents and underground microbial systems.

Plants were major drivers of diversity of the insects that fed on them. The order Coleoptera is by far the most diverse order of animals and has the majority of insect species. The most diverse animal suborder is by far the major suborder of Coleoptera, Phytophaga, and within that suborder is easily the most diverse animal family, the Curculionoidea (weevils). More than 99% of the more than 110,000 described species of Phytophaga are plant feeders. Phytophaga is the oldest and largest radiation of herbivorous beetles. Farrell [1] reconstructed its phylogeny from 115 complete DNA sequences for the 18S nuclear ribosomal subunits and from 212 morphological characters, to interpret the role of angiosperms in Coleoptera diversification. He found several origins of angiosperm-feeding lineages of Coleoptera associated with enhanced rates of beetle diversification, indicating a series of adaptive radiations spurred by coevolution with angiosperms, collectively representing almost half of the species in the Coleoptera. So angiosperms coevolved with and facilitated the adaptive radiation of many herbivorous beetles. Ehrlich and Raven [2] showed that coevolution between plants and their insect herbivores, mainly butterflies and angiosperms, resulted in great diversification of both groups, perhaps even in the origin of the angiosperms, and hypothesized that insect diversity is intimately tied to the rise and diversification of the angiosperms in the Cretaceous.

A substantial portion of the rain in all forests is produced by the trees themselves, via evapotranspiration. The trees take in water from the soil through their roots, and it ascends through their xylem for use in their metabolic processes. Some of this water is expelled through the stomata in their leaves, and evaporates, returning to the air as water vapor. Some humidifies the forest, and some rises and forms clouds, which return the water to the forest as rain when conditions are right. This is an important source of water for the forest. Thus, trees and forests aid their own survival and are self-sustaining through an autocatalytic process. In fact, forests of all types would not have as high diversity as they do if it were not for the rainfall and moisture they create themselves. And amazingly, trees that experience drought produce chemicals that attract water vapor, which adheres to them, increasing water available to the plants.

Tropical rainforests are the most biologically diverse terrestrial ecosystems on the planet. The health of these forests, their high diversity, and even their existence are dependent on high, reliable seasonal rainfall. A large percentage of this rainfall is created by evapotranspiration. If a large area of rainforest is clear-cut, that area can become a desert with very low rainfall. One rainforest tree pumps out about 260 gallons of water per day. In the Amazon rainforest, at least half the precipitation comes from evapotranspiration, while the other half results from water evaporated and blown inland from the Atlantic Ocean [3]. Here, 400 billion trees exude 20 billion tons of water per day in the dry season, when there is actually more evapotranspiration than in the wet season. The trees work as a group. Convection currents carry some of the water from evapotranspiration westward, until some of it is precipitated as snow on the Andes. Some of this snow melts and plays a significant role in replenishing the Amazon River system.

The evapotranspiration also creates a huge "river" of water in the sky, larger than the Amazon River; in fact, it is the largest river on Earth. This flows over the forest to the Andes, which send it south, where it travels 2,000 miles, dropping rain as it goes. It causes ecosystems that would otherwise be deserts to be fertile plains. It provides the water for the Pantanal, making it the largest, most diverse wetland on Earth, with the highest visible wildlife concentration in South America. There are similar sky rivers created by evapotranspiration from the rainforests of the Congo Basin and southeast Asia. These also provide water to distant ecosystems.

All plants carry out evapotranspiration, but trees have the most effect on local rainfall because they are generally bigger than other plants, and hence emit more water. Redwood forests are among the most magnificent ecosystems in the world, with trees of stunning size. These forests need foggy, humid conditions to thrive, and this includes the redwood trees and many of the other plant species in these forests. The redwoods cause rain and moisture through evapotranspiration. In addition, these trees are the tallest in the world, and very thick and massive. The two California species are the Coast Redwood (*Sequoia sempervirens*), which reaches a maximum height of 367 feet, and a maximum diameter of 22 feet; and the Giant Sequoia (*Sequoiadendron giganteum*), which reaches a maximum height of 311 feet, and a maximum diameter of 40 feet. Both of these species, which tend not to occur together, but in forests of one species or the other, are therefore effective wind breaks, preventing wind from building up within the forest, between the trees. Their large size also means any fog or moisture within the redwood forest tends to stay where it is for a much longer time. The result is the redwood trees create and retain a great deal of fog. Their shady branches also hold moisture in the forest, and keep the lower portions of the forest cool by blocking the sun's rays. All of this greatly benefits the redwood forest and the redwood trees in an autocatalytic way. Redwood trees and redwood forests require foggy conditions to thrive.

It appears that trees regulate rainfall in such a way that the optimal amount of rain falls for the specific forest type they are in. However, this has not yet been proven, and needs to be tested further.

Evapotranspiration also cools the environment by converting a great amount of liquid water taken from the soil to water vapor in the atmosphere, which carries latent heat away from its point of origin. This generally helps species of all taxa in both temperate and tropical forests by keeping temperatures from getting too high. And forest canopies can trap water vapor, creating a dense fog that keeps the myriad plant species alive during dry seasons.

Trees also prevent the temperature from getting too cold. Medvigy et al. [13] used a variable-resolution general circulation model (GCM), a model that indicated that deforestation reduces simulated precipitation in the Amazon, but much less than seen in most previous GCM studies. Precipitation was redistributed, with the northwest Amazon becoming drier and the southeast Amazon becoming wetter. Large changes in June–August hydroclimate were also found, with extreme cold events becoming more common. The authors said the changes have consequences for ecosystems and surface hydrology.

Medgivy et al. [14] also showed with a variable-resolution GCM that deforestation in the Amazon induces large changes in the frequency of wintertime extreme cold events. They found big increases in the frequency and intensity of cold events in the western Amazon and, surprisingly, in parts of southern South America, far from the actual deforested area. They propose a possible mechanism for these remote effects, that the temperature changes in the Amazon change the position of the subtropical jet stream.

Forests also regulate temperature and wind patterns through ecosystem engineering by altering the albedo. Forests make large areas of the earth's surface green, which is usually darker than their soil, generally decreasing the albedo of the Earth, causing less energy to be radiated back into space, and more heat to be absorbed. This could offset some of their cooling effects that result from removing CO_2 from the atmosphere. Less often, when the ground is very dark, they increase albedo with respect to the ground. Rainforests are favorable to life in this respect, since albedo affects patterns of atmospheric circulation and rainfall, and the albedo of rainforests generally stabilizes these. One of the many problems of the clear-cutting of rainforest by humans is that it increases the albedo in the tropics. This could potentially adversely affect weather at a distance from the cut rainforest, but more research is needed to show this definitively.

Betts (2000) looked at the relative generally warming effect of albedo change and cooling effect of carbon sequestration caused by planting new forests. He found that new forests in tropical and mid-latitude areas tended to cool, while new forests in high latitudes such as Siberia were neutral or perhaps warming (ibid). This indicates forests could potentially decrease temperature differences between tropical and mid-latitude areas on the one hand and high latitudes on the other, possibly stabilizing climate and decreasing storms and winds that are generated in part by temperature differences in different parts of the Earth.

Boreal forests stabilize local climate by lowering albedo caused by snow cover. Snow does not cover the trees as readily as it does the ground, so winter albedos of treeless areas are 10% to 50% higher than nearby forested areas. This decreases the temperature in treeless areas, and affects atmospheric circulation. This makes recovery slow in areas where forest was removed. Albedo increases can last for 100 years.

The decrease in temperature from the increased reflectivity is deleterious to life because boreal forests are in cold high-latitude areas. Thus, boreal forests decrease albedo from snow cover, keep their temperatures from getting unfavorably low, and possibly help maintain themselves by this mechanism.

In areas where the ground is dark, such as mangrove forests, albedo decreases and temperatures rise if vegetation is removed. When Hurricane Wilma partially defoliated more than 2,400 sq km of a mangrove forest in the Florida Everglades in 2005, it exposed an underlying land surface with a lower albedo. This effectively doubled the warming impact of released CO_2, and affected atmospheric circulation. This shows that mangrove forests regulate and lower local temperature. This may create a more favorable temperature for them in the subtropical heat, so they may help maintain themselves.

Grasslands increase albedo, cooling the Earth, greatly aiding life and increasing global diversity. There is positive feedback whereby grasslands maintain themselves. When grasslands increase albedo, cooling the atmosphere where they occur, this reduces local cloud cover, decreasing local rainfall, favoring grasslands over forests (Hoffman and Jackson, 2000), since grasslands do better in dry conditions and forests in wet ones. They also have less evapotranspiration than forests, making their ecosystem drier, which helps them compete with forests. The dryness also increases fire, which grasses are more resistant to than forests, so this maintains a mosaic of grasslands and forests, instead of only forests. In short, grasslands help maintain themselves by creating conditions that are favorable to grasslands.

Forests also aid cloud formation. Except at extremely low temperatures, moisture in the air does not form into water droplets, which can aggregate into clouds, unless there is a tiny solid core for moisture to aggregate around. This core could be mineral dust, soot, salt, sulfur dioxide from volcanoes, or other particulates. Terrestrial vegetation emits large amounts of volatile organic compounds into the air, which on oxidation produce secondary organic aerosols (SOA). SOA act as cloud condensation nuclei, influencing cloud formation and climate [15]. Scientists have known for a while that the many organically rich particles floating above the Amazon rainforest in a haze act as condensation cores for droplet formation and rainfall there. Now, field studies have shown that the trees of the rainforest provide these cores [16]. They emit microscopic potassium-rich salt particles into the atmosphere, which act as the condensation nuclei for cloud formation. Thus, in addition to evapotranspiration, forests provide the seeds for cloud formation and rain.

Inducing cloud formation and precipitation by this mechanism is not limited to plants. Bauer et al. [17] showed both Gram-positive and Gram-negative bacteria could act as cloud condensation nuclei. Hailstones have thousands of bacteria per milliliter at their cores, supporting the bioprecipitation hypothesis that bacteria stimulate precipitation. Bacteria have protein coatings that cause water to freeze at relatively warm temperatures. They may have evolved to use hail and snow for dispersal. *Pseudomonas syringae* has a cell surface that aligns water molecules in an ordered fashion, and thus acts as a nucleation site, stimulating the formation of ice at temperatures much higher than normally required. It is effective at this. The ice it forms is likely also adaptive for the bacterium in that it can damage plant cell walls, allowing the bacterium to infect and draw nutrients from the cells. The bacteria congregate and form aerosols in forest canopies. The aerosols can ascend during updrafts, eventually stimulating condensation in clouds at temperatures far higher than would be required if soot or dust served as the nucleation sites. The net effect is beneficial to the forest because the bacteria increase precipitation, but not beyond a favorable level. Although this is not an effect by plants, I include it here because it is a local climatic effect by life, and better fits here than in other parts of this book.

Rainforests affect climate and rainfall in areas far from them. Numerical simulation experiments showed deforestation in the Amazon not only locally reduces precipitation, evapotranspiration, and cloudiness, it also reduces rainy season precipitation in several other regions of the world far from the Amazon, including Mexico, Texas, and the Gulf of Mexico [18]. It also results in globally averaged precipitation deficits (ibid). Destruction of forests in Central Africa affects precipitation patterns in the upper and lower U.S. Midwest. Clear-cutting rainforests around the equator has also caused massive storms remote from the cutting.

Further, Avissar and Werth [19] showed that deforestation in the Amazon and Central America severely reduces rainfall in the lower U.S. Midwest during the spring and summer seasons and in the upper U.S. Midwest during the winter and spring, respectively, and that deforestation of southeast Asia affects China and the Balkan Peninsula significantly. Yet, deforestation of any of these tropical forests greatly enhances summer rainfall in the southern tip of the Arabian Peninsula. The combined effect of deforestation of these three tropical regions also causes a significant decrease in winter precipitation in California.

Medvigy [20] investigated simulated effects of deforestation in the Amazon. He found a redistribution of rainfall in the Amazon, accompanied by vorticity and thermal anomalies. In high-resolution simulations, these anomalies resulted in 10%–20% precipitation reductions for the coastal northwest United States and the Sierra Nevada, and declines of snowpack in the Sierra Nevada of up to 50%. But in coarse-resolution simulations, precipitation was not reduced in the northwest United States. He concludes that the deforestation of the Amazon can act as a driver of climate change in some regions in the extratropics, including areas of the western United States.

Another mechanism by which plants aid life is by their effect on the albedo. The albedo is the reflectivity of the Earth's surface. It has profound effects on temperature and weather. In 2005, Hurricane Wilma partially defoliated more than 2,400 sq. km of a mangrove forest in the Florida Everglades, exposing underlying land with a lower albedo than the mangrove forest. This effectively doubled the warming impact of the CO_2 released from mangroves killed by the hurricane [21]. Thus, mangroves have a higher albedo than the land surface, and have a cooling effect, making the temperature more favorable to life. Grasslands also increase albedo, cooling the Earth.

However, forests generally have a lower albedo than their underlying soil, which causes an increase in atmospheric temperature. Forests at mid to low latitudes make large areas of the Earth's surface green,which is darker than their soil, generally decreasing the albedo. On the other hand, water vapor from evapotranspiration in forests can increase albedo when it condenses into clouds. In boreal and other forests at 45 degrees latitude and above, albedo is 10% to 50% higher in deforested areas than areas with intact forest because snow on the ground is highly reflective, and snow does not readily remain on trees,which are darker and have a lower albedo than snow [22]. This warming of high-latitude regions by trees lessens the temperature difference between low-latitude and high-latitude regions, possibly decreasing extreme climate events and thus aiding life, but this is not known for certain. Since recovery is slow in cold regions, albedo increases from deforestation there can persist for 100 years.

The sun is about 30% hotter than when life began, and life kept Earth at temperatures favorable to life by sequestering carbon, cooling the Earth. So one might expect the lowering of the albedo and hence raising of the temperature by forests and other vegetation to be unfavorable to life and counter to the ABH. In fact, the positive effects on life of the decreased albedo induced by forests and other vegetation easily outweigh the negative ones, because this decreased albedo helps ensure adequate local rainfall. An increase of 10% in regional albedo has been associated with a 20% decline in rainfall events connected with thunderstorms. Equivalent reductions in both evaporation and transpiration have also been reported in areas with sudden increases in albedo. We know this from observing the effects on albedo and rainfall resulting from the removal of forests and other vegetation. The greatest changes in albedo occur in regions undergoing deforestation and desertification. Reductions in vegetation can lead to albedo increases of up to 20%. Desertification of the Sahel of Africa during the 20th century caused the albedo to increase from 14% to 25%. This coincided with a 40% decrease in rainfall.

Rainforest trees may play a role in stabilizing the climate globally by decreasing the albedo. Alterations in rainfall and wind far from rainforests may be partially caused by the increase in albedo due to clear-cutting of rainforest. But more research is needed to show this definitively.

The physical presence of vegetation, including its contribution to surface heterogeneity, is another factor that affects heat flux and rainfall. Convection and thunderstorm development tend to occur earlier in the day in heterogeneous landscapes. A little less than 1% of the solar energy reaching Earth that is not reflected into space accelerates the air, generating winds. An equal amount of energy must eventually be lost, or else wind speeds would perpetually increase. Heterogeneities in the geological terrain and in vegetation convert the kinetic energy to heat via friction, controlling wind speeds. Without vegetation, terrestrial systems would have much less friction against the wind, and wind speeds would be almost twice as fast. Marine winds approaching Great Britain are decelerated to about half their original speed because of the friction of the landscape's surface, both from its geology and its vegetation, shortly after the winds make landfall. Excessive wind speeds would negatively impact life. Reduced vegetation and hence reduced surface roughness on the Indian subcontinent has apparently resulted in a weaker monsoon and reduced rainfall, which is highly deleterious to the ecosystems dependent on monsoons.

Plants and their mycorrhizal fungi brought with them rapid modes of biochemical breakdown of rock, breaking rock into smaller pieces, and increasing weathering rates of surface rocks such as granite, basalt, and limestone by an order of magnitude, building soil. Tree roots and their fungi also break up soil particles, aerating the soil, and making passageways for water flow. As soils weather, the dissolution of primary minerals forces plants to recycle nutrients, and they have therefore evolved to do so efficiently. Phosphorus, calcium, potassium, iron, and various other cations must be weathered from rocks before they are incorporated into organisms. Though this is sometimes done abiologically, plants and their mutualistic fungi are important in the weathering that releases these nutrients, making them available to microbes, fungi, and plants, and, through these organisms, to other organisms. Plants are thus essential for ecosystem function for this as well as the other reasons listed in this chapter. Nutrients from this weathering are needed for key biogeochemical processes, from primary production to nitrogen fixation to decomposition, on both the land and in the sea.

Dead trees and plants and fallen leaves decompose, releasing nutrients into the soil, fertilizing and building it. When plants colonized land about 470 mya, it dramatically altered Earth's terrestrial surface. It led to rapid production of soils, including an order of magnitude increase in the rate of clay mineral production. The abundance of clay minerals and the rate of formation of soils increased vastly, providing habitat for more and larger plants and fungi. Plants and their fungal allies have manufactured and improved soil for life ever since their rise. There is a positive feedback whereby plants and fungi produce soil and make habitat for more plant and fungal species, which build yet more soil, and so on, until all ecological space is occupied. As a result of this process, animals, other fungi, other plants, protists, prokaryotes, and viruses diversified, evolving to exploit the niches and resources the forests provided.

Plant roots and their fungal allies also hold soil in place, preventing erosion. Otherwise, soil would be washed and blown away, the plants would die, and the ecosystem would collapse. They also concentrate and purify water, and make pores and air spaces for water and air flow through the soil. Prokaryotes, soil algae, fungi, lichens, invertebrates, and burrowing vertebrates help plants build soil. Soil invertebrates and burrowing vertebrates also create air spaces for air and water passage through soil. Volcanoes also help build soil.

In tropical rainforests, trees on the forest's edge act as a wind break, protecting the other trees from being blown over. As long as the forest is a minimum size, the trees on the edge will not be blown down by the wind. If a rainforest is cut down to the point where it reaches a size lower than a threshold, the wind blows over the trees on its edge, and this process can continue until the forest is gone. Also, the forest dries out and starts to die from the edges if the forest is too small, because a critical number of trees are needed to maintain sufficient moisture for tree survival. So trees on the edge of a rainforest also protect the forest from drying out and gradually disappearing. Rainforests are thus self-maintaining systems as long as they are above a minimum threshold size.

On Easter Island, when the trees were cut, the wind blew away topsoil, and salt spray blown in from the sea increased the salinity of the soil to the point where its ability to support a forest was greatly compromised. This is what generally happens when trees are removed from oceanic islands, where forests protect the soil from wind-blown sea salt in addition to all the standard ways forests protect their soil.

Trees and shrubbery on river banks stabilize them, hold soil in place, and prevent sedimentation of rivers. Sedimentation harms fish and invertebrates, and degrades fish spawning habitat. Suttle et al. [4] experimentally manipulated fine bed sediment in a northern California river and examined responses of juvenile salmonids and the food webs supporting them to determine more precisely how much sedimentation affects them. Increased deposited fine sediments negatively impacted survival and growth of juvenile Steelhead Trout. These declines were associated with a shift in invertebrates toward burrowing forms that were not available as prey. Higher sedimentation increased injury to Steelheads. They found no threshold below which increased delivery of fine sediments was harmless to salmonids. So riverine vegetation keeps rivers clear, aiding survival of invertebrates and fish. It is reasonable to assume this also allows amphibian larvae to thrive (in pools where the river water is still). Trees also shade streams, lowering their water temperatures. Lower temperature allows water to absorb more O_2. This allows more species of animals to live in the stream. Salmonids are an example of fish that need the lower temperature and higher O_2 provided by forest shade.

When plants, especially trees, anchor soils and maintain river banks, this causes rivers to meander in snake-shaped forms and to maintain a good depth and water volume. Without plants, rivers would be like the braided, fan-like alluvial fans at the base of desert mountains and in front of glaciers [12]. These would be shallow and often without water. This would greatly reduce the diversity of freshwater fish, amphibians, invertebrates, other animals, and plants currently in rivers, and in turn land animals that eat any of these species, such as Grizzly Bears and Bald Eagles. Of course, in the cases in which the alluvial fans had no water, there would be no aquatic life.

Forests are very effective at filtering and cleaning water, keeping impurities out of streams, lakes, and ground water [5,6]. Two-thirds of U.S. clean water comes from precipitation that is filtered underground through forest root systems and soil, and then enters into streams [6]. Root systems of trees and other plants, as well as their mycorrhizal fungi for those that have them, keep soils porous, so water can flow through various soil layers which filter out toxins, sediment, and other impurities before entering ground water, streams, lakes, or ponds.

The roots of some herbaceous plants and shrubs growing in rocky mountains in Brazil that have very shallow, nutrient-poor soil use fine hairs and acids to dissolve rock and extract phosphorous [7]. The soil there is poor, with nearly undetectable levels of plant nutrients. Some plants survive on rocky patches with no soil. The poor soil comprises less than 1% of Brazil's land area, yet sustains 5,000 plant species—15% of Brazil's vascular plant species—due to this adaptation. These plants break down rock and help form soil, as lichens do. They thus directly and indirectly help many animal, plant, and microbial species.

Leaf litter provides nitrate for denitrifying bacteria. These undergo aerobic respiration in the presence of O_2, but if enough soil litter is present to block out O_2 (or if the bacteria are deep enough in the soil), they will convert the nitrate to nitrogen gas, releasing it into the air. Nitrate is a useable form of nitrogen for plants, and nitrogen-fixing bacteria produce it in sufficient quantities for plants. But if there is an excess of nitrate in the soil, it gets washed into streams, ponds, and lakes, causing eutrophication. Denitrifying bacteria prevent this. By these mechanisms, trees, denitrifying bacteria, nitrogen-fixing bacteria, and other bacteria in the nitrogen cycle, keep levels of nitrate in the soil at favorable levels for life, using negative feedback.

Trees block sunlight, shading the soil and keeping it from drying out. Soil of cleared tropical forest gets exposed to sunlight, dries out, and can become hard and impenetrable by water. Much of the water evaporates because it is exposed to the sun, and rain is not absorbed by the soil, but runs off, often causing floods. Reduced evapotranspiration from cutting of forests reduces rainfall, causing drought, and exacerbating desiccation of the soil. Dry forests burn more easily. Rainforests normally have only limited fires, but tropical deforestation has led to major fires in rainforests in Mexico, Brazil, Africa, and Indonesia; the latter caused major air pollution. Of course, human-induced climate change also dries out rainforests, and is another cause of these fires. The stress of desiccation and drought can make forests more vulnerable to fungal infections and insect outbreaks. Again, the forest creates a habitat that is hospitable to it. The shade of taller trees helps species of shorter plants and trees that are harmed by too much sunlight, heat, or dryness grow and thrive. This shade benefits many animal species that rest, sleep, and keep cool under these shade trees.

Many species of plants, including essentially all species of trees, provide a three-dimensional habitat, often with platforms, utilized by a multitude of species of squirrels, bats, birds, lizards, insects, other invertebrates, fungi, epiphytes, protists, prokaryotes, and viruses. Ecosystems of tens of thousands of species of invertebrates, especially insects, as well as numerous species of birds, monkeys, sloths, lizards, snakes, frogs, epiphytes, fungi, bacteria, and other taxa, make up the rainforest canopy. In fact, the habitat provided by living organisms that has the most species per unit area is either the rainforest canopy or coral reef. Erwin [11] estimated from his study of a Panamanian rainforest that one hectare of seasonal forest may have over 41,000 species of arthropods in its canopy. He concluded from this study that there could be as many as 30 million tropical, terrestrial Arthropod species extant globally, far exceeding the usual estimate of 1.5 million at the time of the study. Rainforest trees also have diverse, complex ecosystems on their tree trunks and in their roots. Temperate forests house ecosystems in their canopies, on their trunks, and in their roots, of thousands of species of insects and other invertebrates, as well as many species of birds, epiphytes, fungi, microbes, viruses, and mammals, including bats, squirrels, Fishers, and Pine Martins. Temperate, deciduous forests are also habitat for numerous species of springtails (phylum Arthropoda; subphylum Hexapoda; class Entognatha; subclass Collembola), which display niche partitioning with changing assemblages of species as one samples from below the soil to the leaf litter and up the tree to the canopy [8]. In temperate deciduous forests, leaf litter and vegetation typically support 30–40 species of springtails, and in the tropical forests, the number may be over 100 [9]. They are among the most abundant of macroscopic animals, with estimates of 100,000 individuals per square meter of ground [10], and less on tree trunks and in the canopy. They are detritus eaters, and important decomposers and recyclers for the community on the tree. Birds nest and roost in trees in all forests, with tremendous diversity in the tropics. Some nest in temperate climates and migrate to the tropics in winter, occupying trees in both locations. Entire ecosystems live in the root systems of plants, some eating the roots. Cicada nymphs live among and suck the xylem of the roots of oak, cypress, willow, ash, and maple, and the adults also drink tree sap from above ground. Various species of ungulates, cats, and other taxa use tree shade to rest in and keep cool. Various rodent, reptile, and other species burrow under plants or use them for shade. For example, kangaroo rats make burrows under mesquite in the desert.

In fact, the vast majority of full-grown trees in natural environments support small ecosystems. The oak is a good example. In addition to the gall wasps discussed in Chapter 5, any single adult oak tree will support an almost incalculable number of moths, beetles, acorn weevils, and flies that eat various parts of the tree and/or use it as a place to live. Acorns are an especially popular food for invertebrates, many species of which spend their entire lives, generation after generation, on the same tree. These invertebrates are in turn eaten by large numbers and many species of spiders, predatory beetles, and birds. A great diversity of species of mites in great numbers feed on the tree, parasitize insects on the oak, and feed on oak fungus. And a number of species of fungus may grow on the oak. These are found in crevices and cracks in the bark, under and on top of leaves, and in the trees' flowers. Thousands of springtails of several species eat the dead invertebrates on the tree. Several bird species are attracted to the invertebrates as a food source in forests, and many use the tree for shelter or nest sites. At least one and up to many species of each of the following bird taxa eat the invertebrates on oaks: vireos, kinglets, woodpeckers, chickadees, warblers, the Brown Creeper, and several others.

Vertebrates that eat acorns and disperse them in a mutualistic relationship with oaks of various species include Wild Turkeys, woodpeckers, Bobwhite Quail, Mallard Ducks, Wood Ducks, American Crows, jays, Acorn Woodpeckers, Common Tree Squirrels, Fox Squirrels, California Ground Squirrels, flying squirrels, rabbits, mice, voles, raccoons, North American Opossums, wild hogs, White-tailed Deer, Red Foxes, and Gray Foxes. Some of these acorn eaters are eaten by various predators that indirectly benefit from the oaks, such as Coyotes and Mountain Lions.

Golden Eagles (*Aquila chrysaetos*) and various species of hawk and owl nest or temporarily roost in select trees, and eat the small animals such as squirrels and birds that frequent the oaks. Sharp-shinned Hawks hunt small birds that forage in the oaks, such as various species of vireos and kinglets. Bobcats and Gray Foxes climb to rest in the shade of the tree's branches, safely well above the ground, and hunt various bird species there. Turkeys, deer, and small mammals use the oaks for shade and shelter.

Note I discussed biodiversity effects of a temperate zone tree taxon. The effects would generally be just as immense for any temperate tree taxon. For tropical trees, the positive effects on biodiversity would be much higher for essentially any tree taxon, especially considering the great number of arthropods in tropical forest canopies and greater diversity of tropical than temperate forests. Additionally, there are many more species of trees per acre in the tropics than temperate regions, with each tropical tree uniquely aiding a huge number of species.

Many lichens, even some that do not grow on trees, are dependent on the shade of the forest canopy to have a sufficiently cool and moist microclimate to survive. Since lichens provide food and habitat for many species, and some fix nitrogen, trees indirectly aid several species by helping lichens.

The evolution of the plant leaf is a vibrant example of coevolution and life's capacity to generate biodiversity. Leaves cover 75% of the Earth's land surface, and are the key organ for photosynthesis, fueling the entire aboveground terrestrial ecosystem [12]. Dates are imperfect, but about 425 mya, strange, simple vascular land plants lacked leaves, and photosynthesized with their stems. They had a complexity intermediate between mosses and the true vascular plants, which include trees. Various fossil plant finds show this [12]. Over the next 65 million years, from about 425 to 360 mya, during the Paleozoic Era, plants underwent an explosion of innovation and diversification often compared to the Cambrian Explosion in animals [23]. From a simple body plan of but a few cells, proto-land-plants became transformed, developed a blueprint for modern plants, and evolved complex, sophisticated body plans and life cycles. These body plans included roots, but not yet leaves.

As explained in Chapter 4, roots and their fungal allies cause weathering reactions between minerals and CO_2, sequestering CO_2 from the system, helping keep temperatures cool enough to favor life. Roots also hold soil in place, slowing erosion, giving minerals more time to be dissolved by rainwater, increasing weathering, which causes CO_2 sequestration. Even when plants had no leaves, some debris above ground accumulated, forming a moist, acidic environment that weathered soil minerals. By these processes, the plant explosion resulted in plants sequestering tremendous amounts of atmospheric CO_2, which, chemical analyses of fossil soils and fossils show, plummeted by 90% between 400 and 350 mya, corresponding closely to the plant diversification explosion [24,25]. This drop in CO_2 levels is unprecedented in the last 500 million years, and led to such low CO_2 levels and a weakened greenhouse effect that a major ice age occurred, with massive glaciers that spread from the South Pole to the tropics. The low CO_2 stressed plants, since it made photosynthesis more difficult. Stomata are microscopic pores on plants that allow CO_2 to enter plant cells for photosynthesis, and through which water exits the plant. If there is plenty of CO_2, plants can afford to keep the number of stomata low to conserve water, but if CO_2 is low, plants need to increase their number of stomata in order to photosynthesize [26]. In this environment, leaves would be a big advantage, because leaves are flat structures with large surface areas, allowing large numbers of stomata. They also provide a large surface area to catch the sun's light, especially useful when CO_2 is low. So it appears leaves did not appear when CO_2 was high, and appeared as a result of selective pressures for a larger area for photosynthesis and more stomata in low CO_2 air. Low CO_2 also selected for the leaves to increase in size, allowing for a further increase in the area for photosynthesis and in the number of stomata. But larger leaves cannot release their heat in response to wind as well as small ones. Larger leaves create greater friction, slowing the air flow over the leaf, keeping it warmer. Stomata help cool the leaf by expelling water. This was another advantage of the increased number of stomata that became important when the Earth returned to higher temperatures than those in this ice age. Also, the lower temperatures during this ice age due to the reduced atmospheric CO_2 allowed the plants to evolve larger leaves during a time when the need to keep cool was not as much of a problem as at times of higher CO_2 and thus hotter temperatures.

The most important point here is that plants caused their own macroevolution of a key innovation and diversification by lowering CO_2 levels, resulting in selection for the leaf, in a sequential process with feedback leading to its evolution. That is, plants drove their own evolution, resulting in a major key innovation. This led to tremendous plant diversification. Plants did cause a crisis for life and a decrease in species richness by greatly lowering global temperatures, but this was temporary, and the adaptive radiation of plants and their beneficiaries as a result of the appearance of leaves resulted in a great increase in diversity that more than made up for the diversity loss, by a great amount.

This scenario is supported by considering the alternative that CO_2 does not regulate stomata number, and early land plants evolved large leaves with many of these pores, and hence efficient cooling of the leaf, in high CO_2 air. But plants still had primitive roots and transport systems before CO_2 levels dropped. Theoretical calculations show that, in high CO_2 and temperatures, water would exit the stomata-rich leaves of such plants ten times as fast as their primitive roots and transport systems could supply and replenish it [12]. In fact, the demand for efficient water and nutrient uptake and transport in the plant caused by increasing leaf size and stomata number and the need for an efficient cooling system created a selection for better root and shoot systems. So these evolved in tandem with the increase in size and sophistication of leaves and increase in stomata number. Root, shoot, and leaf necessarily coevolved together, each putting selective forces on the others to progress [27]. Once leaves, shoots, and roots evolved to a sufficient level of sophistication, competition between neighboring plants for light and to avoid being shaded out intensified. It is well documented in the fossil record that, in response to this competition, plants got taller and leaves became larger [28,29]. This continued until forests of tall, leafy trees became a dominant feature of the land worldwide by 360 mya, at the end of the Devonian Period and start of the Carboniferous Period. The coevolved set of key adaptations of larger and better functioning roots, shoots, and leaves; increased stomata number; and increased size resulted in tremendous adaptive radiation of higher plants into myriad species of shrubs and trees.

This was followed immediately by the diversification of a great number of new species of animals, epiphytes, fungi, protists, prokaryotes, and viruses that evolved the ability to directly and indirectly exploit the niches created by these new shrub and tree species for food and habitat. And the new species of plants affected animal evolution, atmospheric composition, and the Earth's geology profoundly. Plants from cacti to trees provided a food source and vertical dimension of habitat, often with stable platforms, for a multitude of animal species. This vertical dimension of habitat with platforms was instrumental in the evolution of birds and their flying ability from their theropod dinosaur ancestors. Coevolution with pollinators and seed dispersers, carbon sequestration and as a result keeping the planet at a temperature cool enough to be favorable to life, and the many geoengineering services discussed in this chapter are examples of mechanisms for the tremendous increase in the number of species in essentially all major taxa that resulted from the spectacular story of the evolution of higher plants.

Plants allow animal species to migrate and expand their ranges. Hominins could have traveled through northern Arabia and left Africa as early as 300,000–500,000 years ago, only because it was covered in lush grassland as early as then. Stone tools and probable cut marks support these dates [30]. The presence of animal prey, also dependent on the grass, was also needed for this migration. Animals could not have migrated across Beringia from Asia to the Americas or across the Isthmus of Panama (in both directions) without the presence of plants.

Plants and nonplant photosynthesizers are major components of ecosystems that aid life and increase diversity. Seagrasses, kelp forests, and wetlands such as mangrove forests and salt marshes all capture and hold CO_2 for centuries. They also protect coastal habitats against sea level rise and storms. They act as buffers against wind, rain, and water surges, protecting land. They reduce flood damage by absorbing water. They filter pollutants, cleaning the water in their ecosystems. Wetland plants capture sediment carried by seawater, using it to build habitat, so they can thrive and expand. They support countless animals.

Forests are self-sustaining ecosystems in which trees act as ecosystem engineers and alter the physical environment to their benefit in an autocatalytic process and positive feedback loop. By creating rain, protecting soil and preventing erosion, building and fertilizing soil, protecting the forest from wind, providing food and habitat to animals, regulating greenhouse gasses, harboring nitrogen-fixing bacteria, producing O_2, being mutualistic with root fungi, and other functions, they create conditions that favor them and allow them to thrive. And plants in other ecosystems, such as grasslands and even deserts, are self-sustaining ecosystem engineers that alter their habitat to their benefit by performing many of these functions.

Grasslands are a great example of plant ecosystems that maintain themselves. They are competitors with forests. Grasses reduce rainfall compared to forests, because trees undergo more evapotranspiration than grasses. Droughts cause gaps in forests that allow C4 grasses to grow. Grasses provide combustible fuel, increasing fires that burn grasslands and forests. Grasses are resistant to fire, because their underground roots called rhizomes do not burn, and allow them to quickly grow back. So grasses maintain themselves in a positive feedback loop [31]. Fire also adds huge numbers of aerosol particles to the air. Thousands of water droplets must collide to form a drop heavy enough to fall as rain. Aerosols in fire smoke reduce the size of water droplets to the point where they cannot fall as rain. And the smoke from fires blocks some sunlight from reaching the ground, slowing evaporation from the soil and drying the air. Smoke makes the air darker, so more able to absorb heat and warm up. Hotter air can hold more moisture without dropping it as rain. Thus, the air is drier and yet able to hold more moisture, so precipitation is less over grasslands. So grass-induced fires reduce rain, favoring grasslands in another positive feedback loop. Both grasses and trees sequester carbon, lowering CO_2 levels. Low CO_2 in the air slows the growth of tree seedlings, decreasing the chance that they will reach the height needed for their fire-resistant properties, such as thick bark, to develop. C4 grasses utilize CO_2 more efficiently, hence can survive well in low CO_2. Thus, C4 grasses help themselves thrive, lower CO_2, and conserve carbon. Forests create rain through evapotranspiration, and thus help maintain themselves through positive feedback in the face of these challenges from grasslands. Forests are more diverse than grasslands, but two ecosystems result in higher diversity than only the more diverse forest existing. The result is that life creates a complex mosaic that is more diverse than if either system achieved a monopoly. Here, life involved in a positive feedback loop increases diversity, rather than decreasing it. And a planet with an amount of fire closer to what is optimal for life is created. I will summarize the feedback loops as follows. Grasses make the air drier and decrease rain. This increases grasses. The drier conditions and combustibility of grasses cause more fire, and this helps grasses over forest. More fire makes it drier and decreases rain, favoring more fire and grasses. Trees increase rainfall, helping trees. The mosaic of grasslands and forests is also maintained by herbivores, such as American Bison and elk, which eat tree seedlings, and maintain their grassland habitats. Elephants are most notable for destroying trees and maintaining their grassland habitats. Without such herbivores, there would be much more forests and very few grasslands where these animals live.

For the first three billion years of life on Earth, life in the ocean survived on the slow trickle of rock-derived nutrients arriving via rivers, dust, and volcanic activity [31]. But by 400 mya, there were great numbers of plants on land [32]. Evolving a terrestrial lifestyle allowed plants to access more nutrients from rock-derived nutrients from weathering rocks than their oceanic ancestors could access. The earliest land plants were restricted to wet areas. Plants evolved two mechanisms to be able to colonize the rest of the land. First was the mutualism with fungi whose microscopically fine hyphae could probe into small areas for water and nutrients and thereby accelerate the transformation of the land surface into one that is predominantly mantled by soil [33]. Second were woody root systems that could pry deeply into soils, break apart rocks, and capture water and nutrients. This molded primary minerals into clays, which hold water and nutrients more effectively. These two adaptations resulted in plants playing a major role in the formation of soil. Further, their ecosystem engineering sent nutrients from the soil to the sea, aiding the evolution and diversification of bacteria and plankton in the sea. This aided the evolution and diversification of the rest of the food webs that depend on these organisms. The adaptation of plants and their fungi to obtain nutrients from rocks had secondary effects of releasing nutrients for terrestrial and ocean ecosystems, helping build soil, and sequestering carbon and cooling the Earth. All these benefited life and increased diversity tremendously.

As pointed out in Chapter 2, after angiosperms appeared, they released large amounts of nutrients to the sea, causing the evolution of the four major types of phytoplankton still extant today, These replaced the less complex phytoplankton and diversified. The result was higher plankton diversity and so higher diversity at all levels of the sea's food webs, up to large fish and sharks. Angiosperms also release a large amount of pollen to the sea, where it is used as a nutrient.

Plants make up bout 83% of the total biomass of all life on Earth, about 550 gigatons. Since the soil, geology, rainfall, life-friendly temperature, animals, fungi, and bacteria help and shape plants just as much as plants help and shape all of these entities, there is a mutualistic coevolution between plants and the atmosphere, geology, and biology. The contributions of higher plants to increasing biodiversity is tremendous. And there may well be more mechanisms by which higher plants increase diversity that we are still unaware of. Individual higher plants, especially trees, are complex, interacting, diverse ecosystems intimately interconnected to the biosphere and geosphere, greatly increasing biodiversity, and are not autonomous organisms, individuals, or systems separate from the Earth's biological and geological systems. Plants are ecosystem engineers that are the base of terrestrial food webs, create and stabilize soil, produce O_2, sequester CO_2, provide habitat to tremendous numbers of species, and provide other ecosystem services.

REFERENCES

1. Farrell, B. D. (July 24, 1998). "Inordinate Fondness" explained: why are there so many beetles? *Science* 281 (5376): 555–9. doi: 10.1126/science.281.5376.555.

2. Ehrlich, P. R. & Raven, P. H. (Dec., 1964). Butterflies and plants: a study in coevolution. *Evolution* 18 (4): 586–608.

3. Salati, E. (1987). The forest and the hydrological cycle. In Dickinson, R. E. (ed.), *The Geophysiology of Amazonia*, pp. 273–96. John Wiley & Sons: New York.

4. Suttle, K. B., et al. (2004). How fine sediment in riverbeds impairs growth and survival of juvenile salmonids. *Ecological Applications* 14 (4): 969–74.

5. Barten, P. K. & Ernst, C. E. (April, 2004). Land conservation and watershed management for source protection. *Journal AWWA* 96: 4.

6. Smail, R. A. & Lewis, D. J. (2009). *Forest Land Conversion, Ecosystem Services, and Economic Issues for Policy: A Review*. PNW-GTR-797. U.S. Department of Agriculture, Forest Service, Pacific Northwest Research Station: Portland, OR.

7. Sales, G., et al. (May, 2019). Specialized roots of Velloziaceae weather quartzite rock while mobilizing phosphorus using carboxylates. *Functional Ecology* 33 (5): 762–73. doi: 10.1111/1365-2435.13324C.

8. Ponge, J.-F. (1993). Biocenoses of *Collembola* in Atlantic temperate grass-woodland ecosystems. *Pedobiologia* 37 (4): 223–44.

9. Hopkin, S. P. (1997). *Biology of the Springtails: (Insecta: Collembola)*. Oxford University Press: Oxford, UK, p. 127. ISBN 978-0-19-158925-6.

10. Ponge, J.-F., et al. (1997). Soil fauna and site assessment in beech stands of the Belgian Ardennes. *Canadian Journal of Forest Research* 27 (12): 2053–64. doi: 10.1139/cjfr-27-12-2053.

11. Erwin, T. (1982). Tropical forests: their richness in Coleoptera and other arthropod species. *The Coleopterists Bulletin* 36 (1): 74–5.

12. Beerling, D. (2007). *The Emerald Planet. How Plants Changed Earth's History*. Oxford University Press: Oxford, UK.

13. Medvigy, D., et al. (April, 2011). Effects of deforestation on spatiotemporal distributions of precipitation in South America. *Journal of Climate*. doi: 10.1175/2010JCLI3882.1.

14. Medvigy, D., et al. (June, 2012). Simulated links between deforestation and extreme cold events in South America. *Journal of Climate*. doi: 10.1175/JCLI-D-11-00259.1.

15. Prenni, A. J., et al. (26 May, 2007). Cloud droplet activation of secondary organic aerosol. *Journal of Geophysical Research: Atmospheres* 112 (D10). doi: 10.1029/2006JD007963C.

16. Pöhlker, C., et al. (31 Aug., 2012). Biogenic potassium salt particles as seeds for secondary organic aerosol in the Amazon. *Science* 37 (6098): 1075–8. doi: 10.1126/science.1223264.

17. Bauer, H., et al. (16 Nov., 2003). Airborne bacteria as cloud condensation nuclei. *Journal of Geophysical Research: Atmospheres* 108 (D21). doi: 10.1029/2003JD003545.

18. Werth, D. & Avissar, R. (27 Oct., 2002). The local and global effects of Amazon deforestation. *Journal of Geophysical Research: Atmospheres* 107 (D20): LBA 55-1–LBA 55-8. doi: 10.1029/2001JD000717.

19. Avissar, D. & Werth, D. (April, 2005). Global hydroclimatoloigal teleconnections resulting from tropical deforestation. *Journal of Hydrometeorology* 6: 134–45.

20. Medvigy, D. (Nov., 2013). Simulated changes in Northwest U.S. climate in response to Amazon deforestation. *Journal of Climate*. doi: 10.1175/JCLI-D-12-00775.1.

21. O'Halloran, T. L., et al. (2011). Radiative forcing of natural forest disturbances. *Global Change Biology*. doi: 10.111/j.1365-2486.2011.02577.x.

22. Betts, R. A. (2000). Offset of the potential carbon sink from boreal forestation by decreases in surface albedo. *Nature* 408 (6809): 187–90. Bibcode: 2000Natur.408..187B. doi: 10.1038/35041545.S2CID4405762.

23. Kenrick, P. & Crane, P. R. (1997). The origin and early evolution of plants on land. *Nature* 389: 33–9.

24. Mora, C., et al. (1996). Middle to late Paleozoic atmospheric CO_2 levels from soil carbonate and organic matter. *Science* 271: 1105–7.

25. Ekart, D. D., et al. (1999). A 400 million year carbon isotope record of pedogenic carbonate: implications for paleoatmospheric carbon dioxide. *American Journal of Science* 299: 805–27.

26. Woodward, F. I. (1987). Stomatal numbers are sensitive to CO_2 increases from preindustrial levels. *Nature* 327: 617–8.

27. Raven, J. A. & Edwards, D. (2001). Roots: evolutionary origins and biogeochemical significance. *Journal of Experimental Botany* 52: 381–401.

28. Chaloner, W. G. & Sheerin, A. (1979). Devonian microfloras. *Special Papers in Palaeontology* 23: 145–61.

29. Byfield, L. (1990). *An Oak Tree.* Collins Book Bus: London, UK. Collins Educational. ISBN 0-00-313526-8.

30. Roberts, P., et al. (29 Oct., 2018). Fossil herbivore stable isotopes reveal middle Pleistocene hominin palaeoenvironment in 'Green Arabia'. *Nature Ecology & Evolution* 2: 1871–8.

31. Hoffman, W. A., et al. (2002). Positive feedbacks of fire, climate, and vegetation and the conversion of tropical savanna. *Geophysical Research Letters* 29. doi: 10.1029/20002GL015424.

32. Canfield, D. E., et al. (2010). The evolution and future of Earth's nitrogen cycle. *Science* 330: 192–6.

33. Beerling, D. J., et al. (2001). Evolution of leaf-form in land plants linked to atmospheric CO_2 decline in the Late Palaeozoic era. *Nature* 410: 352–4.

34. Field, K. J., et al. (2015). Symbiotic options for the conquest of land. *Trends in Ecology and Evolution* 30: 477–86.

8 Herbivores Generate Biodiversity

The great amount many herbivores consume makes them important regulators of plant populations. Each adult elephant eats between 149 and 169 kg (330–375 lb) of vegetation per day. Herds usually have about 10–20 elephants. Sometimes several herds merge into super herds of 100 or more. Even a small herd of 10 can eat 1,500 kg in one day! They eat grasses, small plants, bushes, fruit, twigs, tree bark, and roots.

One adult Hippopotamus consumes 150 pounds per day. They mainly eat on land, a diet of almost entirely grasses, with some shrubs and ferns. Large pods can have 100 Hippos. More typically, a pod is one male and about ten females.

One adult rhinoceros eats about 100 pounds a day. The Black Rhinoceros eats 40% grass; for the rest, it strips leaves and other vegetation from bushes and shrubs. The White Rhinoceros eats almost all grasses. Where they occur together, the two species coexist by niche partitioning. They had substantial populations before human impacts.

One grown White-tailed Deer eats about seven pounds per day, or about 2,500 pounds per year. They eat legumes, shoots, leaves, prairie forbs, acorns, mushrooms, and grasses. Twenty deer per square mile could destroy a forest. Natural populations are somewhat less dense than this.

All of these herbivores eat tremendous quantities of food, having enormous ecological impacts. They prevent plant populations from completing competition, which would result in better competitors eliminating poorer ones. The reduced number of plant species would cause a decrease in animal species, because many animals depend on plants for various services, such as food.

The intermediate disturbance hypothesis (IDH) states that species diversity is highest at intermediate levels of disturbance. At low levels of disturbance, interspecific competition decreases diversity, while at high levels, the disturbance drives species locally extinct. The happy medium between the two maximizes species richness. This hypothesis has supporters and critics, evidence for and against it, and needs more research to test it. But when all is taken into account, the majority of the evidence supports it. For a discussion of how diversity can be increased by intermediate disturbance, environmental heterogeneity, systems not at equilibrium, and other ways, see [1].

Herbivores generally tend to disturb and consume at reasonably intermediate levels because they are controlled by predators and disease by negative feedback, keeping their numbers somewhat balanced at an intermediate level. Lubchenco [2] studied the effects of the herbivorous snail, *Littorina littoria*, on the diversity of algae in tide pools on the New England coast. In pools with low snail density, there were as little as one species of alga, the best competitor, the green alga, *Enteromorpha intestinalis*, which eliminated all other algal species. In pools with intermediate snail density, there was the highest algal diversity, with sometimes as many as ten or more algal species. In pools with very high snail density, the snails ate all their preferred algae, leaving only the inedible red alga, *Condrus crispus*, and encrusting corraline algae. This supports the IDH. Since the snail generally grazes at intermediate levels under normal conditions, this supports the idea that herbivores increase the species richness of their prey.

Grazing by the freshwater snail *Physella* resulted in the diversity of its prey algal species being highest at low to intermediate grazing levels, and lowest at high grazer densities [3]. Intermediate levels of grazing by snails resulted in the highest diversity in unenriched and nitrate-enriched environments, while no significant change as a result of grazing level was found in phosphate-enriched environments [4].

Van Klink et al. [5] looked at 24 studies and found support for the IDH and even that intermediate grazing by large herbivores in grasslands increases the variety of abiotic conditions: "In general, heterogeneity in vegetation structure and abiotic conditions increases at intermediate grazing intensity, but declines at both low and high grazing intensity". Of course, a greater variety of abiotic conditions would tend to provide more niches for greater species richness. However, their study found evidence against the Autocatalytic Biodiversity Hypothesis (ABH) in that they found that arthropod diversity was generally negatively impacted by herbivores.

Related to the IDH and herbivores eating more of the best competitors is the Janzen-Connell hypothesis [6,7], a widely accepted explanation for the maintenance of tree species diversity in rainforests, dating from the 1970s. It states that herbivores, seed-eaters, and tree pathogens (fungi, bacteria, and viruses) attack those seeds and young trees closest to the parent, preventing trees of one species from congregating. It states that herbivores, mainly insects, and pathogens that specialize on a tree species make the area surrounding the tree producing seeds inhospitable to seedlings. Seedling survival is thus a function of distance from the parent tree. The farther the seedling is from the parent tree and the lower the density of trees its species, the higher the probability the seedling will escape herbivory and disease. Connell [7] showed that adult trees have a deleterious effect on smaller trees of their species, and that seed predation was greater on seeds near adults of the same species than those near adults of others. He posited that every tree species has host-specific enemies that attack it and any of its offspring which are close to the parent. Thus, herbivores and pathogens are crucial in halting the formation of single-species groves, which is an important mechanism by which one tree species could exclude others by interspecies competition. Thus, herbivores increase tree species richness by consuming seedlings near the parent tree, allowing other tree species to grow there. This hypothesis and observations supporting it also demonstrate the need of seed dispersal for trees. Since animal dispersers are more effective than wind dispersal, the interdependence of species and importance of mutualism are also implied. And herbivores therefore aid plant species richness in an additional way because seed dispersers tend to be herbivores.

The hypothesis was meant to apply to tropical forests, but evidence subsequently was found that showed it applies to temperate forests too. An example that supports it in temperate forests is that Black Cherry Tree (*Prunus serotina*) seedlings closer to their parents are denser, with the result that they have more herbivores and pathogens, such as *Pythioum* [8], and a greater likelihood of being eaten or diseased as a result. Petermann et al. [9] showed soil-borne pests caused results similar to those proposed by the Janzen-Connell hypothesis in temperate grasslands. In addition to Janzen's observations that led to his proposing it and Connell's research, the majority of over 50 studies testing it support the Janzen-Connell hypothesis. For two good studies exemplifying this, see [10,11]. And Bagchi et al. [12] found treating seedlings with fungicides significantly reduced diversity, supporting the hypothesis by suggesting the fungi are increasing diversity by attacking the seedlings closest to the parent plant. The hypothesis is not universally accepted, and there are studies that contradict it, at least partially [13,14]. But most of the evidence and a majority of researchers support it.

It is generally agreed, and there is evidence to support this, that herbivores increase the diversity of plant communities by eating the dominant plants that are the superior competitors [15–18], just as predators do to their animal prey. This is adaptive for the herbivore because it saves it energy to consume the most abundant plant species, and it maintains a greater variety of food sources. Doutt [19] and Janzen [6] have hypothesized that herbivores could be a major reason for the large number of tree species in tropical compared to temperate forests. Janzen argued that herbivores would selectively eat abundant species, allowing less common species to grow. This thesis has support. For example, rubber trees in their native Amazon Basin cannot be grown as dense monocultures because diseases and herbivores attack them. But they grow as dense monocultures in Malaysia, where their natural enemies are not present. Attempts to maintain monocultures of other tree species in the tropics have met with a similar failure. In these cases, it is not only herbivores, but pathogens, that attack the monculture. Both increase tree species diversity.

White-tailed Deer increase herbaceous plant diversity by selectively eating competitively dominant species [20]. They also increase the growth rate of Northern Red Oak (*Quercus rubra*), a canopy tree that supports many species, possibly by increasing nutrient inputs into the soil with their feces [21].

Altieri et al. [22] manipulated the abundance of a dominant herbivorous snail, *Littorina littorea*, in natural tide pools. Seaweed species evenness and biomass-specific primary productivity (mg O_2/h/g) were higher in tide pools with snails because snails preferentially consumed an otherwise dominant seaweed species, *Ulva lactuca*, which can reduce biomass-specific productivity rates of algal assemblages. By preferentially eating the dominant competitor among the seaweed species, the snail increased seaweed productivity and species evenness, but not richness. High and low snail densities similarly increased algal species evenness. Seaweed species identity and thus richness were similar across all levels of snail grazing. But the tide pools with snail grazers had higher species diversity because species diversity is defined as a combination of species richness and species evenness. Productivity was higher where the snail grazed because *Ulva* dominated in its absence, and *Ulva* uses bicarbonate as a carbon source. This raises pH and reduces inorganic carbon levels, causing a fivefold reduction in the photosynthesis rates of the other seaweeds. *Ulva* also inhibits its own photosynthesis and that of other seaweeds because its sheet-like morphology limits light penetration.

Rabbits grazing a small island off England's coast were eliminated by the introduction of myxomatosis virus. The number of plant species decreased dramatically, as the better competitors eliminated the poorer ones without the herbivore to control the former [23].

Mortensen et al. [24] demonstrated that herbivores did indeed maintain plant prey species richness in a tallgrass prairie restoration by limiting temporary pulses in dominance by a single species. Dyer et al. [25] looked at the effects of specialist and generalist herbivores on local seedling diversity in a Costa Rican rainforest with dominant understory *Piper* shrubs. After 15 months, plots in which herbivores had been removed had over 40% less species and over 40% less seedlings on average than control plots with herbivores. However, the plots without herbivores averaged 40% greater seedling evenness. One interpretation of this decrease in species evenness is that high seedling mortality in dominant families due to herbivory allowed the colonization or survival of less common species. This would render the decrease in species diversity due to the decrease in species evenness less meaningful. Increases in herbivore densities on these shrubs resulted in herbivores moving from *Piper* to seedlings of many different plant genera, causing widespread seedling mortality and lower species richness, so high herbivore density decreases plant diversity, supporting the IDH.

Herbivores can also increase plant diversity by eating plants before they reach carrying capacity and complete the competitive process. Charles Darwin did an experiment in which he had his gardener periodically mow his lawn, which had many species of weeds growing in it [26]. The mowing simulated an herbivore. The mowing caused the plant diversity in his lawn to increase by preventing the plants from completing the competitive process, adding evidence to the hypothesis that herbivores increase and maintain plant diversity.

Bakker et al. [27] showed that assemblages that include large herbivores increased plant diversity at higher primary productivity, but decreased it at low primary productivity. Borer et al. [28] found that the loss of plant diversity caused by increased nutrients allowing some plant species to shade others out was alleviated by herbivores. By eating plants that would have shaded others out, thus alleviating light limitation, they increased plant species richness.

Herbivory is crucial in maintaining coral reef health and diversity. Hawkins and Roberts [29] studied six Caribbean islands that ranged in fishing pressure from almost none to very high intensities. They found that algae covered coral most where fishing was highest. Too much algae on coral kills it. Amount of coral and structural complexity were highest on little-fished islands and lowest on those most fished. The results showed that where herbivorous fish were abundant enough to control algal growth on coral, there was higher species richness. The authors said that their results suggest that following the Caribbean-wide mass mortality of herbivorous sea urchins in 1983–1984 and consequent declines in grazing pressure on reefs, herbivorous fishes failed to control algae overgrowing corals in heavily fished areas, but have restricted algal growth in lightly fished areas. Before the sea urchin die offs, herbivorous sea urchins controlled algal growth on corals, increasing species richness on reefs. Seaweeds uncontrolled by herbivores can overgrow and kill corals, often excluding corals from temperate latitudes, where herbivores generally fail to control seaweeds. Hay and Stachowicz [30] showed the omnivorous crab, *Mithrax forceps*, consumes and removes seaweeds and invertebrates growing on or near the coral, *Oculina arbuscular*, allowing it to grow as far north as temperate North Carolina. However, this crab aids the coral not only by acting as an herbivore, but as a predator of predators of coral as well.

Numerous experiments in grasslands show herbivores often, though not always, increase plant diversity by mechanisms additional to those listed above [31]. There is evidence, though it is controversial, that grazing and browsing by herbivores on plants help plants by stimulating their growth. For instance, an herbivore may bite off the terminal bud. This makes lateral buds become more active. This makes a bushier plant with more flowers. Such a plant offers more resource for its herbivores and its pollinators. Pruning by humans mimics this process with similar results.

Ungulate grazers, with their nitrogen-rich urine and saliva, full of essential vitamins, aid higher plant growth. Some argue, however, that the root actually decreases, and thus total plant biomass is unchanged or even decreased due to herbivory. Grasses clearly grow more quickly in response to herbivores. American Bison (*Bison bison*) actually increase the nutritional value of grass when they graze. This result is more nutritious food for them and prairie dogs. Thus, bison both compete with and indirectly aid prairie dogs. Since prairie dogs indirectly aid bison when they help plants by aerating the soil when they dig their burrows, the relationship is both competition and indirect mutualism.

Many herbivores have a positive effect on species richness from their actions as ecosystem engineers. For example, when herds of American Bison graze, paw the ground, take dust baths, or wallow in the mud, they help create fertile prairie mosaics. In winter, snow trails they create open up grazing areas for other species. In spring, wallowing Bison create ponds employed by waterfowl, amphibians, aquatic invertebrates, and microscopic organisms. In summer, their foraging helps prevent catastrophic fires, and aids the growth of shrubs, favoring nesting sites for grouse and sparrows. In fall, their dried-up wallows are habitat for prairie dogs and plovers.

Herbivores help the movement of many animals in grasslands by controlling the height of grasses. Without herbivores, grasses would grow to be so tall and thick that they would impede the movement of insects and other invertebrates, snakes, lizards, turtles, some birds, mice, voles, rabbits, deer, and many other taxa. As a result, many of these would not be able to live in grasslands.

A study found 60% of large herbivores are at risk of extinction. The researchers said this "can have cascading effects on other species, including large carnivores, scavengers, mesoherbivores, small mammals, and ecological processes involving vegetation, hydrology, nutrient cycling, and fire regimes" [32]. Plant diversity will decrease because competition will no longer be lessened by herbivory. This will cause animals dependent on the lost plants to die out, decrease, or go extinct. For example, small mammals will suffer from loss of food and shelter. Carnivores will decrease in diversity due to loss of their herbivore prey. Scavengers will suffer due to loss of food when there are less animals to die. The White-tailed Deer alone supports several scavengers, including vultures, raptors, foxes, and corvids. Large areas of Earth will lack many of the vital ecological services herbivores provide. It will lead to an increase in the number and severity of wildfires because there will be more dense plant life, and such a large amount of fire will reduce the diversity of the system.

Herbivores interact with fire to cause higher species richness and more biofriendly fire regimes. Edwards [33] found American Bison have a pronounced effect on the composition of prairie species by balancing the competition between forbs and the dominant grasses. And he showed that grazing by Bison and Elk (*Cervus canadensis*) in combination with fire maintained the prairie by inhibiting the growth of forest species while favoring prairie species. Land that cattle grazed was invaded by Ponderosa Pines (*Pinus ponderosa*), while the bison grazing areas had few pines. It is of interest that the natural species that coevolved with the system maintained the prairie, and the human-created cow, which never coevolved with the system, did not maintain the prairie. Maintaining the prairie increases diversity because it adds another ecosystem to coexist in a mosaic with the pine forest. Further, Plumb and Dodd [34] found that, although both ate browse species, forbes, and grasses, bison ate more of the competitively superior grasses than cattle, which ate a higher percentage of browse species and forbes, the poorer competitors, than bison. Again, the natural, coevolved species maintained diversity, and the cattle decreased it.

Historically, fire and grazing disturbance regimes interact to create a mosaic of successional areas (thus more habitat types) on the prairie.

Pfeiffer and Hartnett [35] found Bison grazing reduced the density of Little Bluestem, while fire increased it. Thus, a balance between fire and grazing maintained high diversity, keeping Little Bluestem part of the system, but not allowing it to eliminate its competitors.

Pfeiffer and Steuter [36] found that rhizomatous grasses were not affected by burning alone, but were reduced if fire was combined with grazing. Hence the two together controlled this better competitor and maintained species richness.

Hobbs et al. [37] found that less nitrogen was lost from fires in tall grass prairies grazed by cattle. They estimated that grazing conserves 3 to 5 times more nitrogen than is lost in fires in ungrazed areas. Fire temperatures and energy release were reduced by grazing.

Estes et al. [38] reviewed the impact of human-induced loss of large predators and herbivores high in food webs, confirming that their decline has had tremendous, cascading effects on terrestrial, marine, and freshwater ecosystems throughout the Earth. They said this may be humankind's most pervasive effect on nature. They showed that large animals, once ubiquitous across the globe, shaped the structure and dynamics of ecosystems. They showed the loss of large predators and herbivores had profound effects not only on immediate prey species but also on the dynamics of disease, wildfires, biogeochemical and nutrient cycles, carbon sequestration, vegetation growth, invasive species, and soil and water quality. Herbivores control wildfires, preventing them from being too frequent for optimal ecosystem health. Rinderpest, a viral disease introduced by human activity in the late 1800s, decimated populations of wildebeest, buffalo, and other ungulates in the Serengeti in Africa until it was eliminated in the 1960s by a vaccination program. During the time that herbivore populations were low, plant biomass increased and woody shrubs proliferated, fueling large wildfires in the region. After the grazing species returned to normal levels, shrub lands were converted to grasslands, and the frequency and intensity of fires decreased. And wildfire frequency increased following the late Pleistocene/early Holocene decline of megaherbivores in Australia and the northeastern United States. Thus, herbivores cause fires to occur at frequencies and intensities that maximize diversity; without herbivores, there is more fire than optimal for life, and plant diversity declines, causing animals on all trophic levels to do likewise.

Herbivores have caused selection for plant toxins and other antiherbivore defenses. This has resulted in plant-herbivore coevolution (for example, [39]), with herbivores becoming resistant to the toxins, even using them for their own defense against their predators. Increased diversity in herbivores and plants resulted. This coevolution also resulted in many toxic plant chemicals; plant spines, jagged edges, and other physical deterrents to herbivory; aposematic coloration of insects that converted plant toxins to protective agents against their predators; and mimics of the these toxic insects. All of this resulted in great diversification of both herbivores and the plants they consume. Some of this diversification and its mechanisms are discussed in Chapters 4 and 5.

Herbivores have been shown to affect what chemicals roots exude into the soil, how plants allocate their carbon, and, on a longer timescale, root mass and structure. This influences the size and activity of the soil organisms and possibly nutrient supply to the root [40]. While I believe herbivores increase diversity of the soil microbes, and this enhances the health of the plants, more research is needed to fully establish this. More root mass allows greater carbon sequestration while conserving it for the ecosystem, since roots collectively store substantial amounts of carbon.

One form of coevolution between plants and their insect herbivores is an antagonistic pattern termed escape and radiate. This is exemplified by the leaf beetle genus *Blepharida* and its host plant, *Bursera*. When the first *Blepharida* species to eat *Bursera* first appeared, it radiated, since it had no competition from other *Blepharida* species. *Bursera* species soon developed chemical defenses to deter the *Blepharida* species, allowing them to escape and radiate, free from the danger and stress posed by *Blepharida*. However, eventually some *Blepharida* developed a resistance to the chemical defense, allowing them and their progeny to colonize the new *Bursera* species and radiate.

Many herbivores, such as leafcutter ants, termites, the Long-spined Sea Urchin, certain species of parrotfish, some tortoises, gophers, prairie dogs, beavers, and elephants are ecosystem engineers that greatly enhance their habitats, as discussed in appropriate areas in this book. I will discuss a few now in this chapter to emphasize this point.

Large herbivores are important as tillers of the soil. When herds of large herbivores, such as Bison or wildebeest are threatened by predators, they pound the soil with their hooves, then run, digging up the soil. Both these actions break up and aerate the soil, making it much more favorable for plant growth. Here, it is the behavior of predators and their herbivore prey that produce healthy, tilled soil.

Herbivores can make mosaics of habitats. In the African savanna, the larger herbivores, especially the elephants and rhinoceroses, destroy trees. They eat and trample young ones, knock over and trample large ones, eat tree bark, and make room free of excess shade and aerate the soil for the grass species. They dig up and till the soil by walking, and especially by fleeing predators. Without these animals, most to all of the savannah would turn into woodland. Essentially all savannah species would die, including mice, ungulates, and grasses. Like American Bison, elephants and rhinos increase biodiversity by causing two ecosystems—grassland and forest, rather than only forest—to coexist, creating a mosaic of habitats. Each of these ecosystems has several species the other lacks.

Barnosky et al. [41] did an analysis that showed that the Pleistocene mammoths and mastodons maintained grasslands, and hence a mosaic of grassland and forest ecosystems, greatly increasing diversity, because two ecosystems existed instead of but one. The proboscidians ate and trampled trees and ate their saplings, preventing forests from overgrowing grasslands. Also, large herbivores added and distributed nutrients through their movement and defecation, helping the grasslands grow. The grasses were the food of the large herbivores, so grasslands and herbivores were interdependent, and the grasslands and herbivores maintained each other through mutualism. Thus, the herbivores maintained their habitat. Also, megaherbivores kept the savanna from becoming tall grassland, which would have had more frequent and fierce fires. Keeping the grass shorter allowed movement and hence survival of many vertebrate species. When climate and/or hunters killed off the large mammals, the grasslands greatly decreased, causing a great loss of diversity in a cascade of extinction. The late Pleistocene extinction of huge herbivores weighing over 2,200 pounds dramatically changed vegetation patterns, increasing plant species the herbivores had eaten, and decreasing or eliminating their competitors. The loss of megaherbivores is likely why there was a greater loss of plant species at the end of the last Ice Age than at earlier glacial melts. The loss of plants likely caused extinctions and/or emigrations of the many smaller herbivores that ate the plants that decreased or died out.

The migratory behaviors and grazing of wildebeests maintain the open African grassland habitat by inhibiting the growth of woodland and forest. They trample seedling trees, and male wildebeests habitually attack young trees with their horns as an expression of aggression, further inhibiting the regeneration of woodland. They too maintain a mosaic of savanna and forest, increasing habitat types. They fertilize the soil with their urine, feces, and corpses when they die. Their impact is huge because of their numbers.

The Hippopotamus (*Hippopotamus amphibius*) grazes on land 4–5 hours each night, each hippo consuming 68 kg (150 lb) of grass during this time. They defecate in their aqueous habitat, depositing huge amounts of organic matter along the river beds. They move large quantities of nutrients from the nearby land into rivers. Hippo dung provides nutrients from terrestrial material for fish and aquatic invertebrates [42]. This increases abundance and diversity of these taxa. Their feces can be toxic in large quantities, causing downstream fish kills, since its decomposition depletes oxygen in the water [43], although I doubt this occurs often. Over prolonged periods, hippos can divert the paths of swamps and channels.

Additionally, they take the same path to their feeding grounds and back, creating permanent trails and depressing the ground. The trails can be used by other animal species. Over time, they change the courses of swamps and channels, creating habitat for fish and invertebrates. In the distal parts of the Okavango Delta, a large wetland in northern Botswana, their regular movement to feeding grounds, which they may walk a mile to get to, creates incised channels, which they keep clear of vegetation. These act as nodes for swamp expansion. They maintain pathways that lead to the development of new channel systems during avulsion. They make gaps in the vegetation levees that flank channels in the permanent swamps, causing diversion of water and sediment to backswamp areas. Their paths often lead into lakes, and diversion of channels into lakes may occur via these paths [44]. This increasing of the complexity of the swamp system increases the number of amphibian, fish, invertebrate, and microscopic species that can use the microhabitats created. These species, and some reptile and mammal species, use the channels for transportation routes.

Herbivores also increase nitrogen and other nutrients available to roots when they defecate and die and decompose, and the amounts are significant. Small herbivores, such as voles, help increase nitrogen mineralization, aiding nitrogen cycling, as their feces are broken down by microbes to biologically useable forms like ammonium and nitrate, under natural conditions when cattle are excluded [45].

Large herbivores, such as elephants and rhinoceroses, contribute and distribute large amounts of nutrients such as nitrogen on land through their defecation. This greatly aids plant growth and hence ecosystem health. This is also true of large extinct herbivores, whose feces had tremendous positive impacts on Pleistocene ecosystems. Bison, horses, mammoths, giant ground sloths, and llamas played a critical role as nutrient pumps in the last Ice Age, distributing nutrients through eating, and their movements and defecation, across the rangelands that extended across Europe and North America. This helped the grasslands grow. Hippos also add nutrient to freshwater systems, as do large herbivorous fish. Baleen whale defecation adds large quantities of nutrients used by plankton in the ocean.

Humus from the dominant dwarf evergreen shrub *Empetrum*, which releases the allelopathic substance batatasin-III, is infertile for seedlings of local herbaceous plants in the tundra where it grows. Simulated reindeer defecation and wildfire can alleviate this infertility, allowing the growth of herbaceous species [46]. Here, herbivore defecation helps increase species richness within a relatively small plant community.

Herbivores serve as a food source for a great many predator species, and their removal causes catastrophes as devastating as when predators are exterminated, because the predators that eat and depend on them undergo precipitous population declines or extinction when they disappear.

Various species of wildebeest, zebra, gazelle, giraffe, and other herbivores of the African savanna are consumed by Lions, Leopards, Cheetahs, African Wild Dogs, crocodiles, and other predators. All of these predators would greatly decline or go extinct if these herbivores declined or disappeared. This would result in a great loss of diversity in itself. The predators would also switch to other prey that they normally do not eat, devastating their populations. Plants and trees would become less diverse because of increased competition leading to extinction of less fit competitors. Scavengers, including vultures and jackals, and decomposers would be decimated because the loss of predators means a loss of corpses. American Bison were important to wolves on American grasslands before the bison were greatly reduced by overhunting. Just one species, the White-tailed Deer, supported wolves, Mountain Lions, American Alligators, Jaguars, Bobcats, Canada Lynx, bears, Wolverines, and Coyotes, before humans reduced the populations of many of these predators. Herbivorous rodents support many species of birds of prey, snakes, Coyotes, foxes, badgers, skunks, and other species, all of which would decrease without them.

Loss of zooplankton, snails, herbivorous open ocean fish, herbivorous cryptoben-thics in coral reefs, sea urchins, or other herbivores in the sea would cause devastating loss of predators in their respective ecosystems. Herbivorous fish and invertebrates are important in freshwater food webs.

REFERENCES

1. Tillman, D. (1982). *Resource Competition and Community Structure*. Princeton University Press: Princeton, NJ.
2. Lubchenco, J. (1978). Plant species diversity in a marine intertidal community: importance of herbivore food preference and algal competitive abilities. *The American Naturalist* 112 (983): 23–39. doi: 10.1086/283250.
3. Swamikannu, X. & Hoagland, K. D. (1989). Effects of snail grazing on the diversity and structure of a periphyton community in a eutrophic pond. *Canadian Journal of Fisheries and Aquatic Sciences* 46 (10): 1698–1704.
4. McCormick, P. V. & Stevenson, R. J. (1989). Effects of snail grazing on benthic algal community structure in different nutrient environments. *Journal of the North American Benthological Society* 8 (2): 162–72.
5. van Klink, R., et al. (2015). Effects of large herbivores on grassland arthropod diversity. *Biological Reviews* 90 (2): 347–66.
6. Janzen, D. H. (Nov.–Dec., 1970). Herbivores and the number of tree species in tropical forests. *The American Naturalist* 104 (940): 501–28. doi: 10.1086/282687.
7. Connell, J. H. (1971). On the role of natural enemies in preventing competitive exclusion in some marine animals and in rain forest trees. In Den Boer, P.J. & Gradwell, G.R. (eds.), *Dynamics of Populations*. Centre for Agricultural Publishing and Documentation: Wageningen, The Netherlands.
8. Packer, A. & Clay, K. (16 March, 2000). Soil pathogens and spatial patterns of seedling mortality in a temperate tree. *Nature* 404: 278–81. doi: 10.1038/35005072.
9. Pertermann, J. S., et al. (2008). Janzen–Connell effects are widespread and strong enough to maintain diversity in grasslands. *Ecology* 89 (9): 2399–406.
10. Clark, D. A. & Clark, D. B. (1984). Dynamics of a tropical rain forest tree: evaluation of the Janzen-Connell model. *The American Naturalist* 124 (6): 769–88.
11. Mangan, S. A., et al. (8 July, 2010). Negative plant-soil feedback predicts tree-species relative abundance in a tropical forest. *Nature* 466: 752–5.
12. Bagchi, R., et al. (Feb., 2014). Pathogens and insect herbivores drive rainforest plant diversity and composition. *Nature* 506: 85–8. doi: 10.1038/nature12911.
13. Hyatt, L. A., et al. (2003). The distance dependence prediction of the Janzen-Connell hypothesis: a meta-analysis. *Oikos* 103 (3): 590–602.
14. Burkey, T. V. (1994). Tree species diversity: a test of the Janzen-Connell model. *Oecologia* 97: 533–40.
15. McNaughton, S. J. (1985). Ecology of a grazing ecosystem: the Serengetti. *Ecological Monographs* 53: 291–320.
16. Huntley, N. J. (1991). Herbivores and the dynamics of communities and ecosystems. *Annual Review of Ecology, Evolution, and Systematics* 22: 477–503.
17. Belsky, A. J. (1992). Effects of grazing, competition, disturbance and fire on species composition and diversity in grassland communities. *Journal Vegetarian Society* 3: 187–200.
18. Crawley, M. J. (1997). Plant-herbivore dynamics. In Crawley, M. J., (ed.), *Plant Ecology*, pp. 401–74, Blackwell Science Ltd., UK.
19. Doutt, R. L. (1960). Natural enemies and insect speciation. *Pan-Pacific Entomology* 36: 1–13.

20. Royo, A. A., et al. (2010). Pervasive interactions between ungulate browsers and disturbance regimes promote temperate forest herbaceous diversity. *Ecology* 91 (1): 93–105. doi: 10.1890/08-1680.1. PMID 20380200.

21. Lucas, R. W., et al. (2013). White-tailed deer (*Odocoileus virginianus*) positively affect the growth of mature northern red oak (*Quercus rubra*) trees. *Ecosphere* 4 (7): art84. doi: 10.1890/ES13-00036.1.

22. Altieri, A. H., et al. (22 April, 2009). Consumers control diversity and functioning of a natural marine ecosystem. *PLOS ONE.* doi: 10.1371/journal.pone.0005291.

23. Harper, J. L. (1969). The role of predation in vegetational diversity. *Brookhaven Symposium Biology* 22: 48–62.

24. Mortensen, B., et al. (17 July, 2017). Herbivores safeguard plant diversity by reducing variability in dominance. *Journal of Ecology* 106: 101–12. doi: 10.1111/1365-2745.12821.

25. Dyer, L. A., et al. (2010). Herbivores on a dominant understory shrub increase local plant diversity in rain forest communities. *Ecology* 91 (12): 3707–18.

26. Darwin, C. (1859). *The Origin of Species by Means of Natural Selection.* Reprinted by The Modern Library, Random House: New York, N. Y.

27. Bakker, E. S., et al. (2006). Herbivore impact on grassland plant diversity depends on habitat productivity and herbivore size. *Ecology Letters* 9: 780–8. doi: 10.1111/j.1461-0248.2006.00925.x.

28. Borer, E. T., et al. (24 April, 2014). Herbivores and nutrients control grassland plant diversity via light limitation. *Nature* 508: 517–20. doi: 10.1038/nature13144.

29. Hawkins, J. P. & Roberts, C. M. (Feb., 2004). Effects of artisanal fishing on Caribbean coral reefs. *Conservation Biology* 18 (1): 215–26. doi: 10.1111/j.1523–1739.2004.00328.x.

30. Hay, M. E. & Stachowicz, J. J. (Sept., 1999). Mutualism and coral persistence: the role of herbivore resistance to algal chemical defense. *Ecology* 80 (6): 2085–101. doi: 10.1890/0012-9658(1999)080[2085:MACPTR]2.0.CO;2.

31. Olff, H. & Mark, E. R. (1998). Effects of herbivores on grassland plant diversity. *Trends in Ecology and Evolution* 13 (7): 261–5. doi: 10.1016/S0169-5347(98)01364-0.

32. Ripple, W. J., et al. (2015). Collapse of the world's largest herbivores. *Science Advances* 1 (4): e1400103.

33. Edwards, T. (1976). Buffalo and prairie ecology. Midwest Prairie Conference. *Proceedings of a Symposium on Prairie and Prairie Restoration*, Galesburg, IL.

34. Plumb, G. E. & Dodd, J. L. (1993). Foraging ecology of bison and cattle on a mixed prairie: Implications for natural area management. *Ecology Applications* 3 (4): 631–43.

35. Pfeiffer, K. E. & Hartnett, D. C. (1995). Bison selectivity and grazing response of little bluestem in tallgrass prairie. *Jounal of Range Management* 48: 26–31.

36. Pfeiffer, K. E. & Steuter, A. A. (1994). Preliminary response of Sandhills prairie to fire and bison grazing. *Journal of Range Management* 47: 395–7.

37. Hobbs, N. T., et al. (Aug., 1991). Fire and grazing in the tallgrass prairie: contingent effects on nitrogen budgets. *Ecology* 72 (4): 1374–82 doi: 10.1371/journal.pone.0005291.

38. Estes, J. A., et al. (15 July, 2011). Trophic downgrading of planet Earth. *Science* 333 (6040): 301–6. doi: 10.1126/science.120.

39. Ehrlich, P. R. & Raven, P. H. (Dec., 1964). Butterflies and plants: a study in coevolution. *Evolution* 18 (4): 586–608.

40. Bardgetta, R. D., et al. (Dec., 1998). Linking above-ground and below-ground interactions: how plant responses to foliar herbivory influence soil organisms. *Soil Biology and Biochemistry* 30 (14): 1867–78. doi: 10.1016/S0038-0717(98)00069-8.

41. Barnosky, A. D., et al. (2015). Variable impact of late-Quaternary megafaunal extinction in causing ecological state shifts in North and South America. *PNAS USA* 113 (4): 856–61. doi: 10.1073/pnas.1505295112.

42. McCauley, D. J., et al. (2015). Carbon stable isotopes suggest that hippopotamus-vectored nutrients subsidize aquatic consumers in an East African river. *Ecosphere* 6 (4). doi: 10.1890/ES14-00514.1.

43. Dutton, C. L., et al. (2018). Organic matter loading by hippopotami causes subsidy overload resulting in downstream hypoxia and fish kills. *Nature Communications* 9 (1951): 1951. Bibcode: 2018NatCo...9.1951D. doi: 10.1038/s41467-018-04391-6. PMC 5956076. PMID 29769538.

44. McCarthy, T. S., et al. (1998). Some observations on the geomorphological impact of hippopotamus (*Hippopotamus amphibius* L.) in the Okavango Delta, Botswana. *African Journal of Ecology* 36 (1): 44–56. doi: 10.1046/j.1365-2028.1998.89-89089.x.

45. Bakker, E. S., et al. (Jan., 2004). Impact of herbivores on nitrogen cycling: contrasting effects of small and large species. *Oecologia* 138 (1): 91–101. doi: 10.1007/s00442-003-1402-5.

46. Bråthen, K. A., et al. (Aug., 2010). Ecosystem disturbance reduces the allelopathic effects of *Empetrum hermaphroditum* humus on tundra plants. *Journal of Vegetation Science* 21 (4): 786–95. doi: 10.1111/j.1654-1103.2010.01188.x.608. doi: 10.1111/j.1558-5646.1964.tb01674.x.

9 Predators and Prey, Parasites and Hosts

Mutualistic Relationships That Create High Diversity

Predators maintain and increase the biodiversity of their ecosystems. Predation is traditionally viewed as a relationship in which the predator benefits at the expense of its prey, which is negatively impacted. While the predator obviously benefits, the view that predation has only deleterious effects on the prey is overly simplistic. Reality is more complex. Predators are bad for individual prey animals that are consumed. But predators aid many of the prey animals that escape them, by reducing the intraspecific competition they would otherwise face. In fact, the predator-prey relationship is mutualism rather than the predator benefiting at the prey's expense, since the overall effect of predation on the prey population and species is positive, even though some individuals of the prey species are killed or injured. From the viewpoint of the individual prey that is eaten, the interaction is antagonistic, but from the viewpoint of the prey population or species, it is mutualistic. The reasons follow.

The most significant way that its population. This prevents the prey from over-exploiting its environment. Any unchecked population has the capacity to—and in fact will—exhaust its food supply, eating it until there is nothing, or close to nothing, left. This will cause the members of the population to starve to death. The prey will also become overcrowded, with too little space to persist. This tendency of a population to deplete its resource base is fundamental to the ideas of Malthus [1], and is one of the fundamental tenants of Charles Darwin's theory of evolution by natural selection.

Malthus correctly stated that populations grow exponentially and can out-grow their food supply if they are not regulated, An unregulated population will deplete its resources—food, water, land, and so on—from its sheer numbers that result from exponential growth imposed on finite resources. An unchecked population in exponential growth will also pollute its environment with excessive production of metabolic waste products. So any population that lacks regulation will in time decline precipitously or go extinct, from starvation, lack of space, and its waste products. Excessively large populations of organisms degrade their environment in other ways as well. For example, if there are too many hoofed herbivores, they will compact the soil beyond the capacity of other species to undo the damage, with the result that few if any plants can grow. The environmental degradation caused by an excessively high population of any species will greatly negatively impact many other species.

Predators can have tremendous impacts on their prey populations. This is well illustrated by insectivorous bat predators. During the high-energy demand time of late pregnancy, one female Little Brown Bat will eat 5.5 grams of insects, or 61% of her body weight, in one night, including flies, gnats, mosquitoes, moths, crickets, grasshoppers, beetles, and other insects. Their hibernation colonies average about 9,000 individuals and can be as large as 183,000. Mexican Free-tailed Bats (*Tadarida brasiliensis*) consume prey at about the same rate, eating moths, beetles, dragonflies, flies, true bugs, wasps, and ants. They have a colony that spends the warm months in Bracken Cave, north of San Antonio, Texas, with nearly 20 million bats eating 250 tons of insects every summer night. There are about 840 species of insectivorous bats. As a rule, insectivorous bats eat at rates not unlike the Little Brown Bat. Before white-nose syndrome, which would not be very prevalent without human-induced climate change, many of these bat species had very high population numbers. It is clear from these numbers that in natural conditions, bats have a tremendous influence on insect populations, controlling their numbers. Without insectivorous bats, insect populations would explode and deplete their food supply. As a result, many natural ecosystems from tropical to temperate regions would collapse with a precipitous decline in diversity. While not all predators are as impactful as bats, it is generally true that predators prevent catastrophic collapses of their ecosystems by regulating the numbers of their prey. Another example is: nesting murres on Seal Island in the Oshkosh Sea off the coast of Russia eat 100 tons of fish per day. The sea would have a great excess of jellyfish if it were not for predation by Leatherback Turtles. There are numerous other examples of predators that eat enormous quantities of prey, regulating the numbers of insects, rodents, large herbivores, and other prey. If these predators did not provide these ecosystem services, overpopulation of their prey would destroy the numerous ecosystems which they occupy.

Populations of organisms fluctuate due to a myriad of abiotic factors, such as temperature, humidity, rainfall, storms, and so on. Population fluctuations are also caused by biotic factors, such as predation and competition. Density independent factors, so-called because they affect population size and density regardless of population density, and have an effect whose magnitude is independent of it, include all the abiotic factors. Density dependent factors, which increase in their effect as the density of the population increases, include predation and parasitism. Density independent regulators of the population are random with respect to population size and density, and tend to cause the population to fluctuate randomly. A randomly fluctuating population has a much higher probability of going extinct than one that is subject to negative feedback. A randomly fluctuating population will go extinct with a probability on one, if given infinite time.

Density dependent factors such as predation and parasitism enforce stabilizing negative feedback on the population, increasing its likelihood of persistence. The higher the population per unit area, the greater the effect of density dependent population regulators. Predation is a density dependent factor: the denser the prey population, the greater is the dampening effect of predation. This is because high prey populations provide predators with more food, raising their population size, and thus allow more predators to exist and feed on the prey, while the converse is true for low prey populations. Predators and prey form a negative feedback loop, stabilizing each other's populations, and keeping them within reasonably narrow ranges around their mean values.

The predator responds to the prey population, increasing after the prey population increases, and causing the prey to decrease. This causes predators to decrease due to less food, and so the prey then increases. There is thus a cycle of the predator and prey populations, with the predator population lagging behind and following the prey population. This is more stable and has a higher probability of persisting longer than a prey population regulated by density independent factors. The predator population is always lower than the prey, since prey must outnumber predators to support them and keep the system going. Of course, in nature, the fluctuations are not perfectly regular, since other factors, such as weather and disease, affect both populations, but the stability is still present.

So predators make the prey populations fluctuate less randomly, more stable, and confined within fairly narrow limits, increasing the probability of prey persistence compared to randomly fluctuating, density independent prey populations lacking predators.

The time lags in the predator population's response to a change in the prey population and vice versa have the potential to destabilize the system, but this does not normally occur in nature.

Predators also tend to conserve their prey species because they can act as "prudent predators" [2]. This idea states that a predator will itself go extinct if it eats its prey to extinction, so predators tend to eat less of a prey species when the prey population is low or in danger of extinction. This is not done consciously; this behavior of the predator is selected for. The idea has empirical support from field work on tiger beetles [3].

Predators with multiple prey species maintain prey species diversity. This is because a predator will tend to eat the more abundant and successful of competing prey species. This will selectively control the prey that is the better competitor among the prey species, and help its competitor(s). This may result in a dynamic, fluctuating system in which the most abundant prey species continually changes over time, with the predator constantly indirectly aiding the less abundant species by eating the most abundant. This is adaptive for the predator, because the most abundant species is easiest to find and encountered more often, saving the predator time and energy, and because it maintains a greater variety of prey species for the predator.

Research by Paine [4] demonstrated this effect in the intertidal region of a rocky shore habitat in Makah Bay, on the coast of Washington state. The Purple Sea Star, *Pisaster ochraseus*, was the main predator, eating such species in this ecosystem as rock barnacles, gooseneck barnacles, mussels, limpets, and chitons. Paine removed all sea stars in one area, keeping it free of them, while leaving a control area undisturbed. He found after the sea star was removed, the experimental area declined rapidly from 15 to 8 species, while the control area maintained its diversity. The decline in the area where sea stars were removed was mainly because barnacles and the California mussel, *Mytilus californianus*, increased and crowded out other species. These were the best competitors. In the control area, the sea stars preferentially ate these species, controlling their numbers, and keeping high species diversity.

Paine coined the term keystone species as a result of this study. Paine [4,5] also showed that the lower portions of mussel beds smother seaweed and prevent it from growing there. But if sea stars are present, they eat the mussels at the bottom of the mussel bed, allowing the seaweed to grow, which adds diversity to the system. The seaweed is habitat for many fish and invertebrate species. Paine [4] asserted that on a geographic scale, an increased stability of annual production may lead to an increased capacity for systems to support higher-level carnivores, causing higher diversity as carnivores prevent the best competitors among their prey from eliminating their competitors.

Selective consumption of the better competitor is reinforced by search image, a phenomenon whereby the predator actively hunts and eats the most abundant prey species, in response to encountering it or them more frequently. If the most numerous prey population becomes relatively low, the predator switches to other prey species. This is well known in behavioral ecology and results from the predator's behavior. It is particularly common among bird predators. It is in the interest of the predator to seek and to eat the most abundant prey, since it encounters it more, and it is the easiest to find. This conserves the predator's energy. So selective consumption of the more abundant prey and seeking it out by search image are both adaptive for the predator.

Also, the intermediate disturbance hypothesis discussed in Chapter 8, which showed herbivores increase the diversity of the plants they eat, also applies to predators on their prey. Predators can act as intermediate disturbers. And when they keep their prey from reaching carrying capacity, which they sometimes do, competition cannot reach a point where lesser competitors are driven locally extinct.

Predators cause herbivores to stay clumped close together, because each individual herbivore benefits by having many of its kind near it, decreasing the probability that it will be caught and eaten. This concept that animals attempt to reduce their predation risk by congregating in a herd with many individuals of their species is called the selfish herd [6]. In the absence of predators, herbivores spread out much more. The crowding together of herbivores causes their hooves to till the soil very effectively when they move, whereas if they are spread out, the tilling effect is considerably less. This tilling improves the soil's quality by mixing in nutrients, such as feces; breaking up soil into finer particles; and oxygenating the soil. All these allow plants to grow better. The oxygenation also inhibits the growth of anaerobic methanogens, archaea that produce CH_4, which is a greenhouse gas 30 times as powerful as CO_2. This keeps the planet at a cooler temperature beneficial to life. When predators attack, the herbivores run away rapidly and pound the soil with their hooves, greatly increasing the tilling effect.

All trophic levels—producers, herbivores, predators, and decomposers—act together as a system to maximize the health and diversity of ecosystems. Removal of any of these trophic levels can cause the ecosystem to collapse. This is true of predators, which maintain the stability of entire ecosystems. Herbivores move from place to place if they have predators. This is adaptive, because this way, the predators cannot simply keep visiting the same area to predictably find their prey, so have a harder time locating them. The movement of herbivores allows the plants they eat to regenerate. If predators are expirpated from an ecosystem, their prey change their behavior.

They learn they do not have to fear predators, so they stay and eat in one place for much longer time periods, overgrazing and depleting the vegetation they consume locally. Populations of the plants they eat decline precipitously. This causes population crashes of the many species that depend on the plants, including the herbivores that did the overgrazing. Predators also regulate the population size of their herbivore prey, keeping them from overgrazing plants. Without predators, herbivores would decimate the vegetation they eat, causing the local habitat to collapse. The last two paragraphs show that the effect of predators on the behavior of their prey is important in maintaining ecosystem diversity.

The catastrophic disruption of an ecosystem due to the loss of a keystone predator is called a trophic cascade, because it involves multiple trophic levels and becomes amplified with dire consequences for species richness. It illustrates the effects of predators on prey behavior and populations, as well as their entire ecosystems. An example is seen in the eradication and then re-introduction of the Gray Wolf (*Canis lupus*) in the Greater Yellowstone ecosystem. Humans eradicated the wolf from this ecosystem. As a result, Elk populations exploded, and grazed in limited areas without moving around once they did not need to fear wolf predation. Their grazing eradicated such tree species as Quaking Aspen (*Populus tremuloides*), cottonwoods (*Populus spp.*), and willows (*Salix spp.*) in many local areas. The loss of these trees caused the decline of many animals and plants that depend on them, including the Elk themselves. Many species of plants, insects, other invertebrates, amphibians, reptiles, birds, and mammals were greatly reduced or disappeared.

With wolves, Elk clump together, and run when wolves appear, tilling the soil. Without wolves, these behaviors decreased, decreasing soil quality and further negatively impacting plants.

This example is one case of a general rule: predators maintain ecosystem stability, and their loss leads to a great decrease in population size of and even loss of many species. This is truer of some predators than others.

Wolves keep Coyote numbers low because they are competitors and predators of Coyotes. Without wolves, Coyote numbers approximately doubled. The Coyotes also had a much easier time catching their prey without wolves, because with more of the wolf's prey surviving and eating grass, the grass got shorter than when wolves were present. Some birds and rodents use tall grass to hide. Tall grass also provides more food for rodents, insects, and other animals than short grass. And in the presence of wolves, Coyotes are more wary and skittish, so less effective predators. Hence, Coyote's prey populations declined precipitously, as did those of several species that Coyotes did not eat, but which needed the shelter of tall grass from other predators or the food and shade it provided. Coyote prey that suffered greatly include a number of species of snakes, lizards, birds, mice, rats, squirrels, and prairie dogs. Pronghorns suffered declines from increased Coyote predation as well. The loss of reptiles and rodents caused the decline of birds of prey, such as the Osprey, hawks, eagles, and owls; ravens; weasels; foxes; and American Badgers, which also suffered because Coyotes prey on their young. Coyotes also compete with many of these latter animals. The prey of the prey of Coyotes, which includes many insect species, had population explosions that decimated their food sources. This includes various plant species. Coyotes decreased by 80% in areas with wolves when wolves were re-introduced to Yellowstone.

So the return of the wolf caused the increase of mice and rabbits, both of which are prey of Coyotes. This resulted in more weasels, badgers, foxes, ravens, hawks, Osprey, Bald Eagles, and owls. Bobcat populations exploded when the wolf disappeared because wolves compete with them, and that had similar effects to the increase of the Coyotes. Of course, Coyote and Bobcat populations greatly declined after their prey populations collapsed. Since Grizzly Bears and Black Bears steal wolf kill at times, their numbers declined with the removal of the wolf. This caused similar, but less severe trophic cascades with the bears' prey increasing and decreasing the populations of the plants and animals they eat. When trees are scarce because of the lack of wolves and abundance of deer and Elk, the courses of rivers change, because trees stabilize river banks. Without trees, the river banks erode. The resulting changes in the course of rivers is highly detrimental to fish and aquatic invertebrates, causing large declines in their populations. Thus, wolves indirectly affect the courses of rivers and life in them. Also, trees shade rivers and ponds, making them cooler. Without trees, water temperatures increase, which means the water absorbs less oxygen. The higher water temperature and lower oxygen cause some fish and aquatic invertebrate populations to decrease drastically. Multitudes of other species that benefit from the species that were negatively impacted by the loss of the wolf also declined or disappeared as the effects traveled in a domino effect throughout food webs. When wolves were removed, the entire Greater Yellowstone ecosystem collapsed, and it recovered with the re-introduction of the wolf. This trophic cascade illustrates the deep, profound interrelatedness of species in an ecosystem, and interconnectedness of nature.

North American Beavers (*Castor canadensis*) are keystone species that eat some of the tree species that are lost when wolves are removed from the ecosystem. Hence, they decline immensely when wolves are lost. Beavers add diversity by creating a new habitat of ponds in the riparian ecosystem. This creates a mosaic of two habitats, ponds and rivers, instead of just rivers. Without beavers, the ponds created from their dams are not present, and all the species that depend on beaver ponds cannot survive. So the loss of the wolf indirectly negatively impacts the multitude of species dependent on beavers. The benefits of beavers to diversity are detailed in Chapter 5.

The wolf was reintroduced to Yellowstone National Park in 1995/1996 after a 70-year absence. Ripple and Beschta [7] found Elk browsing on aspen decreased from 100%, which they had measured in 1998, to means of less than 25% in the uplands and less than 20% in riparian areas, by 2010. Growth of aspen seedlings and sprouts to above the browse level of ungulates increased as browsing decreased over time in these same stands. Cottonwood recruitment also increased. They found that Elk populations decreased, and both beaver and bison numbers increased, possibly from the increase in available woody plants and herbaceous forage resulting from less competition with Elk.

Another trophic cascade occurred when the Mountain Lion was eliminated or greatly reduced in many parts of its range due to human hunting and habitat destruction. As in the case of the wolf, if Mountain Lions are present, deer, Elk, and other prey species do not forage in one area very long, preventing them from eating out their food species, allowing aspens, willows, and other trees to recover. If the lion is removed, the prey species explode, and they remain and forage until the area is denuded of vegetation. This causes the death of many species of mammals, birds, reptiles, amphibians, and invertebrates that depend on these tree species. Species that depend on these latter animal species that are indirectly impacted also decline in a domino effect. The effect is similar to and the same principle as the case of the wolf. Ripple and Beschta [7] found a trophic cascade in which great reductions in Mountain Lions in Yosemite National Park in California led to an increase in their Mule Deer prey, which ate so many California Black Oak (*Quercus kelloggii*) seedlings that pines and firs replaced and shaded out Black Oaks as the dominant trees. Ferns and grasses, dependent on conifers, replaced various angiosperms, such as evening primroses. Plant diversity decreased sharply, and the diversity of animals aided by the plants that were negatively impacted dropped dramatically. Ripple and Beschta [8] also found that a great reduction in Mountain Lions in Zion Canyon in Zion National Park caused higher Mule Deer browsing intensities, causing reduced recruitment of riparian cottonwood trees (*Populus fremontii*), resulting in increased bank erosion. There were resulting drastic reductions in both terrestrial and aquatic species abundance.

Many scavengers and predators that scavenge that benefit from wolf and lion kills also diminished when these predators were removed. This includes vultures.

Coyotes are also keystone predators. One Coyote can eat up to 1,800 rodents in a year. If Coyotes are removed from their food web, rodents, raccoons, foxes, and, since humans introduced them, feral cats would have population explosions that completely disrupt the ecosystem, depleting various plant and small animal species, causing a trophic cascade. Some pollinators would decline, and decline or loss of certain insect and plant species would cause the species that eat them to decline greatly. Predation, mainly by Coyotes, was responsible for 88% of White-tailed Deer mortality in the Wichita Mountains in Oklahoma [9]. From this, it is clear that ecosystems can crash due to deer overpopulation if Coyotes are removed. They are also crucial controls of Black-tailed Jackrabbit populations, preventing the stripping of vegetation by these lagomorphs. In fact, wolves, Mountain Lions, Coyotes, beavers, aspen, and willows are all keystone species, many of which interact with each other. We see from this paragraph and the one a few paragraphs above it that Coyotes are very beneficial to ecosystems at their natural, intermediate population numbers, and extremely destructive at unnaturally high numbers.

Loss of insectivorous bats, keystone predators, would cause trophic cascades in many ecosystems. Mainea and Boylesa [10] used screens to deny bats access to cornfields at night, removing the exclosures by day, allowing access by birds and other pest-eating species to the corn, to be reasonably sure any effects were due to bats alone. They found 60% more Corn Earworm moth larvae (*Helicoverpa zea*), and 50% more kernel damage. They estimated the suppression of herbivory by insectivorous bats is worth more than 1 billion USD globally on corn alone. The bats also indirectly suppressed pest-associated fungal growth and fungal-produced mycotoxin on corn, because corn weakened by insect attack is more vulnerable to fungal infection. This is not a natural ecosystem, but indicates how much bats control insects and aid plants in natural ecosystems. About 25% of mammal species are bats, and 75% of bat species are insectivorous. One Little Brown Bat typically consumes millions of insects in its 40-year life. The 20 million Mexican Free-tailed Bats that summer in Bracken Cave in Texas eat 250 tons of insects every summer night. Without them, insect populations would explode, plant populations would plummet or disappear, animals dependent on these plants would decline precipitously or go locally extinct, and the ecosystem would crash in a trophic cascade. One colony of 150 Big Brown Bats can protect local farmers from up to 33 million or more rootworms each summer. So bats are highly significant regulators of insect populations. They control moths, gnats, crickets, beetles, locusts, mosquitoes, fruit flies, and many other insect taxa.

Overfishing and other environmental threats to large ocean predators, such as swordfish, sharks, billifish, and tuna, has resulted in an estimated loss of over 90% of all large marine predators. The entire ocean ecosystem is in decline as a result, with jellyfish becoming the dominant animals in some ecosystems with greatly reduced diversity. A specific, well-documented case is the decline of the great sharks off the eastern U.S. coast, including the Bull, Sandbar, Dusky, Blacktip, Scalloped Hammerhead, and Tiger Sharks. This led to a population explosion of their prey, the small sharks, rays, and skates, especially the Cownose Ray, which increased by an order of magnitude to 40 million since the mid-1970s. Combined with fishing pressure, this caused the prey of these species to plummet, notably the Bay Scallop, terminating a century-old industry, but also oysters and other shelled mollusks declined tremendously. All of these shelled species provide habitat and food for myriad other species. The result was devastating to the system's diversity. And many bivalve mollusks feed on plankton, controlling their numbers. When plankton populations explode as a result of loss of these mollusks, they sometimes have large die offs. Bacteria eat the dead phytoplankton, have large population increases, and consume so much oxygen that parts of the sea become anoxic, causing precipitous declines of populations of all animals dependent on oxygen (eutrophication). Large sharks are keystone predators, and there are many instances of ecosystems collapsing due to human-induced loss of sharks.

Even if predators consume a prey species infrequently, the effect on the prey's behavior can be significant. When the predator is present, the prey needs to be wary, and therefore tends not to deplete its food source. Tiger Sharks are only in Shark Bay, Australia, seasonally, but their prey there—sea turtles, dugongs, and dolphins—curtail their foraging there to avoid shark predation, and so do not deplete their prey. Sharks populations were reduced 75% to 90% due to humans in Australian seagrass ecosystems. Sharks only eat small numbers of sea turtles, but they greatly affect their behavior. They affect how many offspring turtles produce, and where they forage and spend their time. Without sharks, sea turtles produced more offspring and spent more time in the seagrass without fear and ate more of it, greatly reducing it. This caused the many species of fish and invertebrates dependent on sea grass to decline. It also negatively impacted coral reefs, which benefit from seagrass (see Chapter 4). Sleeper Sharks in Alaska alter the diving of seals, though the seals are only a small part of the sharks' diet. This keeps the seals' prey at healthy levels.

Dramatic changes in coastal ecosystems have followed the collapse and recovery of sea otter populations. Sea Otters maintain coastal kelp forests by controlling populations of kelp-grazing sea urchins. Loss of otters from overhunting off the west coast of North America caused trophic cascades in which Green Sea Urchins decimated the kelp [11]. After recovery, otters declined off Alaska because Killer Whales switched to eating Sea Otters because humans caused their pinniped food to decline [12].

This caused sea urchins to increase and decimate the kelp there (ibid). Here, loss of prey caused a predator to switch prey with devastating consequences. When Sea Otters decline, myriad fish and invertebrate species, including abalone, dependent on kelp forests for food and/or shelter, disappear or greatly decrease. Sea urchins also decline greatly after they decimate their kelp food. Otters and some kelp are keystone species. There is a quasi-stable state of high kelp and otters and relatively low sea urchins and Great White Sharks, maintained by the otters themselves, which control the urchins, causing the kelp to be abundant. The otters can hide from sharks in the kelp, and thrive and eat more urchins, in a stabilizing negative feedback loop. Female otters use the kelp canopy for nursery habitat. But a decrease in otters causing an increase in urchins leading to a decrease in kelp allows sharks to find otters and eat them more easily, further lowering the otter population, in a destabilizing positive feedback loop. A threshold is crossed, transforming the system form a high-otter, high-kelp, low-urchin quasi-stable state to a high-urchin, low-otter, low-kelp quasi-stable state. The otters thus maintain their own habitat, aiding themselves [13].

Otters also eat crabs, which eat sea slugs, which consume algae that grows on eelgrass. Decrease of otter predation allows high crab populations to decimate the sea slugs, allowing algae to smother and kill eelgrass. Eelgrass reduces erosion in tidal creeks and channels, increases nutrients in estuaries, and provides habitat for various fish and invertebrate species. Its loss would cause the extirpation of many species.

There is a trophic cascade in soil microbiota. Elimination of tydeid mites in a desert ecosystem resulted in a 40% reduction in decomposition, because the mites control cephalobid nematodes, which eat bacteria. Without mites, the nematodes explode and greatly reduce the bacteria that decompose the dead organisms in the system [14]. The result is a great reduction in diversity of both soil life and larger organisms dependent on soil and decomposition.

Estes et al. [15] demonstrated that the loss of large predators and herbivores had profound effects on immediate prey species, as well as on wildfires, disease, vegetation, biogeochemical and nutrient cycles, carbon sequestration, soil and water quality, and invasive species. They showed that top consumers in the food web have tremendous influence on the structure, function, and biodiversity of most natural ecosystems. For example, they found a decrease in Lion and Leopard populations led to an increase in Olive Baboon populations in sub-Saharan Africa, causing an increase in intestinal parasite transmission from baboons to other species, including humans when baboons were forced to move closer to humans for food.

The southern Brazilian wasp, *Agelaia vicina*, is a keystone species as a result of being a major predator, with complex, eusocial colonies, having the largest colony and nest sizes among social wasps, with some colonies exceeding over one million individuals [16]. Their huge numbers, high rate of brood production, and very broad diet allow them to regulate the populations of many arthropod species. After swarming, new nest building is very fast. In most swarming species, nest growth is episodic, but this species continually increases its nest size. Growth rates are so fast that the nest, which gets incredibly large, can double in size in 6 months. Brood production rate is several thousand individuals per day! Without these wasps, any number of small invertebrates could wreak havoc on, or at least negatively impact, the ecosystem, and collectively this wasp's numerous prey species could likely destroy an ecosystem. They mostly devour Lepidoptera and Coleoptera, but spiders are a large part of their prey. They also eat Dermaptera (earwigs), Hymenoptera, Heteroptera (true bugs), Mantodea (praying mantises), Diptera, Neuroptera (lacewings, antlions), Blattodea (cochroaches, termites), and Homoptera (aphids, scale insects, cicadas, leafhoppers) [17]). The degree that they maintain prey biodiversity by selectively attacking more abundant prey species is unknown and needs study.

Predators can easily change their diet according to which prey are available. This accounts for search image, eating the most abundant prey species, and switching prey when a competing predator is eliminated. It also allows predators to survive when their preferred prey is reduced, and promotes predator evolution and speciation. Polar Bears are spending more time on land because climate change is causing sea ice in the Arctic to melt earlier each year. They are encountering less Ringed Seal pups, their traditional prey, and more Snow Goose eggs and Caribou. The time they are on land now corresponds with the Snow Goose nesting period in western Hudson Bay, where the goose population has increased from fewer than 2,500 nesting pairs in the 1960s, to over than 50,000 today. Bear scat samples and video show the bears have greatly increased Snow goose egg predation. They are showing flexibility in diet, and eating what is available. Bears along the coast are ingesting more grass, and those further inland are eating more berries [18–20]. From the late 1800s to the early 1900s, whalers reduced whale populations and krill numbers increased. Penguins switched from eating mostly fish to mostly krill. With less whaling and more krill harvesting by humans in the 1970s–1990s, krill numbers went down, and penguins switched back to eating primarily fish.

Change in diet and dietary flexibility can cause the predator to speciate, usually allopatrically, but perhaps sometimes sympatrically. Killer Whales (*Orcinus orca*) have morphologically differing ecotypes that are not geographically isolated from each other, and which appear to be undergoing sympatric speciation, driven primarily by differing diets. In the northeastern Pacific near Vancouver Island, the resident ecotype eats fish, especially salmon, the transient prefers marine mammals, and the offshore form eats Pacific Halibut and Sleeper Sharks, although the latter's diet is not fully known. A similar divergence of two ecotypes, one eating seals, the other herring and mackerel, is occurring off the northern European coast. There is also divergence happening to five ecotypes in Antarctic waters, with preferred prey for each ecotype being seals, Minke Whales, penguins, Patagonian Toothfish, and Antarctic Toothfish [21–23].

Predator flexibility can also lead to strange, novel adaptations in predators. The European Blue Butterfly (*Maculinea arion*) falls off its host thyme plant to the ground, and produces a mixture of volatile chemicals mimicking the smell of the larvae of the ant species *Myrmica sabuleti*. The ants take it to the nest, mistaking it for their larvae. There, the caterpillar is fed by ant workers, and eventually eats the ants until it pupates and then metamorphoses in the ant nest and emerges as a butterfly [24,25]. This butterfly likely speciated from an ancestor when it acquired this adaptation.

Food webs are very stable, redundant systems. If a predator species is removed, another can substitute for it without disrupting the food web. The presence of predators often prevents their competitors from eating their prey, or at least eating very large quantities of it, with the result that the competitor species must feed on other prey species, adding the prey of the superior competitor to its diet only if the latter is removed. This is the same principle as search image and Paine's work on the sea star food web, because the secondary predator will be responding to increased prey numbers that occur when the primary predator disappears. There are many species of baleen whale that eat krill, and if one goes extinct, others eat more krill. Also, other krill predators that could fill the niche made available when a competing predator is extirpated include various species of penguins, seals, squid, and fish. There could be an increase in the population of any or all of these species, and an increase in the krill they eat per predator if any of the great whale species is reduced or driven extinct. This prevents the collapse of the food web. If krill had a population explosion because of loss of their predators, they would decimate the primary food they eat, phytoplankton, the base of the sea's food web, Earth's major oxygen producer, and a major source of CO_2 sequestration. The result would be destruction of much of the oceanic ecosystem and food webs. Predator diversity eliminates or significantly dampens trophic cascades. The diversity of predators that exists in many ecosystems provides redundancy in the system and is a mechanism that keeps food webs and biodiversity intact, providing evidence for and a demonstration of the Autocatalytic Biodiversity Hypothesis.

The greater predator diversity in open ocean ecosystems with marine mammal predators means trophic cascades are less frequent and pronounced than in ecosystems with less apex predator redundancy, such as Greater Yellowstone, coastal habitats, and seabed fisheries. Trophic cascades are more likely when on land and when megafauna are the prey, because of predator effects on their behavior and their relative lack of predator diversity. The diversity and varied diet and ranging behavior of open-ocean predators make trophic cascades relatively less likely in these ecosystems.

Predators also aid their prey by selectively killing the deleterious mutants, weak, slow, and otherwise less fit, improving the gene pool of the population through normalizing and directional selection. And they strengthen the population by culling the sick and old.

Predators coevolve with their prey. They are a selective agent that favors such traits as speed, strength, armor, poisons, and camouflage. This causes evolutionary change in the prey. The predator responds with adaptive change to counter the prey's adaptation, in an arms race. This coevolution can result in diversification of predator and prey. This general rule is illustrated by bivalve mollusks, which have evolved adaptations to prevent their predators, such as snails that drill into their shells, sea stars, octopuses, and fish, from eating them [26]. This has led to diversification of bivalves, because a variety of differing antipredator adaptations evolved, such as jet propulsion, burrowing, cryptic coloration, powerful adductor muscles, thick shells, and spikes on shells. Of course, the bivalve shell evolved mainly as a result of selective pressure from predators, and is a key innovation that allowed bivalves to enter a novel adaptive zone with many available niches, leading to tremendous adaptive radiation of this taxon. Vermeij [26] believes predators evolve but little in response to the prey adaptations to them, and calls this asymmetry escalation. However, I think predators do evolve in response to prey defense, and coevolution is thus the rule, so predators diversify too. They evolved chemicals to drill holes in prey shells, poisons, strength to pry the bivalves open in the case of sea stars, and so on. In both coevolution and escalation, predators cause their prey to diversify. Beginning in the Early Cretaceous, snail shells became substantially thicker, stronger, and predator-resistant, apparently in response to the evolution of relatively small, powerful, shell-destroying predators, such as teleosts, stomatopods, and decapod crustaceans [27]. Antarctic clams and snails have never been subjected to shell-crushing predators, such as crabs. They have shells so soft, one could crush them with one's hand. This supports the thesis that predation is what caused the shells of these mollusks to get very hard. Macroevolutionary key innovations, such as the development of shells in Gastropods, and strong jaws and other means to puncture shells in predators, likely evolved as a result of selective coevolutionary pressures of predators on prey and vice versa. These led to entry into new adaptive zones with many new niches and thus adaptive radiation, increasing species number.

Fossils from 750 mya show predators, likely unicellular, perforated holes in small eukaryotic prey, and then consumed them [28]. Amoebas of genus *Vampyrella* eat green algae in this way today. Additional evidence of this predation comes from eukaryotic biomineralization, also 750 mya. This predation might have caused unicellular and small multicellular eukaryotes to evolve the beginnings of primitive exoskeletons and the ability to burrow as a defense, huge and significant evolutionary transitions. Both exoskeletons and burrowing would have made many new niches available, and would thus be followed by tremendous adaptive radiations.

The 2-cm long *Habelia optata* of the early Cambrian was a fierce predator that tore apart shells of its prey, such as trilobites [29]. It was a major selective agent in the further evolution of the exoskeleton, resulting in the tough exoskeletons of trilobites, and it evolved mouthparts that were hard appendages similar to the mandibles of insects, to pierce trilobite armor. This coevolution starting over 500 mya was among the first between metazoan predators and prey, and was instrumental in the evolution of protective armor in prey and offensive mouthparts in predators. It aided the tremendous diversification of trilobites, and likely that of early invertebrate predators.

Prey whose major defense is fleeing have coevolved with predators, with both predator and prey becoming much faster. Gazelles of various species in Africa have become fleet of foot because of the speed and cunning of their predators. Predators have coevolved, getting faster. As a result, the Cheetah has become the fastest animal on foot, averaging 64 km/h (40 mph) during a chase, reaching a maximum on a short chase of 112 km/h (70 mph).

There are many other examples of coevolution of predator and prey leading to innovations and radiation in both. Those listed above are sufficient to make the point.

Greig and Pruett-Jones [30] found male Splendid Fairy Wrens of Australia sing a distinct song different from their territorial song after their predator, the Butcherbird, sings its song. The female pays more attention and sings back more often when the male does this, presumably because the predator's song has captured her attention, such that she is now more attentive to the song of the male of her species. The males are using the predator to increase their mating success, in an amazing example of how predator helps prey.

Predators increase species richness by benefiting scavengers. They rarely if ever eat all of their kill, although some predator species are more thorough eaters than others. After a large carnivore makes a kill, a number of scavengers come around to consume a share of the portion that the predator does not eat. An example of this is after a pride of Lions kills an herbivore such as a giraffe, various species of jackal, hyena, vulture, and other species wait until the Lions leave the carcass, and then start to dine on it. Many species of invertebrate and bacteria that are decomposers later benefit from the kill. This is commensalism if one looks at a short time horizon. However, the scavengers recycle the prey, making the nutrients in it available to the ecosystem, and remove carcasses that would otherwise accumulate. This aids many species, including the predator, so the relationship is mutualism if viewed on a longer timescale. Every predator benefits many species of scavenger and decomposer.

Many large predators are major sources of nutrients for their ecosystems. Feces of large predators are generally important in fertilizing the systems they dwell in, helping their prey as well as other species. Baleen whales are really omnivores, since they eat fish, krill, and plankton, which includes phytoplankton, but omnivores are predators. Many whale and pinniped species are ecosystem engineers, and, along with other large animals, play a significant role in the transport of nutrients in global ecological cycles, and are important to the health of the seas. Microbes, zooplankton, and fish are known to be important sources of recycled nitrogen in coastal waters. But cetaceans and pinnipeds contribute more nitrogen through their feces to the Gulf of Maine than do all of the rivers in that system combined [31]. Before commercial whaling, cetaceans and pinnipeds combined distributed three times more nitrogen in the Gulf of Maine than entered it from the atmosphere [31]. Even with current reduced populations, marine mammals are significant fertilizers of oceans where they occur, greatly increasing phytoplankton, and all species in the food web phytoplankton support, including many species of zooplankton, invertebrates, and fish. Species diversity in the ocean is much higher with the nutrient input from cetacean and pinniped feces than it would be without it. All this partially explains why ocean life was so abundant and diverse before whaling and seal harvests became substantial.

Since whales need to surface to breathe, they tend to defecate in shallow water, supplying nitrogen to the upper portion of the water column, and their feces tends to float, and be available to phytoplankton in the photic zone. This reverses the normal downward flow of nutrients, called the biological pump, in the sea, and is termed the whale pump [31]. Whales move nitrogen from deeper in the sea to the surface, keeping it from being buried and lost to the system. Nitrogen is cycled from bacteria and phytoplankton to krill to fish and whales and back in a productive loop.

Following is the nitrogen released in the present time in kg/day of some major baleen whale species: Right, 15.9; Humpback, 9.42; Fin, 15.0; Sei, 8.32; Minke, 2.94. Here is the same data for toothed whales: Pilot Whale, 0.036; Atlantic White-sided Dolphin, 0.15; Common Dolphin, 0.09; Harbor Porpoise, 0.05. It was much higher in the past. So under natural conditions, whale defecation is a major supplier of nitrogen, used by all sea life from bacteria to fish.

Some whale feces sinks, sequestering large amounts of carbon. Defecation by the Southern Ocean's Sperm Whales alone ultimately sequesters about 400,000 tons of CO_2 annually. There are now about 8,000 Blue Whales worldwide, the largest animal ever to exist. There were once more than 200,000 blue whales in the Antarctic Ocean alone. All giant whale species were much more abundant in the past. So the amount of nitrogen their feces added and CO_2 they sequestered was tremendous. The burial of CO_2 helps lower atmospheric temperatures to the benefit of life.

The feces of krill-eating whales is up to ten million times as iron-rich as the surrounding seawater [32], plays a vital role in providing the iron needed for maintaining phytoplankton biomass on Earth, and aids their blooms. This is especially important in the Southern Ocean, which is poor in iron and rich in other nutrients [32]. This benefits entire food webs, since plankton are the sea's main primary producers. It also allows plankton to sequester large amounts of carbon for long time periods. Phytoplankton bury an estimated two billion tons of carbon per year. The iron from the defecation of merely the approximate 12,000 Sperm Whales in the Southern Ocean helps sequester about 200,000 tons of carbon annually by supporting the growth of phytoplankton.

It is noteworthy that a food source of many whales, krill, are also a long-term reservoir of iron in Antarctic surface waters because they store such large quantities of it in their body tissue. The Antarctic krill (*Euphausia superba*) population contains about 24% of the total iron in the surface waters in its range [32]. Thus, krill also play a large role in making iron available to shallow ocean organisms, including, as it cycles, phytoplankton. Therefore, krill also help sequester a great deal of carbon. Krill are mostly omnivores; some are exclusively predators.

Crocodilians are very important fertilizers of the aquatic systems they inhabit through their feces, increasing aquatic plant and fish populations, including those of the fish they eat.

Guano from insectivorous bats (and fruits bats, which are not predators), in caves and below trees, is an important nitrogen source for ecosystems bats inhabit. Guano from sea birds on oceanic islands is also a very important nitrogen source. The feces of whales, bats, and all species discussed in the above paragraphs also nourishes their ecosystems with many other nutrients than nitrogen, such as carbon, phosphorous, potassium, and so on.

Feces of predators support rich communities of prokaryotes, fungi, and invertebrates. Bat droppings in caves support whole ecosystems of organisms, including bacteria that detoxify waste. Likewise, predator carcasses support communities of small organisms. Whale carcasses are eaten by a very diverse set of communities that undergo ecological succession on the carcass.

The vomit and feces of Spinner Dolphins (*Stenella longirostris*) are eaten by 12 species of reef fish from 7 different families, all of which also consume plankton [33].

Most of what I discussed in this chapter about predators is also true of parasites, with some exceptions. Parasites are density dependent population regulators that enforce negative feedback on the host population. As the host population density increases, it is easier for parasites to spread from host to host because hosts are closer to each other, and also hosts are under more stress, weakening their resistance. When the host population is low, transmission is harder because there are less hosts, and host resistance is higher because there is less stress on the host population. Parasites thus dampen host population fluctuations, stabilize host populations, and increase the probability that host populations will persist for long time periods. They prevent host populations from undergoing population explosions and destroying their environment. It is hard to imagine how they could have search images, and more research is needed to know if there are prudent parasites, but there is evidence that they attack the most abundant host. Marine phage maintains marine bacterial diversity by attacking the most abundant bacterial species in a process called "kill the winner". There is no evidence parasites cause their hosts to move and forage at multiple sites. But without them, there could be ecosystem collapses similar to the trophic cascades that occur when predators like wolves are extirpated. While there are not a great deal of experiments involving removal of all parasites in natural animal populations, it is logical to assume that removal of all parasites could result in many host species depleting and destroying their environment and collapsing. We do know that, in the case of humans, the eradication of malaria in Sri Lanka after World War II was followed by a population explosion that devastated the environment, followed by mass starvation, disease, and armed conflict. The fact that parasites are increased when host populations are higher means that they are part of the redundancy of ecosystems, filling in as population regulators when predators of their hosts are eliminated or reduced. As such, they stabilize ecosystems and food webs, and very likely at times prevent ecosystem collapses. Some parasites have multiple hosts and can switch hosts easily, and as such are efficient at replacing extirpated predators when prey populations explode. They thus stabilize ecosystems and prevent disasters by the redundancy mechanism. They do cull the weak and old, strengthening the population. When they kill their host, they aid scavengers. The decomposing host fertilizes the soil. Sometimes the dead host sinks to the sea bottom or is buried on land, sequestering carbon. Like predators, the parasite is commensal with the benefiting scavengers on short timescales and mutualistic with them on long timescales.

Pathogens, including fungi, bacteria, and especially viruses, can jump from one host species to another. Bats are among the more common carriers of viruses that jump to humans and give us diseases. This controls human population as humans invade new habitats, such as rainforests, protecting the habitats and their species. Thus, species such as bats that harbor pathogens help preserve diversity. Whether this mechanism works in controlling nonhuman species that invade new habitats, resulting in preserving diversity, is not known at this point.

Parasites coevolve with their hosts, resulting in key innovations in both host and parasite, followed by radiation in both. For example, the immune system evolved to fight pathogens, which evolved ways to defend against it. Bacteria similarly coevolve in arms races with their phage, resulting in phage-resistant bacteria and phage evolving ways around this, with key innovations and diversification in both. The CRISPR-Cas system is a family of DNA sequences bacteria and archaea genomes derived from DNA fragments of phages that had previously infected them. They act as a procaryotic immune system against foreign genetic elements in plasmids and phages.

Parasites are very diverse. Every cellular species has several species of parasites. Viruses are the largest group of parasites, and all viruses are parasites. Viruses are unique in their ability to infect all types of organisms. Every cellular species has at least ten types (species) of virus, so there are hundreds of millions of types of virus. There are even viruses that parasitize other viruses. There are also viroids, the smallest infectious pathogens known, composed solely of a short strand of circular, single-stranded RNA that has no protein coating. All currently known viroids are inhabitants of higher plants, in which they are pathogens. They are distinct from viruses. All metazoans are infected by bacteria and fungi. Both of these groups have parasitic species likely numbering in the millions. There are multitudes of parasitoid insects. There are 75,000–300,000 species of parasitic worms on Earth.

Parasitic plants can increase the diversity of the system they are in. Avian species richness was found to be about 50% greater in *Eucalyptus camaldulensis* forest plots with the hemiparasitic shrub *Exocarpos strictus* (Santalaceae) than without it in the Gunbower-Koondrook forest in southeastern Australia [34]. The study included comparing *Exocarpus* plots with plots with and without the nonparasitic *Acacia dealbata*. *Exocarpus* provided plentiful fleshy fruits and many microhabitats, and supported a significantly higher arthropod biomass than the *Acacia*. This accounts for the higher bird species richness in the hemiparasitic *Exocarpus* plots. The authors stated that they do not think the result is aberrant, but that the influence of *Exocarpus* on species richness actually relates to its parasitic habit, supporting a hypothesis that states that parasitic plants make resources from their hosts available to a range of trophic levels.

REFERENCES

1. Malthus, T. R. (1798). *An Essay on the Principle of Population As It Affects the Future Improvement of Society, with Remarks on the Speculations of Mr. Godwin, M. Condorcet, and other writers* (1st ed.). J. Johnson in St. Paul's Church-Yard: London, UK.
2. Slobodkin, L. B. (1 Feb., 1968). How to be a predator. *Integrative and Comparative Biology* 8 (1): 43–51. doi: 10.1093/icb/8.1.43.
3. Wilson, D. S. (1978). Prudent predation: a field study involving three species of tiger beetles. *Oikos* 31 (1): 128–36.
4. Paine, R. T. (1966). Food web complexity and species diversity. *American Naturalist* 100: 65–75.
5. Paine, R. T. (1974). Intertidal community structure: experimental studies on the relationship between a dominant competitor and its principal predator. *Oecologia* 15: 93–120.
6. Hamilton, W. D. (May, 1971). Geometry for the selfish herd. *Journal of Theoretical Biology* 31 (2): 295–311. doi: 10.1016/0022-5193(71)90189-5.
7. Ripple, W. J. & Beschta, R. L. (2012). Trophic cascades in Yellowstone: The first 15 years after wolf reintroduction. *Biological Conservation* 145: 205–13. doi:10.1016/j.biocon.2011.11.005.
8. Ripple, W. J. & Beschta, R. L. (May, 2008). Trophic cascades involving cougar, mule deer, and black oaks in Yosemite National Park. *Biological Conservation* 141 (5): 1249–56. doi: 10.1016/j.biocon.2008.02.028.
9. Ripple, W. J. & Beschta, R. L. (Dec., 2006). Linking a cougar decline, trophic cascade, and catastrophic regime shift in Zion National Park. *Biological Conservation* 133 (4) 397–408. doi: 10.1016/j.biocon.2006.07.002.talk.
10. Bartush, W. S. & Lewis, J. C. (1981). Mortality of white-tailed deer fawns in the Wichita Mountains. *Proceedings of the Oklahoma Academy of Science* 61: 23–7 (1981).
11. Mainea, J. J. & Boylesa, J. G. (6 Oct., 2015). Bats initiate vital agroecological interactions in corn. *PNAS USA* 112 (40): 12438–43. doi: 10.1073/pnas.1505413112.
12. Arnold, D. C. (1976). Local denudation of the sublittoral fringe by the green sea urchin, *Strongylocentrotus droebachiensis* (O. F. Muller). *The Canadian Field-Naturalist* 90: 186–7.
13. Estes, J. A., et al. (16 Oct., 1998). Killer whale predation on sea otters linking oceanic and nearshore ecosystems. *Science* 282 (5388): 473–6. doi: 10.1126/science.282.5388.473.
14. Nicholson, T. E., et al. (12 March, 2018). Gaps in kelp cover may threaten the recovery of California sea otters. *Ecography*. doi: 10.1111/ecog.03561.

15. Perseu, F., et al. (June, 1981). The role of mites and nematodes in early stages of buried litter decomposition in a desert. *Ecology* 62 (3): 664–9. doi: 10.2307/1937734.
16. Estes, J. A., et al. (15 July, 2011). Trophic downgrading of planet Earth. *Science* 333 (6040): 301–6. doi: 10.1126/science.120.
17. Zucchi, R., et al. (1995). *Agelaia vicina*, a swarm-founding polistine with the largest colony size among wasps and bees (Hymenoptera: Vespidae). *Journal of the Entomological Society of New York* 103: 129–37.
18. De Oliveira, O. A. L. (2010). Foraging behavior and colony cycle of *Agelaia vicina* (Hymenoptera: Vespidae; Epiponini). *Journal of Hymenoptera Research* 19: 4–11. http://biostor.org/reference/111460.
19. Gormezano, L. J. & Rockwell, R. F. (Sept., 2013). What to eat now? Shifts in polar bear diet during the ice-free season in western Hudson Bay. *Ecology and Evolution* 3 (10): 3509–23. doi: 10.1002/ece3.740.
20. Gormezano, L. J. & Rockwell, R. F. (2013). Dietary composition and spatial patterns of polar bear foraging on land in western Hudson Bay. *BMC Ecology* 13: 51. doi: 10.1186/1472–6785–13–51.
21. Rockwell, R. F. & Gormezano, L. J. (April, 2009). The early bear gets the goose: climate change, polar bears and lesser snow geese in western Hudson Bay. *Polar Biology* 32 (4): 539–47.
22. Riesch, R., et al. (May, 2012). Cultural traditions and the evolution of reproductive isolation: Ecological speciation in Killer Whales? *Biological Journal of the Linnean Society* 106 (1): 1–17.
23. Moura, A. E., et al. (Jan., 2015). Phylogenomics of the Killer Whale indicates ecotype divergence in sympatry. *Heredity* 114 (1): 48–55.
24. Morin, P. A., et al. (Aug., 2015). Geographic and temporal dynamics of a global radiation and diversification in the Killer Whale. *Molecular Ecology* 24 (15): 3964–79.
25. Thomas, J. A. (1995). The ecology and conservation of *Maculinea arion* and other European species of large blue butterfly. In Pullin, A. S. *Ecology and Conservation of Butterflies*, pp. 180–97. Chapman and Hall: New York, NY.
26. Nash, D. R., et al. (2008). A mosaic of chemical coevolution in a large blue butterfly. *Science* 319: 88–90.
27. Vermeij, G. J. (1994). The evolutionary interaction among species: selection, escalation, and coevolution. *Annual Review of Ecology, Evolution, and Systematics* 25: 219–36.
28. Vermeij, G. J. (Summer, 1977). The Mesozoic marine revolution: evidence from snails, predators and grazers. *Paleobiology* 3 (3): 245–58. doi: 10.1017/S0094837300005352.
29. Porter, S. M., et al. (2003). Vase-shaped microfossils from the Neoproterozoic Chuar Group, Grand Canyon: A classification guided by modern testate amoebae. *Journal of Paleontology* 77: 409–29.
30. Aria, C. & Caron, J.-B. (Dec., 2017). Mandibulate convergence in an armoured Cambrian stem chelicerate. *BMC Evolutionary Biology BMC Series* 17: 261. doi: 10.1186/s12862-017-1088-7BMC.
31. Greig, E. I. & Pruett-Jones, S. (2010). Danger may enhance communication: predator calls alert females to male displays. *Behavioral Ecology* 21 (6): 1360–6. doi: 10.1093/beheco/arq155.
32. Roman, J. & McCarthy, J. J. (11 Oct., 2010). The whale pump: marine mammals enhance primary productivity in a coastal basin. *PLoS ONE* 5 (10): e13255. doi: 10.1371/journal.pone.0013255.
33. Nicol, S., et al. (June, 2011). Southern Ocean iron fertilization by baleen whales and Antarctic krill. *Fish and Fisheries* 11 (2): 203–9. doi: 10.1111/j.1467–2979.2010.00356.x.

34. Martins Silva, J., Jr., et al. (2004). Vomiting behavior of the Spinner Dolphin (*Stenella longirostris*) and squid meals. *Aquatic Mammals* 30 (2), 271–4. doi: 10.1578/AM.30.2.2004.271.
35. Watson, D. M., et al. (Aug., 2011). Hemiparasitic shrubs increase resource availability and multi-trophic diversity of eucalypt forest birds. *Functional Ecology* 25 (4): 889–99. doi: 10.1111/j.1365-2435.2011.01839.x.

10 Decomposers Are Indispensable to Their Ecosystems

This chapter concerns all organisms involved in recycling dead organisms and feces. There are three types of these. Decomposers directly absorb nutrients through external chemical and biological processes, and include fungi. Detritivores ingest dead matter and feces internally, and are exemplified by earthworms. Scavengers consume dead animals that they did not kill, and include vultures. For brevity, I will use the term decomposer in this chapter for each of these and all of them collectively, unless I specify otherwise.

Decomposers are necessary for ecosystem function. They recycle minerals and elements such as carbon, nitrogen, and oxygen in dead organisms back into the soil and ecosystem. They do this efficiently, so ecosystems do not waste nutrients, and they make them available to the entire ecosystem. They allow plants to obtain the potassium, nitrogen, and other nutrients they need. Earthworms convert leaf litter and other organic matter into inorganic compounds, helping build soil.

Without decomposers, ecosystems would cease to exist because carcasses, dead trees, leaves, feces, and other organic matter would cover the ground and accumulate into huge piles. This would make movement impossible for many animal species and block oxygen from the soil, killing soil life that requires it. The great majority of plant growth would be blocked. Decomposers thus maintain space for organisms to carry out their functions.

Without decomposers, the soil and ecosystem would not have enough nutrients to persist. There would be too few minerals like calcium, phosphorous, magnesium, and zinc, and too little nitrogen, for ecosystems to survive. Carbon would become too scarce for photosynthesis.

Lignin evolved as a support and for strength in trees, but is very hard to break down. Wood-decay fungi and some bacteria are the only groups that evolved enzymes to break it down. Without them, bark would not decompose.

Decomposers represent a significant portion of each of the various food webs and are in all major phyla. Bacteria have the most species of decomposers in all ecosystems. Fungi are second on land. Mammalian scavengers include hyenas; avian scavengers include vultures. Other terrestrial decomposers are in these taxa: worms, slugs, snails, termites, bark beetles, wood borers, and ants. Ants and some others disperse the nutrient products of decomposition throughout the ecosystem, making them available to a greater numbers of organisms. In the ocean, decomposers include bacteria; archaea; crabs; lobsters; shrimp; hagfish; fish such as sleeper sharks; and other taxa.

Two unintended experiments removed some decomposer species, demonstrating their importance empirically. In the first, Mousseau et al. [1] found that radiation from the Chernobyl nuclear accident negatively impacted decomposers, and that leaf litter accumulated as a result. In control areas with no radiation, 70%–90% of the leaves decomposed after one year. In areas with radiation, loss of leaf litter mass as a measure of leaf litter decomposition was 40% lower in the most contaminated sites relative to sites with a normal background radiation level for the area. The researchers controlled for humidity, temperature, and forest and soil type. Trees were not decomposing as fast as normal. The accumulation of leaf litter and dead trees increased the probability of abnormally large forest fires that could scorch the soil. Nutrients were not efficiently recycled to the soil and the ecosystem. Trees grew much slower than normal, and this is likely because of the lack of soil nutrients. As time passed, the thickness of the forest floor increased with the level of radiation the area was exposed to. The authors state, "These findings suggest that radioactive contamination has reduced the rate of litter mass loss, increased accumulation of litter, and affected growth conditions for plants". The mechanism was the loss of the forest's decomposers.

In the second unintended experiment, many of India's vulture species, especially genus *Gyps*, declined precipitously from the early 1990s until at least 2006. There were up to 80 million White-rumped Vultures (*Gyps bengalensis*) in India in the 1980s, when it was the most numerous raptor species in the world. Now there are only several thousand. This species, and the Indian Vulture (*G. indicus*) and Slender-billed Vulture (*G. tenuirostris*) declined by about 99% in India [2] and nearby countries [3]. Most of India's nine vulture species are endangered now, as a result of a precipitous population decline. The declines rank among the fastest population declines in recorded history, in the same category as the Passenger Pigeon.

Vultures eat dead cattle in India, and it is believed that they declined because of the widespread use of drugs such as diclofenac, commonly used as an anti-inflammatory drug given to livestock during the decline of the vultures. The drug is deadly to vultures. Dogs and rats are eating the carcasses the vultures once ate, and their numbers increased tremendously. Vultures are adapted to their niche as decomposers, and have strong stomach acids that protect them from disease by killing pathogenic bacteria. This means disease does not normally spread in natural and even human-influenced ecosystems when they scavenge. But when other animals that are not adapted to scavenging substitute for them, the abundance of bacteria on carcasses can make them sick. The sick animal becomes a pathogen carrier, and disease can spread widely.

One flock of vultures can consume a pound of meat per minute, so well are they adapted to eating carrion. But if replaced by animals that do not normally scavenge, the carcasses are consumed slowly and can increase, taking up space needed for wildlife in natural systems. Large numbers of slowly decomposing carcasses increase the risk of disease in natural ecosystems, providing a second mechanism for the spread of disease. In India, the much slower decomposition of dead cows is seriously contaminating drinking water and spreading disease. Water contamination negatively impacts wildlife populations as well as humans.

The rat population exploded and rats are not well adapted as scavengers, posing a dire threat of disease to humans and wildlife. The dog population exploded to 18–25 million dogs in India, the largest carnivore population in the world. The dogs have become vectors of rabies, anthrax, bubonic plague, and other diseases in much greater numbers than usual, leading to thousands of human deaths annually. In India, a dog bites a human every 2 seconds, and 30,000 people per year die from rabies, which is one death every half hour. The dogs can also spread the rabies and other diseases to wildlife, leading to deaths in great numbers.

The high dog population has led to Leopards (*Panthera pardus*) preying on the dogs, and Leopards have increased in areas with people, leading to an increase in Leopard attacks on children. Their higher numbers in these areas could also lead to local decreases in their various prey species below healthy levels, and possibly drive some locally extinct.

India banned the use of diclofenac for veterinary purposes in 2006. There is a black market for it, but its use has decreased, and the vulture populations are rebounding.

In a planned experiment, Payne [4] found baby pig carrion kept free of insect decomposers decomposed and dried very slowly. It retained its form for many months. Yet 90% of the carrion that insects were allowed access to decomposed and disappeared in six days.

The unwitting Chernobyl and Indian vulture experiments and those of Payne's on pigs show the prime importance of decomposers to ecosystem health and biodiversity.

Decomposers keep diversity high within their trophic level by the mechanisms of ecological succession and niche partitioning. One decomposer community is replaced by another in a consistent, sequential order. For example, in North America, Turkey Vultures are the first birds to arrive when a carcass appears, because they have the best sense of smell of the scavenger birds that occur there. Black Vultures observe Turkey Vultures to find carcasses, and arrive next. They displace Turkey Vultures, being more aggressive. In some parts of the desert, caracaras arrive last, and displace both vulture species.

This principle generally holds true for scavenger systems. Decomposer species arrive in reverse order of their position on the pecking order, maximizing the species richness of the system. In Africa, what is left of predatory cat kills are eaten after the cat eats its fill of the carcass. The order of arrival of scavengers sometimes varies due to luck and environmental conditions, but a general order of arrival of scavengers after the predator is finished would often be: eagles, vultures, jackals, hyenas. The later arrivals displace the ones that arrived earlier. Other scavenger species are involved, and there are differences in size of prey desired between these species as well. The mammals use the vultures to locate carcasses by observing them above the carcasses. Even later-arriving bird species use the sight of earlier-arriving bird species to find the carcasses more quickly. Bacteria, fungi, and insects arrive largely in that order before the vertebrates, and continue decomposing while the vertebrates eat.

A great example of this niche partitioning in decomposers is seen in how they break down whale carcasses in an ecological succession. If the carcass ends up on the beach, a succession of bacteria, archaea, protists, fungi, isopods, shrimp, other invertebrates, birds, raccoons, bears, and others consume and recycle it, in a process that takes up to ten years, usually with some of the nutrient going to land, some to sea. That which is recycled on land goes into the soil and the plants, supporting terrestrial ecosystems. If the carcass remains in the sea, sharks and other fish are the first to the carcass. Many sharks are scavengers as well as predators. It is believed that some sharks and fish are specialized to feed exclusively on dead whales.

If the carcass sinks in the sea, it can last up to 50 years because of the low temperatures, supporting entire communities. Whale carcasses are a major transmission vector for nutrients to the seafloor, an otherwise static, nutrient-poor part of the ocean. They support high diversity by providing immense quantities of organic matter. There is a complex microbial succession, with archaea, bacteria, fungi, and protists of many species succeeding one another.

There are three successional stages of animal decomposers that feed on the sunken whale carcass, each differing from the others in species composition [5]. The first stage takes from four months to five years. Sleeper sharks come first, followed by hundreds of hagfish. As the tissue fragments decrease in size, they are consumed by successively smaller scavengers. Rattail fish, then amphipods, and finally copepods eat the remaining meat. This stage ends when all the soft tissue, which is nearly 90% of the whale's wet weight, has been removed. The larger scavengers of this stage disperse a great deal of nutrients when they defecate.

The second stage lasts about two years. It starts when only bone and scraps, surrounded by rich organic sediments, remain. Large and dense populations of bottom dwellers, including sea anemones specific to whale carcasses, polychaete worms, gastropods (notably snails), juvenile bivalves, some crustaceans, and more taxa colonize in and around the whale bones. Polychaete zombie worms grow their roots into the bones and feed on lipids. Snails and shrimp feed on both the sediment and bone.

The third stage is the most diverse, and the most diverse ecosystem of all hard substrates in the deep sea. A very large, trophically-complex community thrives on the skeleton. Anaerobic microbes produce hydrogen sulfide, which is used as an energy source by sulfophilic bacteria. These serve as an energy source for tube worms, clams, mussels, limpets, and other species. And they break down lipids embedded in the bone. Bacterial grazers, bone lipid consumers, predators, and other organisms tolerant of high sulfide levels also occupy the skeleton. This stage can last 50 to 100 years.

On land or sea, the whale can support entire ecosystems because of its size. The average local species richness of 185 macrofaunal species on single whale skeletons approaches levels of the global cold-seep macrofaunal species richness of 229 species, and exceeds the richness of the most speciose cold-seep vent field known (121 species) [6]. Twenty-one species in five phyla are known to live exclusively on whale fall. Decomposers on dead whales follow a sequential succession that maximizes their diversity and makes it tremendously high. They recycle nutrients, supplying a tremendous amount of them that nourish a myriad of species.

Scavengers and decomposers create high diversity. Without them, dead animals and plants would build up so much that animals would not be able to move freely from place to place; disease would increase in wild animals tremendously; and nutrients would cease to be available in ecosystems except in low quantities. Decomposers close the loop of energy flow, connecting all trophic levels to the primary producers.

REFERENCES

1. Mousseau, T. A., et al. (May, 2014). Highly reduced mass loss rates and increased litter layer in radioactively contaminated areas. *Oecologica* 175 (1): 429–37.
2. Prakash, V., et al. (2007). Recent changes in populations of resident *Gyps* vultures in India. *Journal of the Bombay Natural History Society* 104 (2): 129–35.
3. Baral, N., et al. (2005). Population status and breeding ecology of White-rumped Vulture *Gyps bengalensis* in Rampur Valley, Nepal. *Forktail* 21: 87–91.
4. Payne, J. A. (1 Sept., 1965). A summer carrion study of the baby pig *Sus scrofa* Linnaeus. *Ecology.* doi: 10.2307/1934999.
5. Smith, C. R. & Baco, A. R. (2003). Ecology of whale falls at the deep-sea floor. *Oceanography and Marine Biology: An Annual Review*: 311–54.
6. Baco, A. R. & Smith, C. R. (2003). High species richness in deep-sea chemoauto-trophic whale skeleton communities. *Marine Ecology Progress Series* 260: 109–14. doi: 10.3354/meps260109.

11 Eight New Proposed Principles of Ecology and Evolution

I will now introduce eight ideas that have never been proposed as precisely and with as much claimed generality as will be presented here before. Their general applicability has not been sufficiently established to call them laws, so I call them principles of ecology and evolution. I hope in time they will be tested and found to be generally valid to the point where they can be called laws. The First Principle of Ecology and Evolution can also be called the Principle of Beneficial Species. It states that in a natural ecosystem, situation, and conditions, over a time interval sufficiently long to accurately represent the effect of a species on its ecosystem, every species has a net positive effect on the species it interacts with, its ecosystem, and biodiversity. Said another way, in a natural ecosystem, the net effect of each species over time is to maintain and even increase the diversity of its ecosystem. Species do this by various mechanisms, some of which were discussed in this book. These include providing food, providing habitat, regulating prey populations, selectively attacking the superior competitor, regulating temperature and CO_2, increasing and regulating oxygen, fixing nitrogen, enhancing habitat, and so on. The law is supported by the observation that the removal of any species from an ecosystem generally reduces the number of other species, or at least reduces the populations of some species. A natural ecosystem, situation, or condition is one which was not interfered with by humans. Human alterations of natural ecosystems and the climate, and the introduction of exotic species are not natural situations. The chytrid fungi, *Batrachochytrium dendrobatidis* and *B. salamandrivorans*, have caused the extinction and reduction of so many amphibian species because of the unnatural situations of anthropogenic climate change and their being introduced to widely separated environments that they are not native to by humans. Herbivores denude areas of vegetation and reduce diversity greatly as part of trophic cascades only when humans remove their natural predators. Mule Deer and Elk can destroy forests and be key players in trophic cascades, but only after the removal of Grey Wolves by humans. Introduced species greatly lower diversity and harm ecosystems only because of their unnatural introduction to new ecosystems by humans. The principle requires that the time interval being considered is sufficiently long to be a representative sample of the impact of the species in question. For example, migratory locusts in Africa periodically denude large areas of natural vegetation. However, these events are short-lived, and the ecosystem recovers. There is no long-term, lasting negative impact on diversity or the ecosystem. The time interval of a locust outbreak does not fairly represent the overall, long-term impact of the migratory locust on its ecosystem. Given sufficient time, the overall effects of the migratory locust on its ecosystem will be positive and diversity-promoting in natural conditions. When oxygenic photosynthesis evolved, it increased levels of O_2, which combined with the potent greenhouse gas CH_4 and thus lowered Earth's temperature. This may have caused the Huronian glaciation and snowball Earth, 2.4 to 2.1 bya. The sun gave off much less heat at that time. But in time life helped thaw the Earth from the snowball state. The O_2 produced by the early photosynthesizers was toxic to the life, which was only prokaryotes. This reduced diversity for a period of time. But prokaryotes evolved enzymes to protect themselves. In time, they evolved the ability to use O_2 through respiration, and O_2 became the major energy source for the evolution of multicellular life. In the long term, there was a great increase in diversity as a result of oxygenic photosynthesizers. In another example, sulfate-reducing microbes produced toxic hydrogen sulfide (H_2S) that had several negative effects on life and caused die-offs that decreased diversity for long time periods, including during the Permian-Triassic mass extinction. But in time, photosynthesis evolved, and the oxygen produced greatly limited the anaerobic sulfate-reducing microbes and hence H_2S. Moreover, sulfate-reducing microbes increased diversity in the long term because they have diversified into about 69 genera within nine phyla in the bacteria and 37 genera within two phyla in the archaea (Florentino, 2016). They are ubiquitous in anoxic environments, where they have an important role in both the sulfur and carbon cycles (Muyzer and Stams, 2008). They helped oxygenate the atmosphere. They are in symbiotic consortia with many species with methanotrophs and filamentous sulfur bacteria, mainly in anoxic marine sediments, reducing the amount of CH_4 released into the air, thus regulating temperature. In some marine areas, the process produces H_2S, but it is used by commensal filamentous sulfur bacteria. The consortia help many other species by producing nutrients for other prokaryotes and some animals. In some cases, sulfate-reducing bacteria live in animals, such as clams and tube worms, with sulfide-oxidizing bacteria that convert the H_2S to a form of sulfur useable by the animal hosts. There are other examples of life reducing diversity for a period of time, but organisms always in time stop or limit the action that is causing the loss of diversity, and institute some form of ecosystem engineering that causes a net increase in diversity, even when the reduction in diversity is taken into account. One can argue that the time period of deleterious effects on diversity was long in the case of oxygen and hydrogen sulfide production by life, but the effects actually ended up increasing diversity in the long run in both cases. If one thinks the time that diversity is negatively impacted by oxygenic photosynthesis and sulfate-reducing microbes in unreasonably long for this principle to have meaning, the principle can be stated to say that that every species *almost always* has a net positive effect on other species, its ecosystem, and biodiversity. If one accepts the time qualification even if it is very long, the principle can omit "almost always", and be stated: In a natural ecosystem, situation, and conditions, over a time interval sufficiently long to fairly and accurately represent the effect of a species on its ecosystem, every species has a net positive effect on other species, its ecosystem, and biodiversity. A positive effect on other species means the species benefits other species and helps them survive; on the ecosystem, it means it helps the ecosystem survive and persist; on biodiversity, it means the species maintains or increases biodiversity. The *net* effect is the weighted sum of all effects—positive, negative, and neutral—that the species has on the species it interacts with, its ecosystem, and biodiversity. There are likely negative impacts from every species on at least some individuals of some other species. Some or all species have some negative effects on biodiversity and/or their ecosystem. For example, interspecific competitors can negatively impact each other. But the *overall* net impact of each of these competitors on the diversity of their ecosystem will be positive in a natural system over sufficient time. The law states that the overall net effect on other species, diversity, and the ecosystem is positive when all impacts of the species in question are added together in a weighted sum, when all of its effects are taken into account considering the relative importance of each. This law may work by the species regulating its prey, providing food to its predators, being mutualistic and commensal with other species, and so on. But it could additionally increase the diversity of its ecosystem over time by its evolution and its effects on the evolution of other species. For example, it could coevolve a mutualism with another species. This could then lead to the diversification of both mutualists and the evolution of new species that exploit the niches created by the many new mutualist species. Hence, I call it the First Principle of Ecology and Evolution, not just of Ecology. Since all of the principles potentially involve evolution as well as ecology, I call all of them principles of ecology and evolution.

The Second Principle of Ecology and Evolution, also called the Principle of Ecosystem Engineers, states: All species are ecosystem engineers. Even viruses alter the environment when they reduce the population of their host. Every species of soil bacterium in the nitrogen cycle benefits from chemicals produced by others and manufactures chemicals used by species in the subsequent step of the cycle. This too is ecosystem engineering. The behavior of the genome includes sexual recombination, viral transduction, transposable elements, polyploidy, hybridization, epistasis, and other things. These are mechanisms of the ABH, of organisms increasing diversity. This genomic behavior may be considered a form of ecosystem engineering. If one accepts this, behavior of the genome can be incorporated in this Second Principle. If one thinks considering genomic behavior a form of ecosystem engineering is too much of a stretch, the behavior of the genome can be considered a separate mechanism of the ABH and not part of the Second Principle. Either interpretation is acceptable. One cannot objectively determine which interpretation is better.

Combining the first two principles above, we get what I call the First Two Principles of Ecology and Evolution, also called the Principle of Beneficial Ecosystem Engineers, which states: In a natural ecosystem, every species is an ecosystem engineer whose actions, over a sufficient time period to accurately represent its effect on its ecosystem, causes a net positive effect on species it interacts with, its ecosystem, and biodiversity. Thus, in a natural ecosystem, all species alter their environment in a way that is on the average favorable to life, other species, and diversity.

To illustrate these first two principles and their combined form, let us consider one species, the American Badger (*Taxidea taxus*) of North American grasslands. It regulates populations of pocket gophers (Geomyidae), ground squirrels (*Spermophilus*), moles (Talpidae), marmots (*Marmota*), prairie dogs (*Cynomys*), pika (Ochotona), woodrats (*Neotoma*), kangaroo rats (*Dipodomys*), deer mice (*Peromyscus*), voles (*Microtus*), skunks (*Mephitis* and *Spilogale*), ground-nesting birds, snakes, lizards, amphibians, fish, insects (including bees and honeycomb), some plants, sunflowers

(*Helianthus*), and mushrooms and other fungi. It regulates sunflowers by eating their seeds. Ground-nesting birds it eats include Bank Swallows (*Riparia riparia*) and Burrowing Owls (*Athene cunicularia*). Badgers are considered the most important predators of rattlesnakes in South Dakota [1]. Any of its prey could potentially reach destructive population levels in the absence of badger predation, especially the animals. The prey animals are controlled by other predators, so a catastrophic trophic cascade might not occur, but increases in its prey populations would occur and be deleterious to the ecosystem, if badgers were removed. Badgers are also scavengers that recycle carrion. Young, small badgers provide food to Golden Eagles (*Aquila chrysaetos*), Coyotes, and Bobcats. The main predator of adults is the Mountain Lion, and they are also eaten by bears (*Ursus* spp.) and occasionally Gray Wolves. Abandoned badger burrows are commensal homes to California Tiger Salamanders, California Red-legged Frogs, Burrowing Owls, foxes, and skunks. It hunts as a mutualist with Coyotes. Badgers till and aerate the soil by digging their burrows. They support a diverse microbiome. Their fur is home to a number of insect species. Yet there is nothing unique about badgers with respect to their effects on diversity. This principle could have been illustrated with any species, from species of viruses to protists to snakes to trees to fungi.

Considering larger taxa than species illustrates the magnitude of collective effects, both positive and negative, of higher taxa on their ecosystems. So the principle will be further illustrated with a discussion of diatoms, looking especially at the genus *Rhizosolenia*. Diatoms produce 20%–50% of the O_2 on Earth annually, contribute almost half of the organic material found in the sea, and absorb over 6.7 million metric tons of silicon per year from the waters they dwell in [2]. Silicon is deleterious to and limits the growth of phytoplankton if in concentrations that are too high, so this ecosystem engineering is beneficial to phytoplankton and all the species in the food web they support. Yet, silicon is needed by life, and they produce 9% to 13% of silica globally [3]. Thus, they keep silicon at levels close to optimal for sea life, neither too high nor too low. The shells of dead diatoms reach as far down as 800 m on the seafloor, where they provide nutrients to the food webs there. Every year, 2.7 million tons of fossil diatoms blown from the Sahara Desert fertilize the entire Amazon basin, Caribbean islands, and the Everglades (see Chapter 4). *Rhizosolenia* play a significant role in the carbon, silica, and nitrogen cycles in the oligotrophic seas. They excrete nitrogen in the forms of nitrate, nitrite, and ammonium, all of which play an important role in the nitrogen cycle in the ocean's photic zone. They migrate down to depths of 300 m to obtain nutrients, some of which they excrete in the photic zone after they return there, where other aquatic organisms use them [4]. Their living and dead shells are eaten as food by bacteria and benthic animals. *Rhizosolenia* species are known to form huge blooms of 400,000 cells/L between May and June in the English Channel and in the Bay of Brest. These blooms are the primary producer level of foods webs, directly and indirectly supporting countless species at all trophic levels beyond this level.

Some diatoms sink to the seafloor when they die, and this way they get buried there in great amounts. They remove great amounts of carbon from the system by this mechanism. This regulates Earth's temperature, keeping it cool and favorable to life. Some *Rhizosolenia* species lower numbers of other phytoplankton species by competing for nutrients. Some *Rhizosolenia* species sometimes kill fish by clogging their gills with their hard silica shells and from the depletion of O_2 after their blooms. Their negative impacts are far outweighed by their positive effects when averaged over sufficiently long time periods.

The Third Principle of Ecology and Evolution, or the Principle of Niche Provision, states that all species provide niches for other species. It is also true that every species uses other species as niches. There is not one species that does provide at least one other species with a niche and use at least one other species for a niche. This is one reason why diversity is autocatalytic. It is an important mechanism by which organisms create more diversity. After an adaptive radiation, there is always a second wave of diversification because the new species that originated as a result of the radiation provide niches for other species that evolve to fill them.

The Fourth Principle of Ecology and Evolution, also called the Principle of Interconnectedness, states: No gene, organism, or species exists or evolves as an autonomous, independent entity, or in isolation. Genes, organisms, and species evolve and interact with several other like entities in an interactive ecosystem and environment. They all exist and evolve in concert with a network or system of other genes, organisms, species, and a physical and chemical environment that they are connected to, affect, and are affected by. They are all linked. Every species requires other species for its survival, and its ecology and evolution affect and are affected by other species, as well as the physical environment. Species need other species to obtain energy, for habitat, for removal of their wastes, and so on, and need the physical environment for oxygen, nitrogen, habitat, and so on. It follows from this principle that, since all species in an ecosystem are directly or indirectly connected and interacting, and affecting and affected by other species, all species are part of one unitary system that they are inseparable from. Since ecosystems interact, the entire planetary system is one unified system, each organism, species, and part of which is connected to one whole. The system is as real as the individual gene, organism, or species. The concept of an individual organism or individual species is not wholly valid. There are aspects of organisms and species that fit the definition of the individual, but there are equally aspects that do not. The most realistic interpretation is that organisms exist only partly as individuals, and species exist only partly as independent units. They are less like autonomous individuals or entities and more like non-autonomous, inter-connected parts of a greater whole, system, or ecosystem. Ecosystems and the Earth possess a profound, unified interconnectedness. Pollination and horizontal gene transfer are examples of this principle. The microbiome shows the extraordinary veracity of the principle.

The Fifth Principle of Ecology and Evolution, which can also be called the Principle of Filled Niches, states that no niche remains empty (unoccupied) for a very long time. Said another way, no unused resource will remain unused for a long time. There are no empty niches for very long. The evidence for this principle is discussed in Chapter 2.

The Sixth Principle of Ecology and Evolution, which can also be called the Principle of Coevolution, states: All biological evolution is coevolution. I say biological evolution to specify evolution between organisms, species, genera, and so on, as opposed to evolution driven by the nonbiological environment, such as the effect of climate on evolution. This specification is required because life does not always coevolve with the abiological environment. Organisms often coevolve with it. Of course, the physical-chemical environment affects life and its evolution. But life often affects the nonbiological environment—the atmosphere, climate, soil, and so on. Thus, they often coevolve, and this is an important aspect of the ABH. But they do not always coevolve, so the principle cannot state that all evolution is coevolution, and the word "biological" is needed to assure the general truth of the principle. Obvious examples of this principle are predators and prey, and species of pollinators that coevolve with the species of angiosperms they pollinate. But each species has at least some coevolution with any species it interacts sufficiently strongly with for a sufficiently long time.

The Seventh Principle of Ecology and Evolution, also called the Principle of Diversity and Stability, states: The number, quality, and distribution of connections (relationships) between species in an ecosystem are of fundamental importance in determining its stability. From this it follows that the number of positive connections (relationships) between species in an ecosystem is fundamentally important in determining its stability. An ecosystem's stability is partly determined by the number of species it has. However, an ecosystem in which all species are at the same trophic level (for example, all predators) is highly unstable, even if it has a million species. Species richness is correlated with diversity, because most natural ecosystems have a balance of trophic levels and a predominance of positive connections, and adding more species tends to keep the balance and add stability. But species richness is only one variable that correlates with stability.

Having more connections is correlated with greater stability. If a species is extirpated and there are a great number of connections between the surviving species, they are less affected by the loss of the one species. The number of connections correlates with the number of species, and this is another reason why species richness is correlated with stability, and another reason why the number of connections correlates with stability. The nature, or quality, of the connections or relationships between species is important to the stability of the system. Mutualism is a positive connection, and more positive connections makes for a more stable system. I already pointed out that a balance between trophic levels is stable (a system with only herbivores is unstable). The distribution of the connections refers to whether it is the case that one or a few species are important nodes with almost all the connections, or all species have about the same number of connections, or there is some situation in between these extremes. An even distribution of connections is relatively more stable, all other factors being equal. If almost all the species in the system benefit from one species, the system is more unstable than a system of more evenly distributed connections because the loss of that one species will cause many species to decline and destabilize the system. Thus, one can distill the seventh principle into a succinct statement featuring diversity: coevolved diversity is correlated with ecosystem stability. This is because coevolved systems have more positive connections between species and a healthy balance of trophic levels. They do not tend to be all predators, for example. It is useful to visualize species as nodes that are linked together by lines representing their relationships. The links can be strong or weak, and positive on both species (mutualism, predator-prey), negative on both (competition), or positive on one and neutral on the other (commensalism), and so on. If a link is positive, it is more stabilizing if strong; if negative, it is more destabilizing if strong. A system with many links, a high percentage of which are positive, and whose links are evenly distributed is the most stable, all else being equal. A system with few links, a high percentage of which are negative, with links unevenly distributed, is the least stable.

A high number of species and links provides stability through redundancy. If a predator is removed, a trophic cascade can be prevented because another predator can switch its food source and replace it in a redundant system. When a species or a link is lost in a system with more of them, a lower percentage of the total species or links is lost, so such a system is more stable. Such systems are also better at homeostatically maintaining themselves. They are more able to maintain themselves and return to an equilibrium state when disturbed by changes in weather, fires, insect outbreaks, or mild human intervention such as selective cutting of trees. Monoculture farms are unstable and need constant input of fertilizers and pesticides to persist.

The Eighth Principle of Ecology and Evolution, also called the Principle of Beneficial Chemistry in Nature, states: Nature does not make any compound that cannot be broken down and recycled. Thus, no chemicals build up to levels that become severe environmental problems in nature. And all compounds are broken down, and the breakdown products are used as nutrients or building blocks for nutrients useful to at least some species. Thus, no chemicals in nature persist long enough to accumulate and cause problems to life, and every natural chemical is converted to one or more other chemical(s) useful to life, including to species other than the species that created it.

This book has discussed various specific mechanisms by which species promote diversity and concerning how the ABH works, such as effects of mutualism, commensalism, competition, plants, herbivores, predators, and so on. I will now discuss the two underlying general, fundamental, unifying mechanisms by which this occurs.

The most comprehensive unifying concepts of the ABH are the first two Principles of Ecology and Evolution combined, that all species are ecosystem engineers with a net positive effect on diversity, other species, and their ecosystems in natural systems over time. A predator acting as an ecosystem engineer in controlling its prey by negative feedback, the prey supporting the predator, species in closed-loop systems like the nitrogen cycle, any benefactor species in commensalism, and any two mutualists—these are all ecosystem engineers. It is not hard to see why all of these species do the engineering they perform. A beaver building a dam creates a pond habitat that benefits the beaver. It is not surprising that this pond also benefits many additional species, or that adding a habitat in addition to the river habitat increases diversity. Since many species have similar needs, it is not surprising that other species benefit from the beaver pond, or that other species benefit from an elephant digging a water hole. Each of these is a secondary effect of the work of the ecosystem engineer. It is not surprising that mutualism benefits many species. The flower evolved to aid the pollinator in the engineering task of pollination. The flower is an engineer that molds the pollinator's behavior. The flower and pollinator are ecosystem engineers that each benefit from the other, so it is no surprise they both diversified once pollination evolved. Nor is it a surprise that many species evolved to exploit them, since they provide benefits to these species.

An important example of ecosystem engineering by species is negative feedback between species at adjacent trophic levels. If a prey population increases, its predator has more to eat, and so increases and causes the prey to decrease. If the prey decreases, the predator decreases and the prey numbers increase. Similarly, more predators mean less prey, and hence fewer predators. The same mutual negative feedback applies to parasites and their hosts. And if a species becomes numerous to the point that it is a threat to its environment, its abundance means there is more of it to consume, it is a readily available and abundant resource, and there is selection for more predators and parasites to attack it. It is thus regulated and does not disrupt the ecosystem for very long. The predator-prey and host-parasite systems are mechanisms by which life generates diversity and maintains stability by ecosystem engineering that are easily seen.

In a nutrient cycle such as the nitrogen cycle, in which the waste product of one set of organisms is a nutrient for another set, compounds are kept at near-optimal levels by a negative feedback system. This is a mechanism that employs ecosystem engineering and that promotes diversity and stability.

And when there is a major die off, there are more carcasses, so there is more food for decomposers. So decomposers increase and recycle more dead organisms, as needed by the system. A situation with too many carcasses is unstable and does not last long. In fact, it produces more nutrients. This is ecosystem engineering by decomposers, and, again, it is easy to understand how the decomposers benefit from performing this service.

Less easily comprehended are secondary effects that have no apparent adaptive reason to help other species. Plant roots are adaptive because they absorb nutrients and water. There is no obvious reason why they should weather rocks, resulting in the building of soil and the sequestering of carbon. This is ecosystem engineering that appears to be a fortuitous secondary effect.

It is also not obvious why a waste product of photosynthesis, O_2, would be exploited, after some evolution, as the major energy source that made higher life possible. Why photosynthesis and respiration are opposite reactions allowing a fundamental mutualism in nature is an interesting question. It is partially answered by the fact that every waste product is a nutrient for any species that evolves to utilize it. But the perfect complementarity of the two most important reactions for higher life is a fortuitous occurrence allowing fundamental ecosystem engineering that is worth exploring.

The other general, unifying mechanism of the ABH is natural selection. This includes life's ability to adapt and its tendency to diversify. Natural selection is one of several mechanisms by which the ABH works. Organisms have the ability to adapt to environmental challenges, adapt to extreme environments, and evolve spectacular adaptations. Organisms often are causative agents of speciation and macroevolutionary breakthroughs, populations often have a tendency to speciate, and sexual selection drives diversification. Natural selection is a mechanism by which life increases diversity (Chapter 3). This unifying mechanism includes the tendency of the genome to provide the variability that selection can act on and produce macroevolutionary breakthroughs with, by such mechanisms as sexual recombination, gene duplication, and transposable elements, including those that jump between species. It includes the fact that the amount and type of genetic variability in a population is subject to natural selection. It includes the fact that after a key innovation, taxa greatly diversify.

It includes the mechanisms of evolution of altruistic behavior and cooperation within species, such as kin selection.

Since general unifying mechanisms are a major goal of science, I am excited to have discovered two such mechanisms for the ABH. However, a comprehensive, unifying mathematical model of the ABH—of how life generates biodiversity—is lacking and is a necessary and important goal. I hope this book inspires other researchers to work on this laudable pursuit.

The null hypothesis is life does not affect diversity. We have seen that species drive their own speciation and biology is a major cause of macroevolutionary innovations that lead to adaptive radiations. Mutalism, commensalism, competition, plants, herbivores, predators, and decomposers all increase diversity tremendously. Without these effects of biology, life would be limited to simple prokaryotic communities. The null hypothesis must be rejected and the ABH is supported.

REFERENCES

1. Klauber, L. M. (1997). *Rattlesnakes: Their Habits, Life Histories, and Influence on Mankind.* Volume 1. Second edn., p. 1076. University of California Press: Berkeley, CA. ISBN 0520210565.
2. Treguer, P., et al. (1995). The silica balance in the world ocean: a reestimate. *Science.* 268 (5209): 375–9. Bibcode: 1995Sci...268..375T. doi: 10.1126/science.268.5209.375. PMID 17746543.
3. Shipe, B., et al. (July, 1999). *Rhizosolenia* mats: An overlooked source of silica production in the open sea. *Limnology and Oceanography* 44 (5): 1282–92. https://doi.org/10.4319/lo.1999.44.5.1282.
4. Singler, H. & Villareal, T. (2005). Nitrogen inputs into the euphotic zone by vertically migrating *Rhizosolenia* mats. *Journal of Plankton Research* 27: 545–56.

Index